Horst Kisch, Robin Perutz
Light

I0027578

Also of interest

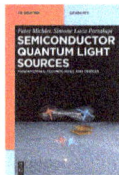

Semiconductor Quantum Light Sources.
Fundamentals, Technologies and Devices
Peter Michler and Simone Luca Portalupi, 2024
ISBN 978-3-11-070340-5; e-ISBN (PDF) 978-3-11-070341-2

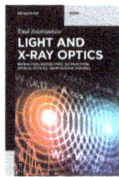

Light and X-Ray Optics.
Refraction, Reflection, Diffraction, Optical Devices, Microscopic Imaging
Emil Zolotoyabko, 2023
ISBN 978-3-11-113969-2; e-ISBN (PDF) 978-3-11-114010-0

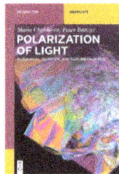

Polarization of Light.
In Classical, Quantum, and Nonlinear Optics
Maria Chekhova and Peter Banzer, 2021
ISBN 978-3-11-066801-8, e-ISBN (PDF) 978-3-11-066802-5

3D Printing with Light
Edited by: Pu Xiao and Jing Zhang, 2021
ISBN 978-3-11-056947-6, e-ISBN (PDF) 978-3-11-057058-8

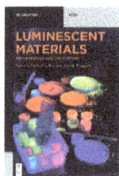

Luminescent Materials.
Fundamentals and Applications
Edited by: Mikhail G. Brik and Alok M. Srivastava, 2023
ISBN 978-3-11-060785-7, e-ISBN (PDF) 978-3-11-060787-1

Horst Kisch, Robin Perutz

Light

From Photochemistry, Photocatalysis and Photobiology
to Visual Arts

DE GRUYTER

Authors
Prof. em. Dr. phil. Horst Kisch
Genglerstraße 18
91054 Erlangen
Germany
horstkisch@gmail.com

Prof. Robin N. Perutz
Department of Chemistry, University of York
YO10 5DD York
UK
robin.perutz@york.ac.uk

ISBN 978-3-11-102934-4
e-ISBN (PDF) 978-3-11-102937-5
e-ISBN (EPUB) 978-3-11-102987-0

Library of Congress Control Number: 2024937890

Bibliographic information published by the Deutsche Nationalbibliothek
The Deutsche Nationalbibliothek lists this publication in the Deutsche Nationalbibliografie;
detailed bibliographic data are available on the Internet at http://dnb.dnb.de.

© 2025 Walter de Gruyter GmbH, Berlin/Boston
Cover image: lena5/iStock/Getty Images Plus
Typesetting: Integra Software Services Pvt. Ltd.

www.degruyter.com

Prologue

Life on Earth depends on light from the sun. Without it, there would be no photosynthesis and consequently no oxygen to breathe and no food to eat. Worship of the Sun as a Godhead is common to many religions. The Pharaoh Akhenaten (also written as Echnaton) abolished worship of multiple gods in ancient Egypt around 1350 BCE and decreed that the Sun (Aton) is the one and only God. In religion, light acts as a metaphor for enlightenment, redemption and purity, in philosophy for knowledge and truth. The history of humanity has been hugely influenced by light, whether in art or science. In nature, light supplies the energy required for plants via photosynthesis. That in turn exploits carbon dioxide and water to make the oxygen and carbohydrates without which animal and human life would not be possible. The impressive selectivity of this remarkable process is brought home when you consider that air contains only 0.04 volume percent of carbon dioxide, but 21% of the much more reactive oxygen. Light from the sun enables not just our vision, but also steers our hormonal balance and therefore the rhythm of life. We can think of the colour of plants and animals as light particles, or photons as they are called, converted into camouflage and communication. Light also drives the formation and destruction of ozone in the atmosphere, whether it is in the upper atmosphere where ozone is protective or in the lower atmosphere where it is harmful.

There are other much weaker sources of light in Nature apart from the sun that are mostly observable only in darkness. Chemical reactions occur in many organisms that generate energy in the form of light. The resulting glow is called bioluminescence and is essential to life in the depths of the ocean. The deep-sea region below 1000 m forms the biggest ecosystem on Earth. Only through the light of bioluminescence can a partner be attracted or an enemy deterred. These observations lead us to the conclusion that light is the most used carrier of information in Nature.

Goethe experienced that fascination with sunlight, as we can read in the first verse of the *Prologue in Heaven* in Faust:[1]

> *The Sun sings out, in ancient mode,*
> *His note among his brother-spheres,*
> *And ends his pre-determined road,*
> *With peals of thunder for our ears.*
> *The sight of him gives angels power,*
> *Though none can understand the way:*
> *The inconceivable work is ours,*
> *As bright as on the primal day.*

For almost 50 years, scientists and engineers have worked intensively to exploit the energy of the sun as a means of ending the reliance on coal, oil and gas – the fossil

---

1 Translation from https://www.poetryintranslation.com/

https://doi.org/10.1515/9783111029375-202

fuels – with the expectation of slowing down climate change. The heat from the sun generates the wind that we use in wind turbines, while the light from the sun is used to generate electrical energy from photovoltaics in solar panels and to generate heat with solar thermal systems. There is the further possibility of converting solar energy into chemical energy. An ideal process would split water into hydrogen and oxygen as Jules Verne predicted in his book *The Mysterious Island* back in 1874:

> *Yes, but water decomposed into its primitive elements . . . . and decomposed doubtless, by electricity . . . Yes, my friends, I believe that water will one day be employed as fuel, that hydrogen and oxygen which constitute it, will furnish an inexhaustible source of heat and light*[2]

Since water is regenerated when hydrogen is burnt, it would provide an inexhaustible fuel so long as the sun shines. Electrolytic splitting of water using electricity from photovoltaic panels and wind turbines is performed on a commercial scale. The competitive one-step splitting of water by light absorption at a semiconductor powder surface is on the way to commercial application. The idea of artificial photosynthesis for the production of carbohydrates and other commodities from the components of air, carbon dioxide, nitrogen, and water is being pursued enthusiastically by many scientists. This concept that light absorption can lead to chemical reaction underlies the subject of photochemistry.

Artificial light also plays a critical role in everyday life. It's not just used for lighting but for information transfer (fibre optic cables), it stores books and music as tiny changes in DVDs and might one day be used by humankind as a common language by analogy with deep-sea communication by fish. Absorption of light of one colour (wavelength) can lead to emission of light of another colour – photoluminescence. The use of europium (one of the rare earth elements) as a luminescence agent in Euro banknotes is not a gimmick, but a security device. Emission of light can also be achieved by electrical stimulation – electroluminescence – and acts as the basis for many displays in mobile phones and televisions.

The UN General Assembly named the year 2015 as the "International Year of Light and Light-based Technologies" because of the fundamental role of light for humankind. Following the great success of this initiative, UNESCO introduced May 16 as an annual "International Day of Light" with the aim of making people more aware of the central role that light plays.

Light acts as an essential tool in chemistry, physics, medicine and engineering and the role of light absorption and emission is becoming ever more important in science. A series of 14 Nobel Prizes in chemistry, physics and medicine have celebrated these developments over the last 35 years:

3D structure of the photosynthetic reaction centre (1988)

Formation and decomposition of ozone in the atmosphere (1995)

---

2 www.gutenberg.org/files/1268/1268-h/1268-h.htm

Development of methods to cool and trap atoms with laser light (1997)
Studies of transition states using laser (femtosecond) spectroscopy (1999)
Development of laser-based precision spectroscopy (2005)
Discovery and development of green fluorescent protein (2008)
Development of super-resolved fluorescence microscopy (2014)
Invention of blue light-emitting diodes (2014)
Design and synthesis of molecular machines (2016)
Discoveries of molecular mechanisms controlling the circadian rhythm (2017)
Groundbreaking inventions in laser physics (2018)
Experiments with entangled photons, pioneering quantum information science (2022)
Development of attosecond laser spectroscopy of ultrafast electron movements (2023)
Discovery and synthesis of quantum dots (2023)

In undergraduate courses, it is rare to find an interdisciplinary course on the properties and effects of light – this book is designed to fill this gap. We aim to outline the fundamental principles and to stimulate interest in the many physical, chemical and biological consequences and applications of the interaction of light – especially light absorption – with both inanimate and living matter. The book should act as a primer for students of physics and life sciences who receive little or no training in these areas as well as for chemists who may be more familiar with some aspects. We complete the book with a chapter on the cultural aspects of light, from religion and philosophy to art and architecture.

This book represents a translation and substantial expansion of "Licht: eine Einführung für Chemiker, Physiker und Lebenswissenschaftler" written by one of us (HK) and published by De Gruyter in 2021. It has been made possible through the longstanding collaborations with all those who have worked with us on light-driven reactions. We owe a great debt to them and to those who have taken the subject forward. Thanks also go to Dr Terry Dillon for reading the section on atmospheric chemistry and to Vivien Perutz for advice on art history. Last but not least, huge thanks to our families for their support and patience not just with this project but the many photochemical projects that preceded it.

Contents

Glossary of acronyms

Acronym	Full term
A	Acceptor
ATRA	Atom transfer radical addition
BDE	Bond dissociation energy
BET	Back-electron transfer
BLUF	Blue light using flavin
BNAH	1-Benzyl-1,4-dihydronicotinamide
bpy	2,2′-Bipyridine
CB	Conduction band
CDOM	Chromogenic dissolved organic matter
CFC	Chlorofluorocarbon
CFP	Cyan fluorescent protein
CT	Charge transfer
CTTS	Charge transfer to solvent
D	Donor
DMF	Dimethylformamide
DNA	Deoxyribonucleic acid
DOM	Dissolved organic matter
dppz	Dipyridophenazine
EDTA	Ethylenediaminetetra-acetic acid
EnT	Energy transfer
EPR	Electron paramagnetic resonance
ET	Electron transfer
eV	Electron volt
FAD	Flavin adenine dinucleotide
FLIM	Fluorescence lifetime imaging spectroscopy
FMNH	Flavin mononucleotide
FRET	Förster resonance energy transfer
GFP	Green fluorescent protein
HAT	Hydrogen atom transfer
HER	Hydrogen evolution reaction
HOMO	Highest occupied molecular orbital
IC	Internal conversion
IFET	Interfacial electron transfer
IPCT	Ion pair charge transfer
IR	Infrared
ISC	Intersystem crossing
LC	Ligand centred
LEC	Light-emitting electrochemical cell
LED	Light-emitting diode
LLCT	Ligand-to-ligand charge transfer
LMCT	Ligand-to-metal charge transfer
LSPR	Local surface plasmon resonance
LUMO	Lowest unoccupied molecular orbital
MC	Metal centred
MeV	Mega electron volt
MLCT	Metal-to-ligand charge transfer

https://doi.org/10.1515/9783111029375-204

MMCT	Metal-to-metal charge transfer
MO	Molecular orbital
MOF	Metal organic framework
NIR	Near infrared
NMR	Nuclear magnetic resonance
OER	Oxygen evolution reaction
OERS	One-electron reduced state
OET	Optical electron transfer
OLED	Organic light-emitting diode
PALM	Photoactivated localisation microscopy
PAN	Peroxyacylnitrate
PAS	Photoacoustic spectroscopy
PC	Photocatalyst
PCET	Proton-coupled electron transfer
PDT	Photodynamic therapy
PET	Photo-induced electron transfer
PET	Polyethyleneterephthalate
Photo-CORM	Photochemical CO-releasing molecule
ppy	Deprotonated phenylpyridine
PS	Photosystem
PS I	Photosystem I
PS II	Photosystem II
QD	Quantum dot
RISC	Reverse intersystem crossing
ROS	Reactive oxygen species
SAD	Seasonal affective disorder
SC	Semiconductor
SET	Single electron transfer
SPR	Surface plasmon resonance
STORM	Stochastic optical reconstruction microscopy
TADF	Thermally activated delayed fluorescence
STED	Stimulated emission depletion spectroscopy
TEOA	Triethanolamine
THF	Tetrahydrofuran
TOF	Turnover frequency
TON	Turnover number
TTA	Triplet-triplet annhilation
UC	Photon upconversion
UV	Ultraviolet
UV-A	Ultraviolet 315–380 nm
UV-B	Ultraviolet 250–315 nm
VB	Valence band
VE	valence electrons
VOC	Volatile organic compound
VR	Vibrational relaxation
YFP	Yellow fluorescent protein

1 Fundamental principles

1.1 Energy, colour, absorption and emission

Light is a child of the sun that is about 150 million kilometres away. Light escapes from the outer layer of the sun that has a temperature between 5200 °C and 6800 °C and begins its lightning-speed journey to Earth [1]. When it reaches the Earth's atmosphere, it loses much of its ultraviolet radiation through absorption by nitrogen and oxygen gas. After about 8.5 min, the light arrives at the Earth's surface and ends its short life, where it supplies us with heat and, through photosynthesis, with food to eat and air to breathe.

Light absorption can lead to chemical reactions or to emission of light of a different colour – those were two of the messages of the prologue. The consequences and applications range from how we see with our eyes, to the formation of ozone in the atmosphere, to the speeding up of chemical reactions. Recent Nobel prizes have celebrated both the laser tools we use to study these processes and their applications in imaging, in understanding the basis of our daily rhythms, in new inventions such as synthetic molecular motors and in many other phenomena. In this chapter, we will delve into the underlying principles, keeping the mathematics to the bare minimum. Some of you will be familiar with what follows – if so, skip to the next section.

Light consists of electromagnetic radiation with a mixture of different wavelengths. The idea that it is composed of a stream of tiny particles called photons provides an alternative description. Light is generated from hot objects, whether stars or flames or incandescent light bulbs. It can also arise from the conversion of electrical energy (diodes) or by emission from energy rich (*excited*) atoms and molecules (luminescence). In the sun, hydrogen nuclei (protons) are fused together to form helium nuclei with concomitant emission of positrons and neutrinos. Energy is released corresponding to the difference in mass of the helium atom compared to the four protons (about 0.7%). The amount of energy can be calculated by Einstein's famous equation $E = mc^2$ and is 26.72 Mega electron volts (MeV) for each helium nucleus generated (Equation 1.1). This energy serves in part to heat the sun while the rest is released into space as photons and energetic particles.[1]

$$4\,^{1}\mathrm{H}^{+} \rightarrow {}^{4}\mathrm{He}^{2+} + 2\,\mathrm{e}^{+} + 26.7\,\mathrm{MeV} \tag{1.1}$$

In this process, the heat of the sun is converted into light, just as when burning a candle or heating iron in a furnace. The iron glows dark red at 700 °C, light orange at 1200 °C and white above 1300 °C. When atoms are heated or absorb light, they form

---

[1] 1 eV per particle corresponds to 96.5 kJ mol^{-1} or 23.0 kcal mol^{-1}, so 26.7 MeV corresponds to 2580 million kJ mol^{-1}.

https://doi.org/10.1515/9783111029375-001

excited states that emit light. The colour and energy of this emitted light are character-istic of the element and can be used to identify the atoms. This is how astrophysicists determine the composition of stars (spectroscopic analysis).

Most electromagnetic waves, such as infrared or radio waves, cannot be detected by our eyes but play a huge role in everyday life. Remote controls, fibre-optic cables, lasers for surgery and DVDs all use infrared (IR) radiation. The field of photonics de-pends on infrared radiation for information transmission, while radio communica-tion is all pervasive. IR radiation corresponds to wavelengths from about 700 nm to a few cm, while radio waves may be from tens of cm to a few hundred metres. The wavelength range that our eyes can detect runs from about 400 nm to 700 nm (1 nm = 10^{-9} m). The photons elicit a response in our eyes that we perceive as different col-ours. At shorter wavelengths than 400 nm lies the ultraviolet region which we again cannot perceive. We should not assume that other vertebrates see as we do – many birds and fish can see into the ultraviolet as far as about 300 nm [2].

Light travels at a speed, c, of 3×10^8 m s^{-1} in a vacuum. The frequency of light, ν, corresponds to the number of wave-crests passing per second and is related to the wavelength (Figure 1.1) λ, by Equation 1.2:

$$\nu = c/\lambda \tag{1.2}$$

so we can calculate that the frequency of light of wavelength 400 nm is 7.5×10^{14} s^{-1}.

wavelength

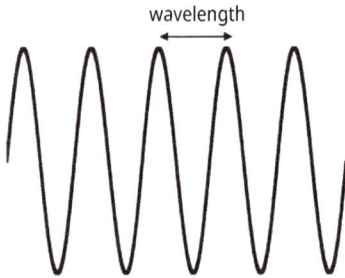

Figure 1.1: Light is a sinusoidal electromagnetic wave.

Sometimes, we use a wavenumber, $\tilde{\nu}$, in place of a frequency. This quantity is the in-verse of wavelength and is often quoted in cm^{-1}, so the wavelength of 400 nm corre-sponds to 25000 cm^{-1} (Equation 1.3):

$$\tilde{\nu} = 1/\lambda \tag{1.3}$$

It was Max Planck who first used the concept of light particles, photons, to relate the frequency of light to the energy of the photon by Equation 1.4:

$$E = h\nu \tag{1.4}$$

In other words, light particles are indivisible units or quanta with an energy that corresponds to their frequency and wavelength. The value of Planck's constant, h, is incredibly small, 6.63×10^{-34} J s, so we can calculate that the energy of the 400 nm photon is $6.63 \times 10^{-34} \times 7.5 \times 10^{14} = 4.97 \times 10^{-19}$ J. Chemists and biologists measure the energy of chemical bonds for each mole of substance: 1 mole (abbreviated mol) corresponds to the Avogadro number, $N = 6.02 \times 10^{23}$ molecules, enabling us to calculate the energy of 1 mole of photons (Equation 1.5):

$$E = Nh\nu \tag{1.5}$$

For our 400 nm photon, the energy is $6.02 \times 10^{23} \times 4.97 \times 10^{-19} = 299$ kJ mol^{-1}. For comparison, the bond energies of H_2 and Cl_2 are 432 and 240 kJ mol^{-1}, respectively. This calculation provides some startling information. The 400 nm photon has enough energy to break the bond of the chlorine molecule into its constituent atoms, but not enough to do the same for hydrogen. Table 1.1 provides some energy conversions and shows the energies of different chemical bonds.

Table 1.1: Wavelengths, wavenumbers and photon energies in comparison to bond dissociation energies (BDE).

	λ (nm)	$\tilde{\nu}$ (10^3 cm^{-1})	E (eV)	E (kJ/mol)	E (kcal/mol)	BDE (kJ/mol)
UV	200	50.0	6.20	598	143.0	H-CCH (523), O_2 (494),
	300	33.3	4.13	399	95.3	H_2 (432), H-CH$_3$ (460)
	350	28.6	3.54	342	81.7	C-O (360), C-C (350)
	400	25.0	3.10	299	71.5	Cl-CCl$_3$ (295)
VIS	500	20.0	2.48	239	57.2	Cl_2 (240)
	600	16.6	2.07	199	47.7	
	700	14.3	1.77	171	40.8	Br_2 (190)
	750	13.3	1.65	159	38.1	
NIR	800	12.5	1.55	150	35.7	I_2 (148)
	1000	10.0	1.24	120	28.6	
	5000	2.0	0.25	24	5.7	

The light reaching Earth from the sun consists of about 3% ultraviolet (UV), 47% visible (Vis) and 50% infrared (IR) radiation. The distribution of light intensity, i.e. the number of photons per second, is shown schematically in Figure 1.2. Also shown are the positions of the primary colours, blue, green and red which correspond to wavelengths of about 450, 510 and 640 nm, respectively. All other colours can be obtained from mixtures of these three primary colours. The complementary colours of yellow, magenta and cyan are obtained by combinations of two of the primary colours in equal proportions.

Different materials and substances absorb different wavelengths of light. The colour of a substance that we perceive is complementary to the colours that it absorbs. If the substance absorbs blue, it appears yellow; if it absorbs green, it appears magenta, and if it absorbs red, it appears cyan. We see the substance as the primary blue

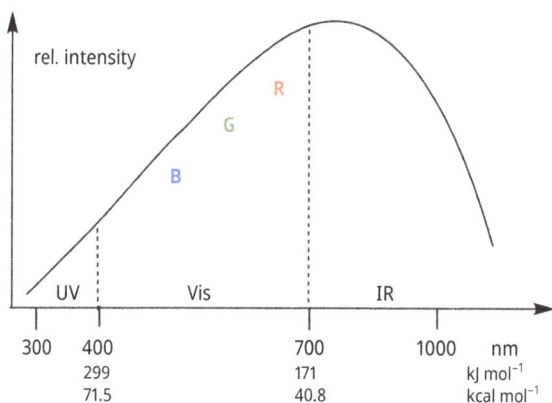

Figure 1.2: Approximate intensity distribution of the sunlight reaching the Earth's surface. R = red. B = blue, G = green. The energies of photons are shown below the x-axis.

if it absorbs red and green. If it absorbs all three primary colours, it appears black. The colour wheel (Figure 1.3) represents the primary and secondary colours. To find out why a substance absorbs a certain colour, we must turn to the way the electrons behave in an atom or molecule. Quantum theory tells us that the energy of the electrons does not vary continuously – the energy can only take certain values. Absorption of light shifts the molecule from its minimum energy state, called the *ground electronic state,* to higher states, called *excited electronic states.* The light absorbed by the substance corresponds to the difference in energy between the ground state (**G**) and the excited state (**G***) (Figure 1.4A). We can interconvert the energy of the quantum, $h\nu$, and the wavelength of light via Equation 1.2. Use of Equation 1.5 allows conversion to molar quantities. Light absorption is not the only way to generate colour. Light scattering makes the sky blue, while interference and diffraction effects create

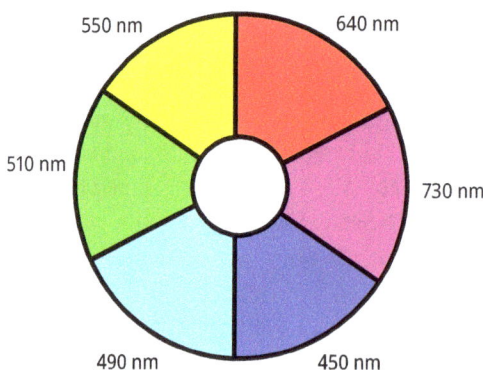

Figure 1.3: Colour wheel.

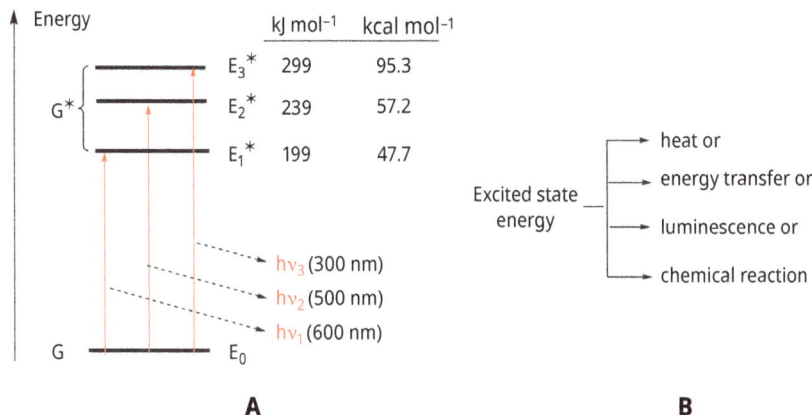

Figure 1.4: A: Energetic consequences of absorption of light by a hypothetical molecule **G** showing the ground electronic state (E_0) and the excited states $E_1{}^*$, $E_2{}^*$, $E_3{}^*$. Excited states are often labelled with an asterisk. The energies are shown for 1 mol of photons. B: Four possible fates of excited state energy.

the colours of soap films and many butterflies. These mechanisms will not be pursued further in this book.

The concept of the electronic excited state is central to this book. Now that our molecule is in an excited state, it has to dispose of its extra energy and there are four important ways (Figure 1.4B). One way is to turn this energy into heat; a second is to transfer the excitation energy to another molecule (*energy transfer*). A third way is to emit light of a different and longer wavelength than the incoming light – this *spontaneous emission* of light is also called *luminescence*. For instance, the europium compound in the Euro banknotes mentioned in the prologue absorbs UV radiation and emits in the visible region, returning the europium compound to its ground state. Since the emission is at longer wavelength than the incoming radiation, some of the energy is also converted to heat so that energy conservation is maintained. In summary, *light-in leads to absorption, light-out is emission*. A fourth way is to use the energy of an excited state for chemical reaction – this is called a photochemical reaction. For instance, ozone undergoes a photochemical reaction generating an oxygen molecule and an oxygen atom (Equation 1.6).

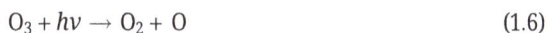

$$O_3 + h\nu \rightarrow O_2 + O \tag{1.6}$$

Most of this book will be concerned with the twin phenomena of luminescence and photochemical reactions. Looking back at what we've learnt, a substance acquires its colour through light absorption, but just because it's coloured it doesn't mean that it will luminesce or undergo photochemical reaction.

Light absorption is not the only way of generating excited states and hence luminescence. Other phenomena include chemical reactions, pressure, temperature and application of an electrical potential each of which can result in luminescence. The terminology for these different sorts of luminescence appears in the textbox below.

Luminescence Terms

Cause	Effect
Light absorption	photoluminescence
Higher pressure	triboluminescence
Higher temperature	thermoluminescence
Electrical potential	electroluminescence
Chemical reaction	chemiluminescence

The triboluminescence of sugar was described more than 400 years ago (see textbox) but the explanation requires much more recent concepts [3].

Sparkling sugar lumps

The English scientist Francis Bacon reported back in 1605 that – after the eyes have adapted to complete darkness – a cold blue-green glow can be observed on scraping sugar lumps with the blade of a knife. This *triboluminescence* can also be seen when two sugar lumps are rubbed against each other in a darkened room. The glow is caused by emission from nitrogen molecules in the air that have been ionised and promoted to their electronically excited state. When crystalline solids are fractured equal and opposite charges are generated on the surfaces that have been pulled apart. The resulting electric field is evidently strong enough to form nitrogen molecule ions, N_2^+ in their excited state that luminesce on converting to their ground state.

There are also numerous compounds that become coloured or change colour on irradiation – they are described as *photochromic*. Similarly compounds that become coloured on heating are *thermochromic* and those that become coloured on application of an electric potential are *electrochromic*. Azobenzenes have been used in numerous applications for their photochromism and their colours can be tuned by substitution (Scheme 1.1) [4]. Applications of photochromism include spectacles that darken on exposure to sunlight, while electrochromism is used in displays. Many conducting organic polymers and metal oxides show electrochromic behaviour. An example is poly(3,4-ethylenedioxythiophene) (PEDOT) that changes colour from red to blue on oxidation.

(a)

(b)

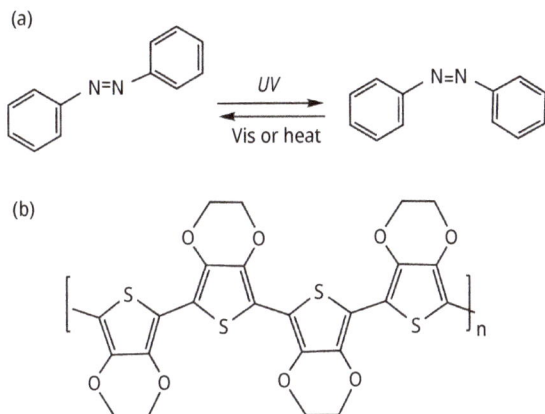

Scheme 1.1: (a) A photochromic compound, azobenzene (b) an electrochromic conducting material poly-(3,4-ethylenedioxythiophene) (PEDOT).

1.2 Absorption and emission spectra

1.2.1 Spectra of liquids and gases

We measure the wavelengths of light that are absorbed in an absorption spectrometer and the wavelengths that are emitted in an emission spectrometer. How these spectrometers work is explained shortly, but the result is shown schematically in Figure 1.5 for a hypothetical molecule **G**. This figure shows an electronic absorption spectrum with three maxima corresponding to the existence of three excited states (see Figure 1.4). In many cases the position and absorption coefficient provide important indications of the electronic structure of **G**. The lowest energy absorption bands can be traced back to features of a larger molecule since there are bands characteristic of different chromophore groups[2] such as C=C, C=O, N=N, or ring structures such as benzene or porphyrins. Similar principles apply to transition-metal complexes. This section mainly concerns liquids (usually solutions) and also applies to gases. Solids are addressed separately in Chapter 3.

Another important part of Figure 1.5 is the emission maximum shown in red which usually lies to lower frequency (longer wavelength) than the absorption maxima. Sometimes, we don't just want to measure the emission spectrum, but need to check the intensity of luminescence on excitation at different absorption wavelengths. This is called the excitation spectrum and is shown in blue in Figure 1.5. In simple cases, the excitation spectrum tracks the absorption spectrum. If the absorption spec-

---

2 Chromophore groups are elements of the molecular structure which are principally responsible for the absorption characteristics.

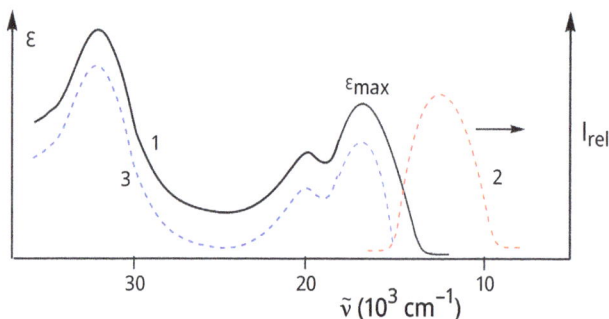

Figure 1.5: Molecular electronic spectra: (1) absorption shown in black, (2) emission in red, (3) excitation shown in blue. The y-axis on the left refers to the absorption coefficient, ε; the y-axis on the right refers to the relative intensity of emission, I_{rel}.

trum and the excitation spectrum, do not track one another, this may be a sign of impurities present in the sample. When we measure an excitation spectrum, we are finding out which excitation wavelength causes luminescence.

The layouts of typical absorption and emission spectrometers are shown in Figure 1.6, while the different measurement procedures are shown in the textbox. This figure shows the layout for solution measurements, but it is equally possible to measure in reflection or scattering of the incoming light. The light source and detector usually operate over a wide range of wavelengths. However, a conventional white-light source may be replaced by a tuneable laser which can be adjusted to emit a very precise wavelength (the results are known as *laser-induced fluorescence,* LIF). Absorption spectrometers may operate in a single-beam mode (as shown) or a double-beam mode where one beam (the reference beam) goes through a cuvette containing solvent only and the other passes through a cuvette with dissolved sample (the sample beam). For absorption measurements, the cuvette has a rectangular or square section with two parallel polished sides, while for emission, it has four polished sides.

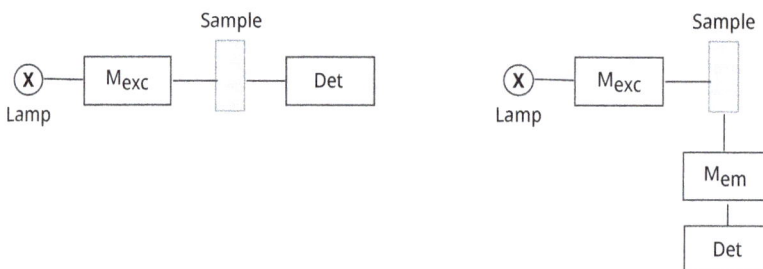

Figure 1.6: Layout of typical absorption (left) and emission (right) spectrometers for dissolved or gaseous samples. M = monochromator, exc = excitation, em = emission, Det = detector. Note that there are many variants on these layouts.

Spectroscopic measurements	
Absorption spectrum	Measure absorbance as a function of excitation wavelength, λ_{exc} Measure sound intensity as a function of excitation wavelength, λ_{exc} (see Chapter 3)
Emission spectrum	Measure emission intensity, I_{em}, as a function of emission wavelength, λ_{em}, at constant excitation wavelength, λ_{exc}
Excitation spectrum	Measure emission intensity, I_{em}, as a function of excitation wavelength, λ_{exc}, at constant emission wavelength, λ_{em}

When measuring a sample in solution, the output is usually shown as an absorbance. This quantity is related to the intensity of transmitted light logarithmically (Equation 1.7). Thus an absorbance of 1.0 corresponds to 10% transmission, 2 corresponds to 1% transmission, etc. The absorbance is related to the concentration of the sample and the pathlength of the cuvette (sample container) via the Beer-Lambert law (Equation 1.8) where ε is called the molar (decadic) absorption coefficient and is characteristic of the substance. The units of molar absorption coefficient are usually written as dm^3 mol^{-1} cm^{-1} or M^{-1} cm^{-1}. These are spectroscopists' units and can be converted to SI units of m^2 mol^{-1} (1 m^2 $mol^{-1} = 10$ dm^3 mol^{-1} cm^{-1}). Since m^2 corresponds to an area, absorption coefficients are described as an area in the literature of solids.

$$(I_0/I) = 10^A \text{ or } A = \log_{10}(I_0/I) \tag{1.7}$$

where I_0 is the intensity at the detector without sample and I is the intensity with sample and A = absorbance.

$$\text{Beer-Lambert Law} \quad A = \varepsilon\,cl \tag{1.8}$$

where ε is the molar absorption coefficient, c the concentration and l the pathlength.[3]

Example: a dissolved sample of concentration 0.1 mol dm^{-3} with absorption coefficient 2×10^3 dm^3 mol^{-1} cm^{-1} has an absorbance of 2 in a pathlength of 0.01 cm, meaning that only 1% of light is transmitted. For a practical measurement with a 1 cm path cuvette, a much lower concentration is needed – a concentration of 5×10^{-4} mol dm^{-3} will give an absorbance of 1. Molar absorption coefficients vary over at least 8 orders of magnitude, but the dynamic range of the spectrometers is much smaller. Consequently, the concentration and pathlength must be adjusted to obtain accurate measurements.

Emission measurements are more sensitive than absorption measurements, but this can be put to advantage to monitor trace impurities by their emission spectra. For this reason, it is important to use samples and solvents of optimum purity when characterising their spectra. Comparison of absorption and excitation spectra can sometimes reveal impurities.

---

3 In earlier literature, *absorbance* may be described as *optical density* and *absorption coefficient* as *extinction coefficient*.

1.3 Emission lifetimes and imaging

Imagine yourself in a clean white shirt at a disco. Under the UV illumination, your shirt glows violet with luminescence. The lights are suddenly turned off. How long does the luminescence of your shirt take to fade? We call that quantity the luminescence (or emission) lifetime. In other words, luminescence has a time dimension as well as a wavelength dimension. In practice we use a flash lamp or a pulsed laser to excite the luminescence and a suitable detector to monitor its decay. The decay of luminescence is exponential when reactive collisions or similar events are absent (Figure 1.7). Such exponential decay corresponds exactly to radioactive decay or first order kinetics. Photochemists usually quote the decay time as a lifetime, τ, which is defined as the time to decay to $1/e$ times the initial intensity I_0, or the inverse of the first order rate constant, k (e = 2.7183, Equation 1.9). Figure 1.7 shows an exponential decay with lifetime of 25 ms. In practice, lifetimes vary over 12 orders of magnitude from a few picoseconds to seconds. In the next chapter, we will encounter the effect of collisions on the luminescence decay.

$$I = I_0 \exp - (t/\tau) \text{ or } \ln_e(I/I_0) = -t/\tau \tag{1.9}$$

$$k = 1/\tau$$

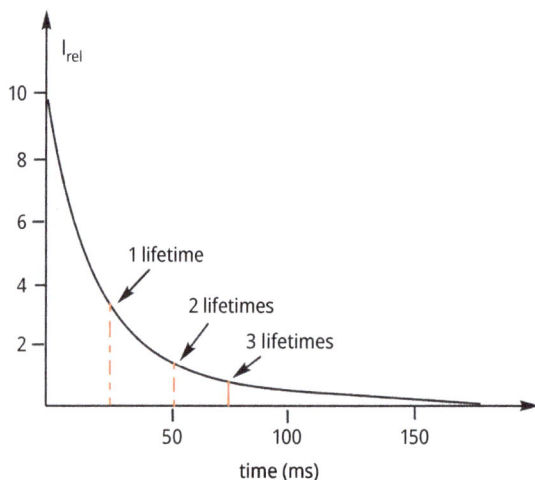

Figure 1.7: Exponential decay of luminescence shown schematically for a lifetime of t = 25 ms. The red broken lines indicate 1, 2 and 3 lifetimes. After 3 lifetimes, the intensity is ca. 5% of its initial value.

Under the UV lamp, the Euro banknote shows the parts that contain the luminescent material – the *luminophore*. In Figure 1.8, the europium luminophore glows red and shows a pattern of stars: we see an image. Luminophores are used extensively in biological microscopy to image cells and organelles. Biological samples often exhibit short-lived luminescence which can obscure what the observer wants to see under the microscope. If the sample is labelled with a dye with a long luminescence lifetime, the background emission can be blocked out by using a pulsed light source and only beginning measurement after a delay. This is called gating. It is also possible to obtain images where the intensity of each pixel is based on its emission lifetime – this is called fluorescence lifetime imaging microscopy (FLIM).

Figure 1.8: A €50 banknote under UV illumination. The red stars contain a europium compound [5].

The luminescence discussed until now occurs without further stimulus once an excited state has been formed – it is *spontaneous emission*. The light is emitted in all directions and consists of many wavelengths (*polychromatic* light). The different waves have no phase-relationship with one another (*incoherent* light). These features contrast with those for stimulated emission and laser light discussed in the next section.

1.4 Stimulated emission and lasers

It was Einstein who recognised that the emission process described above (spontaneous emission) is not an exact parallel of absorption. He argued that absorption of light results from the action of an incoming photon. There ought to be an analogous emission process where an incoming photon causes emission – *stimulated emission*. The three processes, absorption, spontaneous emission and stimulated emission are shown in eqs 1.10–1.12 for an excited molecule A^*.

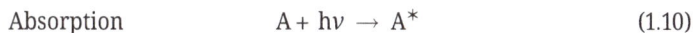

$$\text{Absorption} \qquad\qquad A + h\nu \rightarrow A^* \qquad\qquad (1.10)$$

$$\text{Spontaneous emission} \quad A^* \rightarrow A + h\nu \qquad\qquad (1.11)$$

$$\text{Stimulated emission} \quad A^* + h\nu \rightarrow A + 2\,h\nu \qquad\qquad (1.12)$$

Stimulated emission is the basis of the *laser* (Laser = Light Amplification by Stimulated Emission of Radiation). Laser action requires more molecules to be in the excited electronic state than in the ground electronic state. This situation is called a *population inversion* and contrasts with the behaviour of molecules at equilibrium where populations are governed by the Boltzmann distribution. Excited molecules generated by the typical processes of light absorption also represent a very small proportion of the total. Population inversions can be achieved very briefly by use of an intense flash lamp and exploitation of selection rules (see below) or by an electric discharge. An incoming photon can stimulate this extremely unstable state to emit an avalanche of photons that are:

- identical in wavelength – *monochromatic*
- *directional*
- in phase – *coherent*

Lasers can generate a short flash (pulsed laser) or act as a continuous light source (continuous wave, CW). Population inversions are usually generated by an indirect route involving two or more excited states. A simplified diagram (Scheme 1.2) shows how this works for a neodymium-doped yttrium aluminium garnet (Nd-YAG) laser which generates an intense pulse in the near IR at 1064 nm.

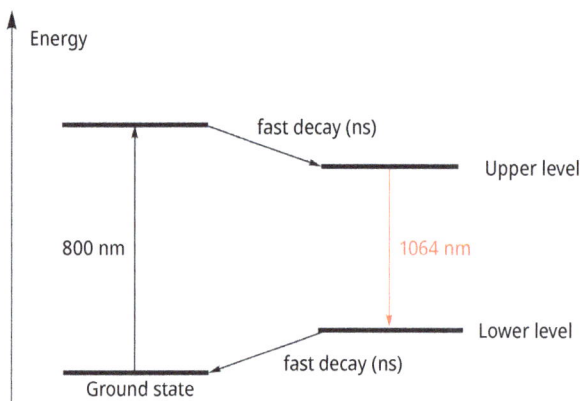

Scheme 1.2: Simplified energy level scheme for a Nd-YAG laser. The flash lamp causes absorption at 808 nm. Rapid decay populates the state labelled *upper level* which undergoes stimulated emission at 1064 nm, usually as a pulse lasting a few ns.

The special properties of laser radiation enable it to be manipulated in many ways. The frequency of the light can be doubled or tripled, the frequency can be reduced through a tuning mechanism, the pulse length can be reduced etc. Pulsed lasers are key to many imaging techniques (super-resolution imaging, Section 1.5) and essential for many luminescence measurements and for time-resolved spectroscopy (Section 1.8). They are available for a wide range of the electromagnetic spectrum from far UV through visible to far

IR. The monochromatic nature of lasers, often with very narrow range of emission wavelengths, allows very selective excitation of luminescence, leading to single molecule spectroscopy (Section 1.5). The high intensity enables absorption of two photons at once (2-photon spectroscopy).

In addition to their uses in science, lasers are important in surgery, welding and cutting of metal, distance measurement, 3D-printing, CD players, light displays, as pointers, barcode readers and many other applications.

1.5 Single molecule fluorescence and super-resolution imaging

In an experiment with a standard luminescence spectrometer (Figure 1.6), the sample contains trillions of molecules (1 mL at 10^{-6} M contains 6×10^{14} molecules) and the measurement depends on exciting numerous molecules and measuring their emission – this is called ensemble averaging. If, instead of detecting the emission of many molecules once each, we detect the emission of one molecule many times, we would be able to study single molecules. Amazingly, this is indeed possible. *Single molecule fluorescence* requires chromophores that withstand multiple excitation without bleaching, a laser to excite them, extremely dilute samples, a powerful microscope to reduce the sample volume and exceptionally sensitive detectors[4] [6]. As a result, it has been possible to observe how individual molecules embedded in a solid or polymer differ from one another, how molecules change with time in a catalytic process, and to create images with nanometre resolution. Single molecule fluorescence has become an essential tool in the life sciences and has uncovered new phenomena in optical behaviour. It has also led to the development of super-resolution imaging – see below. Even more importantly, it has enabled one of the biggest revolutions in science affecting biology, archaeology, anthropology and medicine – "next generation" DNA sequencing. In the Solexa/Illumina method developed by Balasubramanian and Klenerman,[5] fluorescent tags are attached reversibly to the DNA nucleotides with different colours for each of the four bases. The sequence is read, one base at a time, by the single molecule fluorescence of the tags. Several other versions of this method have also been developed more recently.

---

4 *Fluorescence* is defined as luminescence with no change in electron spin state; see Chapter 2.1. The chromophore that causes the fluorescence is the *fluorophore.*

5 Shankar Balasubramanian and David Klenerman (Cambridge, UK) invented this method of DNA sequencing in 1998; its impact led to the Millennium Technology Prize in 2021 and the Breakthrough Prize in 2022. The idea for the method came about at a meeting in the beer garden of the pub near the Department of Chemistry in August 1997.

The resolution of conventional optical microscopy is limited by diffraction of visible light according to the Abbe limit of $\lambda/2(NA)$ (NA = numerical aperture, λ is wavelength of light) in practice about 200 nm. This limit can be overcome by super-resolution imaging methods, all of which depend on luminescence, and some of which interrogate single molecules[6] [7]. Here we summarise two methods.

The first method is called Stimulated Emission Depletion Spectroscopy (STED) and can achieve resolution of 20–50 nm. Here many fluorophores are distributed through the sample and are excited by one laser beam. A second laser beam of longer wavelength illuminates the sample in a ring shape so fluorophores in the centre receive far less light from this beam than those further out. This beam causes stimulated emission (see above) effectively switching off the emission from the molecules in the ring and leaving only the molecules in the centre emitting (Figure 1.9A). A STED image of the nuclear pore complex is shown in Figure 1.9B; this structure controls transport across the membrane of the nucleus in eukaryotic cells.

The second approach called single molecule localization microscopy depends on the ability to switch fluorophores on and off photochemically. The fluorophores on the sample, (e.g. a biological structure or a polymer) are excited with a continuous laser until they cease emitting because they have been converted to an isomeric structure – switched off. A pulse from a second laser at a different wavelength switches a few of them back on and the image is collected showing a few bright spots. This process is repeated many times until the final image is obtained comprising many emission spots as can be seen in the image of actin filaments in Figure 1.9C (actin is one of the proteins in muscle). The density of fluorophores is low enough that the spots don't overlap. The method can be adapted to use dyes emitting at different wavelengths marking different positions in the structure. It can also be employed to obtain 3-dimensional images. Two of the popular versions of single molecule localization microscopy go by the acronyms PALM and STORM.

---

[6] Three of the pioneers of super-resolution imaging, Eric Betzig, Stefan Hell and William Moerner, were recognised by the 2014 Nobel Prize. Moerner also was the first to demonstrate single molecule fluorescence in condensed phases.

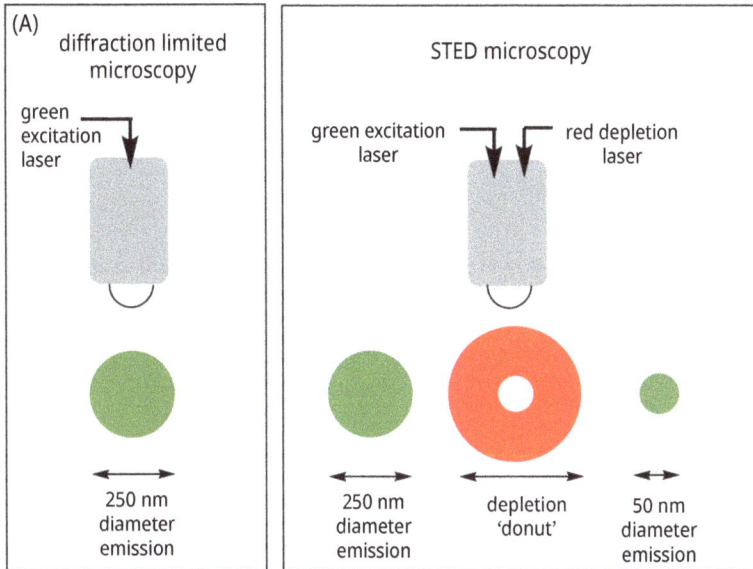

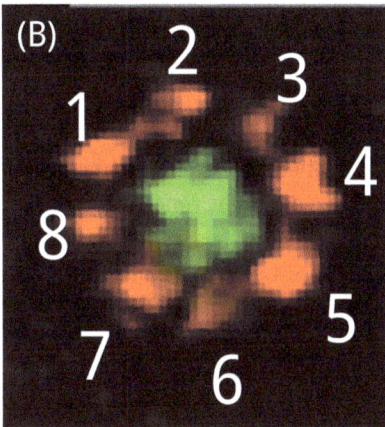

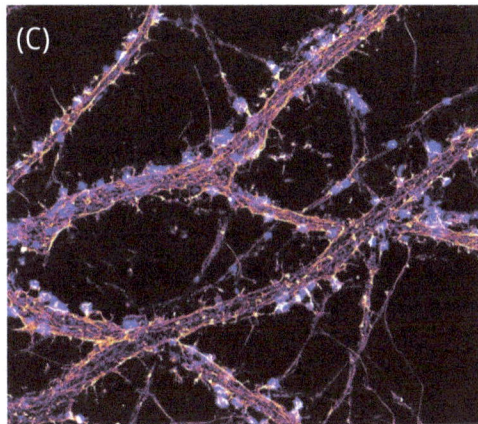

Figure 1.9: (A) Principles of STED. (B) STED image of nuclear pore complex showing 8-fold symmetry. (C) Actin filaments imaged by single molecule localization microscopy (A and C adapted from ref [8], B from ref [9], reprinted from *Biophysical Journal*, **105**, L01, Göttfert et al, Copyright 2013 with permission from Elsevier).

1.6 The laws of photochemistry, quantum yield and action spectra

The laws of photochemistry (textbox) are so simple that some textbooks omit them entirely, but they are essential to understand what happens after light absorption.

Laws of Photochemistry

(1) Only light absorbed causes photochemical change (Grotthus-Draper Law)
(2) One molecule is excited when one photon is absorbed (Einstein-Stark Law)
(3) The absorption of a photon takes place instantaneously
(4) The absorption is so fast that there is no change in molecular geometry during absorption (Franck-Condon Principle)

Quantum Yield (Φ)
Number of times event occurs per photon of wavelength λ absorbed in a standard volume.

$$\text{For photoreaction: } \Phi_p = \frac{\text{rate of reaction in s}^{-1}}{\text{no of photons absorbed per second }(I_a)}$$

$$\text{For luminescence: } \Phi_L = \frac{\text{no of photons emitted per second}}{\text{no of photons absorbed per second }(I_a)}$$

When these laws are combined with the energy of the photon ($E = h\nu$, Equation 1.4), we see that light absorption leads to instantaneous formation of an excited state of energy $h\nu$ as shown in Figure 1.4. It is the frequency of light that determines the energy of the excited state, not its intensity. This parallels the principles of the photoelectric effect – no wonder Einstein was involved in both. In the photoelectric effect and in photoelectron spectroscopy, the photon energy, $h\nu$, supplies the energy to generate the cation and to provide the kinetic energy of the electron released.

What is the probability that a molecule luminesces after light absorption or undergoes a particular chemical reaction? This critical question leads to the concept of *quantum yield* (also termed *quantum efficiency*). Quantum yield (Φ) is defined as the number of times a particular event occurs per photon absorbed. For most events, the quantum yield is less than or equal to one (or 100%) as a direct consequence of the laws of photochemistry. It is best understood by remembering that there is a choice of four classes of events and Φ is the probability of any one event: release of heat, emission of light, energy transfer, photochemical reaction. Quantum yields depend on the wavelength of light absorbed, so must be measured at a specific wavelength. The quantum yield is usually measured as a rate of the relevant event divided by the rate of photon absorption (see textbox). For light emission, the relationship between the intensity of emission (I_{em}), the quantum yield and the intensity of light absorbed, I_a is given by Equation 1.13 where each quantity is dependent on the wavelength λ. Equation 1.13 can be combined with eqs 1.7 and 1.8 to yield equation 1.14.

$$I_{em}(\lambda) = \varphi_{em}(\lambda) \times I_a(\lambda) \tag{1.13}$$

$$I_{em}(\lambda) = \varphi_{em}(\lambda) \times I_0(\lambda)\ln_e(10) \times \varepsilon cl \tag{1.14}$$

Quantum yields for reaction can exceed one if light absorption initiates a chain reaction, for instance by generating a radical that proceeds to cause polymerisation in the dark (see photocatalysis). In the stratosphere, chlorofluorocarbons undergo photoreactions producing chlorine atoms that proceed to destroy ozone catalytically. The chlorine atoms are regenerated through the reactions shown in Equation 1.15.

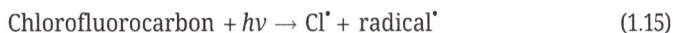

$$\text{Chlorofluorocarbon} + h\nu \rightarrow \text{Cl}^{\bullet} + \text{radical}^{\bullet} \tag{1.15}$$

$$\text{Cl}^{\bullet} + \text{O}_3 \rightarrow \text{ClO}^{\bullet} + \text{O}_2$$

$$\text{ClO}^{\bullet} + \text{O} \rightarrow \text{Cl}^{\bullet} + \text{O}_2$$

When we perform a photochemical reaction, we want to measure which absorption wavelengths cause reaction. This is called an action spectrum and is the equivalent of the excitation spectrum for luminescence. Figure 1.10 compares the absorption spectrum of chlorophyll with the action spectrum, measuring the rate of photosynthesis with wavelength.

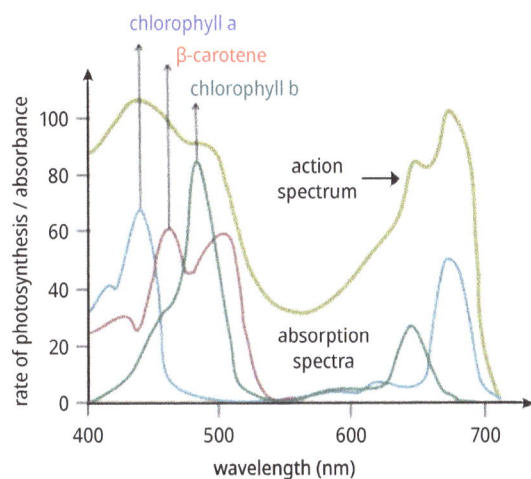

Figure 1.10: Comparison of action spectrum for photosynthesis with absorption spectra of chlorophyll a, chlorophyll b and β-carotene. The action spectrum measures the rate of photosynthesis against the wavelength of light absorbed. Adapted from ref [10].

1.7 Ground states, excited states and electron configurations

The Periodic Table of the elements is built-up by filling the *orbitals* 1s, 2s, 2p, 3s, 3p, etc. with electrons in turn as the charge on the nucleus increases. This Aufbau Principle leads to the electron configurations of the gas-phase atoms in their ground electronic states that may be familiar to you. Each s orbital accommodates two electrons of opposite spin. There are three p orbitals of equal energy that can fit six electrons in total (orbitals of equal energy are termed *degenerate*), while d orbitals have degeneracy of five and f orbitals degeneracy of seven, fitting in 10 and 14 electrons, respectively. Light absorption allows excitation of *one electron* to a higher lying (less tightly bound) orbital following a set of *selection rules* that limit the possible *allowed* transitions. For example, the sodium atom has a ground electron configuration $(1s)^2(2s)^2(2p)^6(3s)^1$. For sodium, absorption transfers the 3s electron to a 3p orbital, while emission does the reverse. Thus the excited electron configuration is $(1s)^2(2s)^2(2p)^6(3p)^1$. The bright orange colour of a sodium flame or sodium flame arises from this transition that shows as a very narrow pair of lines at 589.0 and 589.6 nm in their spectra. (Scheme 1.3). The concept of the ground and excited *electron configurations* provides a very useful approximation to the ground and excited *electronic states* that we will use frequently in this book. It is not the whole story since electron configurations can interact in different ways. The sodium atom provides an example of this complication since there are actually two very closely spaced transitions, hence the pair of lines, arising from interaction of the spin and orbital contributions of the excited electron.

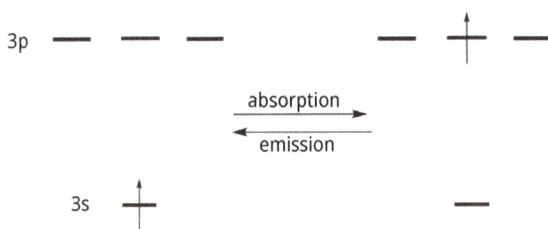

3p — — —   — ✟ —

                       absorption
                ⟶
                ⟵
              emission

3s ✟   —

Scheme 1.3: Ground and excited electron configurations of the sodium atom.

This book is not concerned with atoms in the gas-phase but with molecules in a variety of phases. Some of the concepts needed for atoms can be transferred to molecules but first we need to introduce *molecular orbitals* (MOs) of covalent molecules. These MOs are formed by overlap of the atomic orbitals (AOs) of the constituent atoms, making use only of the highest occupied shell (the valence orbitals). Thus carbon forms molecular orbitals by overlap of 2s and 2p orbitals, silicon, by overlap of 3s and 3p orbitals, etc. We refer to the inner electrons (for instance 2s and 2p in the case of silicon) as the core and they play a negligible part in our story. The number of MOs must be identical to the number of AOs from which they formed. When two AOs overlap, two MOs are formed, one bonding in which the AOs are in phase (plus overlaps with plus in orbital diagrams) and one antibonding in which the AOs are out of phase (plus overlaps with minus). This principle can be seen for the C–C bonds of ethene in Figure 1.11 – notice that the orbitals mostly tightly bound to the nucleus lie lowest in an MO diagram. In ethene, we see that two electron pairs are present in C–C bonding orbitals and none in antibonding orbitals so we have a double bond. The MO formed by overlap along the C–C axis is called a sigma (σ) bond, while the MO formed by overlap in a plane perpendicular to the molecular plane with zero electron density along the C–C axis is called the pi (π) bond.

If an electron is promoted from a full orbital of ethene to an empty orbital, the MO diagram (Figure 1.11) shows us that the lowest energy transition is from the highest occupied molecular orbital (HOMO) to the lowest unoccupied molecular orbital (LUMO), in this case from π to π*. Since that removes an electron from a bonding orbital and places it in an antibonding orbital, the C–C bond will be much weaker in the excited state. When formed initially, there can be no change in the bond length (Franck-Condon Principle), but at equilibrium the excited state will have a longer bond than the ground state and internal rotation about the C–C axis will no longer be restricted as in the ground state. It is not just the molecular geometry that changes in the excited state, but the chemical reactivity changes too. One consequence of the lack of restricted rotation is that substituted alkenes isomerise from cis (Z) to trans (E) configurations and vice versa: this isomerisation forms the basis of our vision.

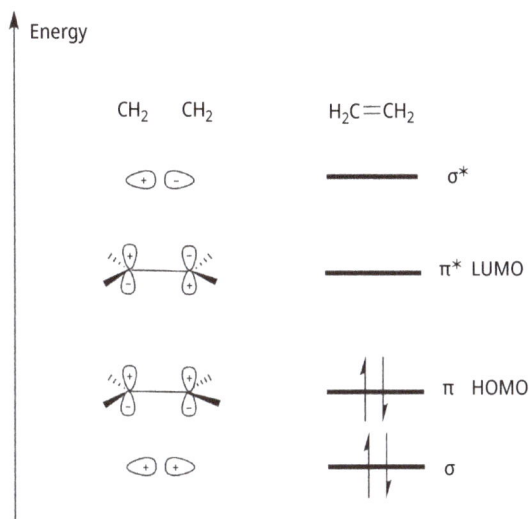

Figure 1.11: Molecular orbital scheme for the C-C bond of ethene. Left: overlap of atomic orbitals. Right: Resulting MO diagram showing occupancy of MOs in ground electronic state.

The spin of an electron is designated by the quantum number $s = \frac{1}{2}$ and can adopt two orientations in a magnetic field. The component in the direction of the field is labelled m_s and can adopt values $+\frac{1}{2}$ or $-\frac{1}{2}$. In the ground electronic state of ethene, all the MOs are either completely filled or completely empty and the electrons are all paired. The total electron spin, S, corresponding to the sum of all the m_s contributions is zero. When the electron is promoted from the π to π* orbital of ethene, there are two possible orientations of the electron spin in the excited configuration, spins opposed or parallel. When the spins are opposed, the total electron spin, S, remains zero, but when they are parallel, S = 1. Consequently, two possible electronic excited states can be formed, although both have electron configurations $(\pi)^1(\pi^*)^1$. Confusingly, we do not refer to these states by the value of S, but the value of (2S + 1), called the *spin multiplicity*.

For ethene, the two possible excited states are distinguished by their spin multiplicities which are either zero (singlet) or one (triplet state) – see Figure 1.12 and textbox. When two states have identical electron configurations, the state of higher multiplicity usually lies lower in energy. Thus the triplet state of ethene lies lower in energy than the singlet. Sodium gas (Scheme 1.3) has one unpaired electron and therefore a doublet state.

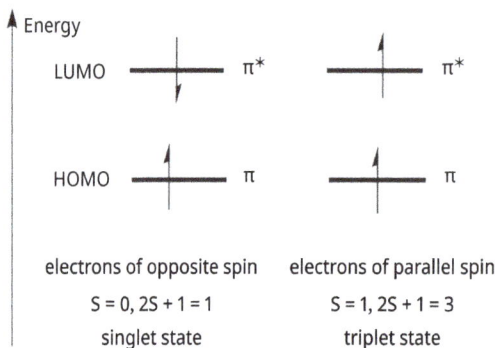

Figure 1.12: Excited electron configurations of ethene.

Spin states

Spin of a single electron s = ½
Total electron spin = S
Spin multiplicity = 2S + 1

No unpaired electrons	S = 0	2S + 1 = 1	Singlet state
One unpaired electron	S = ½	2S + 1 = 2	Doublet state
Two unpaired electrons	S = 1	2S + 1 = 3	Triplet state, etc.

Spin selection rule: $\Delta S = 0$
$\Delta S = 0$, absorption coefficient ε much greater than for $\Delta S \neq 0$
Luminescence with $\Delta S = 0$ – *fluorescence*, emission lifetime τ short
Luminescence with $\Delta S \neq 0$ – *phosphorescence*, emission lifetime τ long

---

The vast majority of molecules, like ethene, have singlet ground states (S = 0), but important exceptions occur when there are degenerate orbitals or when orbitals lie close in energy. Doubly or triply degenerate orbitals occur when molecules have high symmetry – linear, pyramidal, trigonal bipyramidal, square-based pyramidal and octahedral structures all have 3-fold or higher axes (i.e. rotation by 120° or less leaves the molecule indistinguishable). Dioxygen, O_2, has two electrons in its doubly degenerate π^* orbital which have parallel spin in its ground state. It therefore has a triplet ground state. [Cr $(NH_3)_6]^{3+}$ has an octahedral structure and three electrons in the triply degenerate component of its d orbitals (d_{xy}, d_{xz}, d_{yz}) (Scheme 1.4). In the ground state, the electrons must have parallel spin according to Hund's Rule so it has a quartet ground state.

We also find molecules with unpaired electrons in their ground electronic states when orbitals lie close in energy but are not degenerate – *high-spin* molecules. Car-

Energy

π orbitals of O_2 d orbitals of $[Cr(NH_3)_6]^{3+}$

π*

π

$d_{x2-y2}\ d_{z2}$

d_{xy}, d_{xz}, d_{yz}

ground states

S = 1 (triplet)   S = 3/2 (quartet)

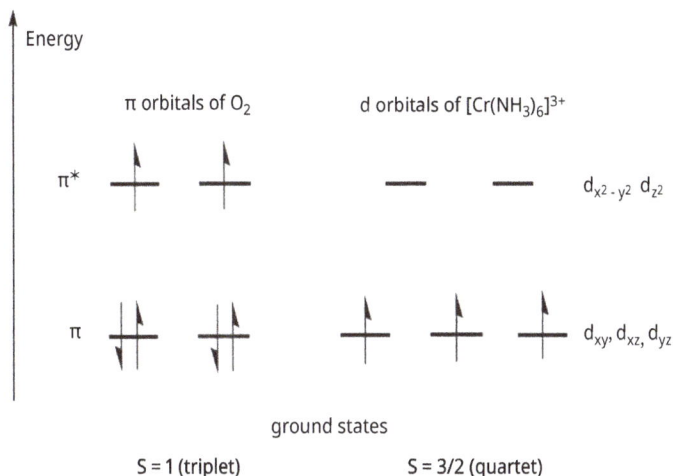

Scheme 1.4: Partial occupation of degenerate orbitals, illustrated for the π* orbitals of O_2 and the d orbitals of $[Cr(NH_3)_6]^{3+}$.

benes act as reaction intermediates and form an interesting example. Methylidene, CH_2, has two unpaired electrons in its ground state, while the electrons in the related dichlorocarbene are all paired. Many complexes of first row transition metals such as $[Mn(H_2O)_6]^{2+}$ also adopt high-spin ground states.

Transitions between ground and excited states (and vice versa) are governed by a set of *selection rules*. In reality, these rules dictate the probability of an event. A useful analogy is the law – selection rules are obeyed no more rigorously than the law. For example, speeding offences are recorded about 3000 times as frequently as murder. Minor selection rules are broken frequently, just as speed limits. Major selection rules, such as the rule that there is no change of spin during a transition, are very rarely broken, at least for light atoms like carbon or oxygen – compare to murder! Consequently, the absorption coefficient, ε, for transitions involving no change of spin ($\Delta S = 0$) have absorption coefficients that are many orders of magnitude larger than those for a change of spin ($\Delta S \neq 0$). Correspondingly, the emission lifetimes, τ, are many orders of magnitude shorter when there is $\Delta S = 0$ than when $\Delta S \neq 0$. This difference leads to names to distinguish these two types of luminescence: *fluorescence* when $\Delta S = 0$, *phosphorescence* when $\Delta S \neq 0$ (see textbox, above [11]).[7] For heavy atoms like platinum, the differences in ε and τ remain, but they are not so marked.

---

7 George Gabriel Stokes (1819–1903) was a polymath of Irish origin who became Lucasian Professor of Mathematics in Cambridge, England. He carried out experimental work on luminescence of quinine sulfate and discovered the luminescence of chlorophyll. He coined the word "fluorescence" which he named after the fluorescent mineral fluorspar. His first paper on fluorescence was 100 pages long [11].

In the case of O_2, MO theory predicts that there should be two excited singlet states formed simply by rearranging the electrons in the π^* orbital. One of these is of great importance in photochemistry (Chapter 4.2.5). The absorption bands for direct excitation from the ground state triplet to these excited state singlets are extremely weak because of the spin selection rule. Famously, Gerhard Herzberg, the Nobel laureate spectroscopist, could not detect one of these absorption bands until he looked at the spectrum of the setting sun, thereby achieving an exceptionally long pathlength. The spin forbidden transition of Cr^{3+} formed the basis of the ruby laser, the first laser (ruby consists of Cr^{3+} in Al_2O_3).

The excited states of ethene (Figure 1.12) have a weaker CC bond than the ground state of ethene. Excited states of other molecules may behave differently: there are some where the bond is weakened so much that the molecule dissociates (example: Cl_2), others where the bond is strengthened (example: XeF used in excimer lasers), others where the electrons are redistributed so as to generate a strong dipole (example: $HO\text{-}C_6H_4\text{-}NO_2$). Remarkably, it is possible for the excited state of a molecule to behave both as a strong reducing agent (electron donor) and a strong oxidising agent (electron acceptor) (example: $[Ru(bpy)_3]^{2+}$, bpy = 2,2'-bipyridine). We will see examples of each type in later chapters.

1.8 Practical photochemistry, how to stimulate and monitor a photochemical reaction

Photochemical reactions usually need an intense light source, preferably at a specific wavelength or narrow range of wavelengths since the reaction may be wavelength-sensitive. Moreover, the reaction product may also be photosensitive. For synthetic purposes, the source should be continuous but for studies of reaction mechanism a pulsed source is usually needed. Monochromatic and white (multi-wavelength) light-emitting diodes (LEDs) have supplanted tungsten lamps for excitation with visible or near-UV radiation. For shorter wavelength UV excitation, mercury arc lamps are commonly used. Alternatively, CW lasers offer high intensity sources. For study of solar energy conversion, white-light sources are available that mimic the output of the sun (solar simulators). Some reactions are so light-sensitive that room lights cause reaction – consequently, the substances must be stored in the dark. The desired wavelength can be obtained through judicious choice of laser. The wavelength ranges of light sources for photochemistry are shown in Figure 1.13. When changing the excitation wavelength, it is important to pay attention to the absorption of the solvent. Typical cut-off wavelengths are as follows: water 185 nm, methanol, 205 nm, benzene 280 nm, acetone 330 nm.

Pulsed lasers are used for time-resolved spectroscopy (see below) and generate a flash lasting from tens of femtoseconds (1 fs = 10^{-15} s) to tens of nanoseconds (1 ns = 10^{-9} s). Since vibrational spectroscopy tells us that nuclei take $\sim 10^{-13} – 10^{-14}$ s to move, femtosecond spectroscopy can, in principle, probe any chemical reaction that involves

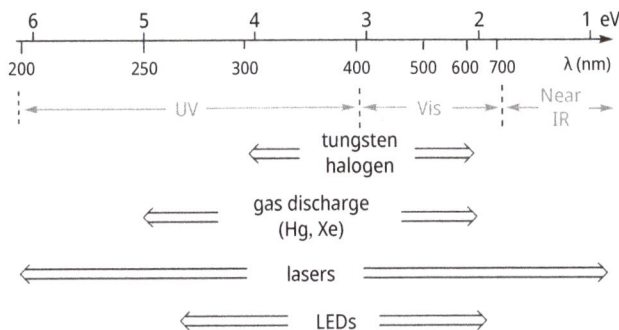

Figure 1.13: Wavelength ranges of light sources used in photochemistry.

nuclear motion. If we want to study changes in electron distribution without motion of nuclei we need to go even faster and that is becoming possible with lasers that produce pulses in the ~100 attosecond range (1 as = 10^{-18} s) [12].

The Beer-Lambert law (Equation 1.8) has major consequences when designing a photochemical experiment: if the solution is too concentrated or the pathlength is too long, the radiation will never reach much of the sample. Heat is also an issue: if the light source is too intense or too tightly focused, the sample will overheat. Another factor is the presence of oxygen from the air. A sample may be air-stable in the dark but air-sensitive in the light. Additionally, oxygen has unpaired electrons that can influence the course of reaction (see Chapter 2.2.2). These issues can be overcome even when carrying out a reaction at large scale either by using flow reactors or by irradiating solutions as thin films (Figure 1.14) [13]. In Chapter 5 we will illustrate an example of how a flow

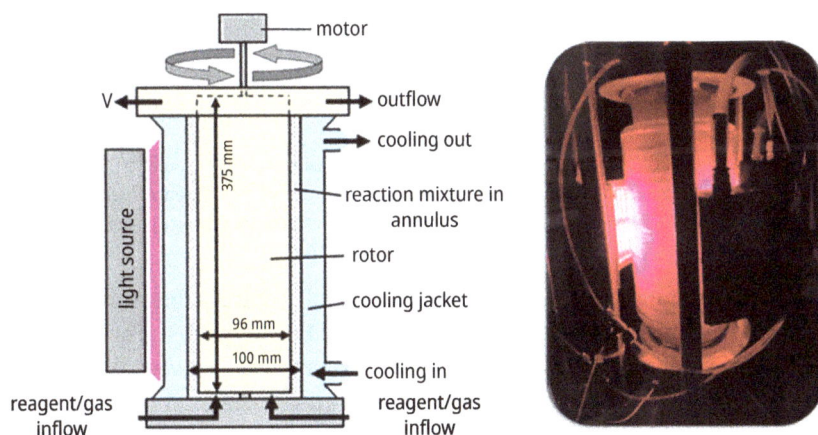

Figure 1.14: Vortex photochemical reactor allows synthesis on a kilogram scale. Left: diagram showing rotor at centre with spinning fluid. Right: reactor in action. Reprinted with permission from *Org. Process Res. Dev.*, 2020, 24, 201–206, copyright 2020, American Chemical Society, ref [15].

reactor with an array of LED light sources can be used to run a photocatalytic reaction at a scale of 500 g/h [14].

When a photochemical reaction is performed, it is essential to monitor how it is progressing. A wide variety of spectroscopic methods can be used for this purpose: UV/ visible absorption, IR spectroscopy, electron paramagnetic resonance, (EPR), nuclear magnetic resonance (NMR), X-ray spectroscopy, etc. It is often informative to display spectra as a difference between the spectrum before and after irradiation, with the result that the product appears positive and the reactant appears negative (Figure 1.15).

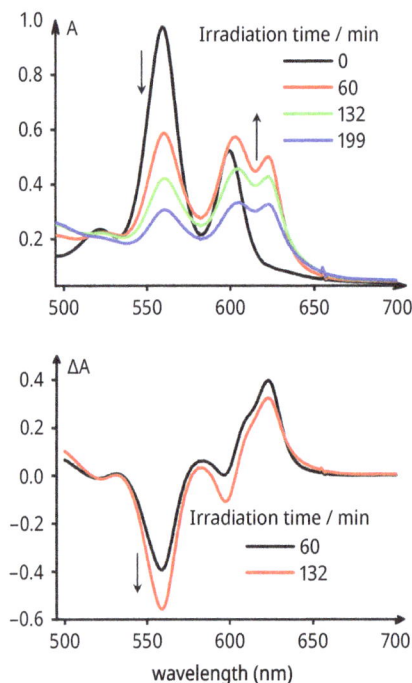

Figure 1.15: Above: visible absorption spectra of a zinc porphyrin measured before irradiation and after 60, 132, and 199 min irradiation. Below: the 60 and 132 min spectra displayed as difference spectra relative to the initial spectrum. Now the reactant peaks appear negative and the product peaks positive.

Measurement of quantum yields is clearly important and requires a measurement of the rate of an event and the number of photons absorbed per second, I_a. For luminescence, there are several standards available allowing comparison between the sample and the standard directly on the emission spectrometer. For chemical reactions, the rate of a reaction can often be measured more accurately than I_a. The value of I_a is measured with a chemical or physical actinometer or radiometer (radiation detector): in the chemical case, this is a standard photochemical reaction (see Chapter 2.3), in the physical case it is a calibrated silicon diode. Complications arise when the product absorbs at the same wavelength as the reactant, so the product acts as a filter (inner filter effect). For heterogeneous samples used in semiconductor catalysis (Chapter 3), measurement is even harder because of reflection and scattering of light. The uncertainty in quantum yield measurements is rarely less than 10%.[8]

Try heating a reaction and it goes faster, while cooling it slows it down. Those are the familiar principles that lead to the concept that thermal reactions need to overcome a barrier, the activation energy. What happens if you cool a photochemical reaction? The answer is that the majority of photoreactions have no barrier, at least in the initial stages, so they can still occur even at very low temperature. The photochemical step is temperature independent. This principle can be used to observe reaction intermediates that would otherwise have a very fleeting existence. The example in Scheme 1.5 shows the formation of a metal complex of methane in low temperature solution, formed by photodissociation of a carbonyl ligand in the presence of methane gas. Observation of this reactive complex depends on the low temperature as well as the fluorinated solvent and counterion; the light source is switched on throughout the observations [16].

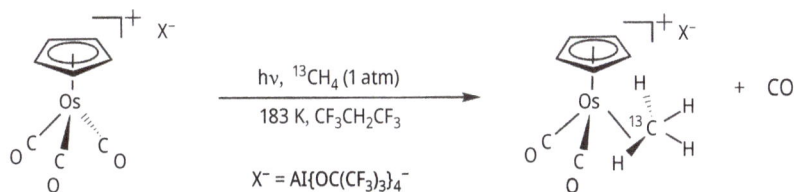

Scheme 1.5: Formation of a methane complex in low temperature solution that is observed by NMR spectroscopy. Such complexes have been postulated as intermediates in many reactions.

Measurements in solution at low temperature capture spectra of reaction intermediates that are moderately reactive but unsuitable for the most reactive species such as CH_2. Just as an insect embedded in amber lasts for millions of years without decay, a highly reactive molecule will stay fixed in an inert solid such as a frozen noble gas, so long as it's kept cold. Moreover, frozen noble gases are completely transparent across the UV, visible and IR spectrum. In this technique, called matrix isolation, a precursor compound is embedded in solid neon, argon, krypton or xenon at temperatures between 4 and 70 K and irradiated to stimulate a photochemical reaction. The product is observed spectroscopically. Matrix isolation stabilises reaction intermediates, but not excited states. It has been used to characterise many intermediates mentioned in organic chemistry textbooks such as benzyne and carbenes, CH_2 and CCl_2.

The chlorine oxide Cl_2O_2 has a stable structure $ClClO_2$, but is isomerised in the matrix to ClOClO and ClOOCl; all three isomers prove important in ozone destruction in the stratosphere (Scheme 1.6) [17]. This example illustrates *wavelength-selective*

---

8 The inner filter effect can be circumvented by measurement of initial rates.

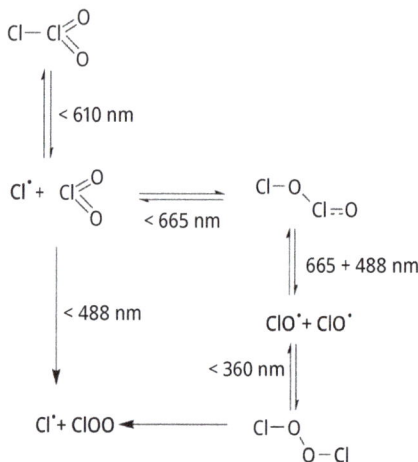

Scheme 1.6: Photochemistry of Cl_2O_2 in argon matrix at 11 K [17].

photochemistry – different isomers are formed by selecting a wavelength where one isomer absorbs strongly but another only weakly or not at all. The initial idea of matrix isolation was that the noble gas matrix would be inert. Sometimes, the noble gas matrix proves reactive leading to remarkable noble gas compounds such as HArF, the first neutral compound of argon, formed by irradiation of HF in solid argon [18].

Photochemistry offers another distinctive feature compared to thermal chemistry. In a thermal reaction, individual molecules react at random times during the course of a reaction. In contrast, if a photochemical reaction is started by a short-lived flash of light, the flash acts like the starting gun for a sprint race. All the molecules start in synchrony,

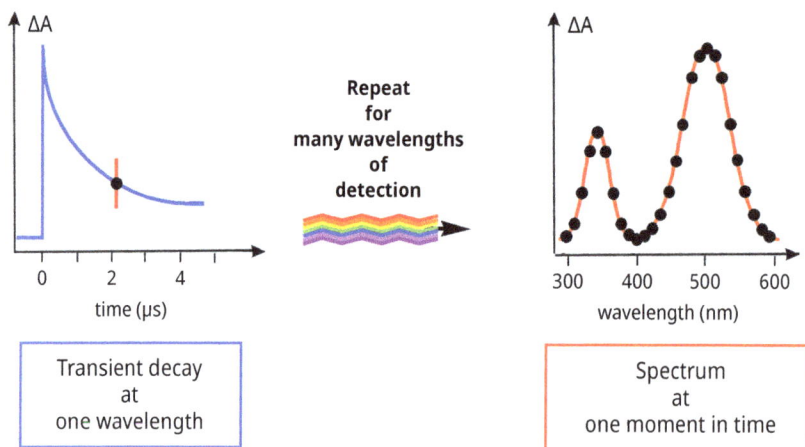

Figure 1.16: Schematic for time-resolved spectroscopy showing time (kinetic) and spectroscopic dimensions.

forming a transient excited state or reaction intermediate. Like the runners in the race, the transient molecules do not maintain their synchronisation after the flash is over. This principle forms the basis of *time-resolved spectroscopy* (formerly called laser flash photolysis). The reaction is started by a flash from a laser (the *pump beam*) lasting from a few nanoseconds to a few femtoseconds according to the laser. This is followed by a *probe beam* at a selected delay time to monitor the reaction product which may be pulsed or continuous. Different types of spectroscopy may be used for the probe: UV/vis absorption, UV/vis luminescence, IR absorption, Raman, EPR, X-ray, etc. Time-resolved spectroscopy therefore provides a spectroscopic and a time dimension. We usually display a spectrum at one moment in time or a kinetic plot measured at one wavelength (Figures 1.16 and 1.17).

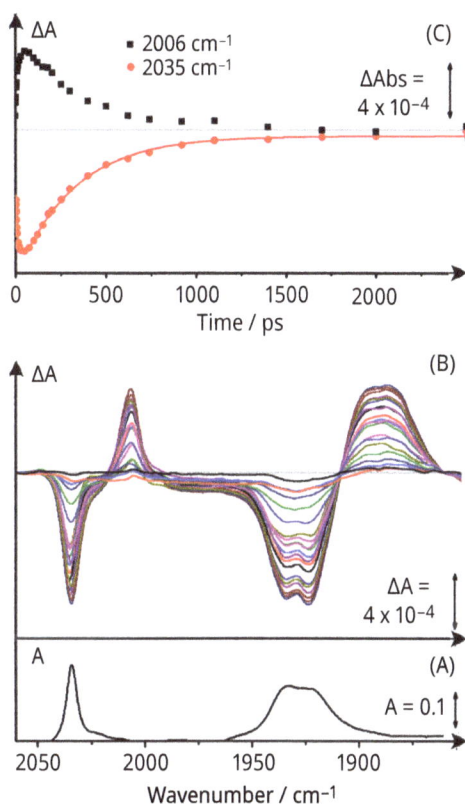

Figure 1.17: Time-resolved IR spectra of a metal carbonyl-porphyrin complex. (A) IR spectrum before irradiation, (B) time-resolved spectra measured at intervals of a few picoseconds, (C) kinetic plots: black rise and decay of product; red: loss and regeneration of reactant. Reproduced from Ref [19] Copyright 2015 Royal Society of Chemistry.

Time-resolved spectroscopy may be used to monitor excited states and reaction inter-mediates. It is used in all areas of photochemistry in order to understand reaction mechanisms. It is also possible to follow reactions by time-resolved X-ray diffraction or electron diffraction. An example of time-resolved IR spectra of a metal carbonyl-porphyrin complex used for photoreduction of carbon dioxide is shown in Figure 1.17. The spectra are measured in the CO stretching region and are shown as difference spectra. Negative peaks are due to loss and regeneration of reactant, positive peaks are due to formation and decay of product. The product is formed with a risetime of 8 ± 1 ps and decays with a lifetime of 320 ± 15 ps.

The major principles described in this section are summarized in the textbox below.

Practical Photochemistry
Designing a photochemical reaction: light source, concentration, pathlength, excluding O_2
Monitoring reaction; choice of spectroscopy, difference spectra
Lack of temperature dependence of primary photochemical process
Use of low temperature photochemistry
Wavelength-selective photochemistry
Time-resolved spectroscopy
Pump-beam and probe-beam experiments

Questions

1. What are the wavelength regions of UV, VIS, and IR light?
2. (a) What is the approximate energy (kJ/mol) of one mol of light quanta (1 Einstein) of a wavelength of 500 nm? (b) Two peaks in a spectrum are at 400 and 500 nm. What is their difference in energy in cm^{-1}.
3. Formulate and explain the Beer-Lambert Law.
4. Explain the difference between absorption and emission spectrum. How are they measured?
5. A dissolved compound exhibits two maxima in its emission spectrum. By what experiment can it be decided if both belong to one pure compound or one of them belongs to an impurity?
6. What is the difference between spontaneous and stimulated emission?
7. What is the meaning of *coherent* light, and how is it produced?
8. Define the terms Frank-Condon Principle, Grotthus-Draper Law, Einstein-Stark Law, and Quantum Yield.
9. Formulate the spin states of a molecule with zero and two unpaired electrons.
10. Draw the LUMO – HOMO electron occupation (including electron spin) of O_2.

References

[1]   L. Green, *"15 Million Degrees"*, *Viking*, *London*, 2016.
[2]   T. H. Goldsmith, *Scientific American* **2006**, *295*, 68–75.
[3]   https://www.sciencedirect.com/topics/physics-and-astronomy/triboluminescence. Accessed on March 4th, 20024.
[4]   H. D. Bandara, S. C. Burdette, *Chemical Society Reviews* **2012**, *41*, 1809–1825.
[5]   https://upload.wikimedia.org/wikipedia/commons/4/44/050euro-uv.jpg. Accessed on March 4th, 20024.
[6]   W. E. Moerner, *Angewandte Chemie International Edition* **2015**, *54*, 8067–8093.
[7]   S. W. Hell, *Annalen der Physik* **2015**, *527*, 423–445.
[8]   https://www.technologynetworks.com/neuroscience/articles/what-is-super-resolution-microscopy-sted-sim-and-storm-explained-328572. Accessed on March 4th, 20024.
[9]   F. Göttfert, C. A. Wurm, V. Mueller, S. Berning, V. C. Cordes, A. Honigmann, S. W. Hell, *Biophysical Journal* **2013**, *105*, L01–L03.
[10] https://www.vedantu.com/question-answer/the-red-and-green-lines-in-the-graph-shown-below-class-11-biology-cbse-6080f860d2b6191bfed4a3a8. Accessed on March 4th, 20024.
[11] a). R. Ceredig, *Philosophical Transactions. Series A, Mathematical, Physical, and Engineering Sciences* **2020**, *378*, 20200105; b) G. Stokes, *Philosophical Transactions of the Royal Society* **1852**, *142*, 463.
[12] K. Ramasesha, S. R. Leone, D. M. Neumark, *Annual Review of Physical Chemistry* **2016**, *67*, 41–63.
[13] a). D. Cambie, C. Bottecchia, N. J. Straathof, V. Hessel, T. Noel, *Chemical Reviews* **2016**, *116*, 10276–10341; b) S. Y. Park, S. Lee, J. Yang, M. S. Kang, *Advanced Materials* **2023**, *35*, e2300546.
[14] E. B. Corcoran, J. P. McMullen, F. Levesque, M. K. Wismer, J. R. Naber, *Angewandte Chemie International Edition in English* **2020**, *59*, 11964–11968.
[15] D. S. Lee, M. Sharabi, R. Jefferson-Loveday, S. J. Pickering, M. Poliakoff, M. W. George, *Organic Process Research & Development* **2020**, *24*, 201–206.
[16] J. D. Watson, L. D. Field, G. E. Ball, *Nature Chemistry* **2022**, *14*, 801–804.
[17] J. Jacobs, M. Kronberg, H. S. Mueller, H. Willner, *Journal of the American Chemical Society* **1994**, *116*, 1106–1114.
[18] L. Khriachtchev, M. Räsänen, R. B. Gerber, *Accounts of Chemical Research* **2009**, *42*, 183–191.
[19] C. D. Windle, M. W. George, R. N. Perutz, P. A. Summers, X. Z. Sun, A. C. Whitwood, *Chemical Science* **2015**, *6*, 6847–6864.

2 Fundamental principles II

2.1 Primary photophysical processes

2.1.1 Singlets and triplets

The concept of the excited state dominated the previous chapter and will continue to play a major role in this chapter as we explore more about how excited states behave. Chapter 1 introduced the laws of photochemistry and quantum yields. The excitation of ethene (or other alkenes) was described in terms of molecular orbitals (MOs, Figure 1.11). The idea of singlet and triplet states was developed showing the influence of electron spin.

The ground electronic state is normally a singlet and is labelled S_0. Light absorption leads to the excited configuration and two possible excited states, the first excited singlet, S_1, and the corresponding triplet, T_1 (Scheme 2.1, left). At the right of Scheme 2.1, we see a diagram of the energies of the states. The allowed transition is from S_0 to S_1, which may convert to T_1 with release of heat. Conversion between states of different spin multiplicity is called *intersystem crossing* (ISC).

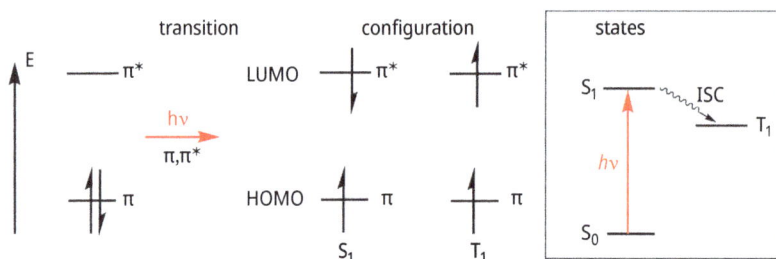

Scheme 2.1: Electronic configurations and states of ethene (or an alkene), showing transition and intersystem crossing (see Figure 1.11).

Control of the magnitude of the singlet-triplet splitting, ΔE_{ST}, is extremely important in applications of luminescent materials for photonics (Chapter 2.2.3 and 2.2.5). The stability of the triplet state relative to the singlet is recognised in Hund's Rule which depends on the quantum mechanical stabilisation gained by exchange of electrons with parallel spin. The main determinant of ΔE_{ST} is the exchange integral which contains the operator $1/r_{12}$, where r_{12} represents the spatial separation of the electron distributions for the two exchanging electrons. Assuming that the electronic states can be simplified to refer to the HOMO and LUMO of S_0, we deduce that ΔE_{ST} is small when the spatial region of the LUMO is very different from that of the HOMO. This principle explains why ΔE_{ST} is much greater for the π,π^* transition of an alkene than

https://doi.org/10.1515/9783111029375-002

for the n,π^* transition of an imine (Figure 2.1a,b).[1] The lone pair on the imine nitrogen always lies perpendicular to the 2p orbitals forming the C=N π-bond and hence well-separated from them [1, p.64], whereas the π and π^* orbitals of the alkene occupy the same spatial region as one another. In both excited states, the occupation of antibonding orbitals weakens the double bond and allows rotation about this bond leading to geometric isomerisation (cis to trans or vice versa, Figure 2.2). This principle of spatial separation of HOMO and LUMO is used in luminescent compounds that contain an acceptor moiety (LUMO) lying perpendicular to a donor moiety (HOMO), resulting in very small values of ΔE_{ST} (Figure 2.1c).

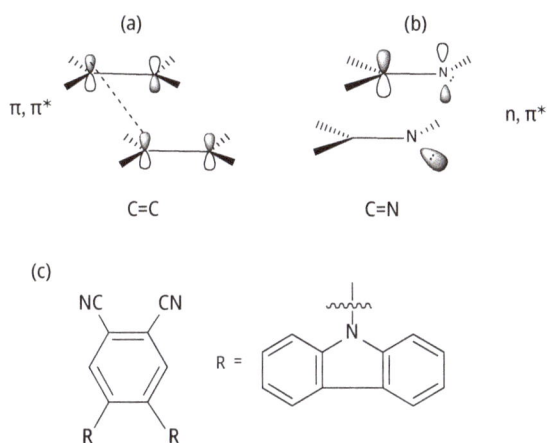

Figure 2.1: (a) π,π^* excited state of alkene and (b) n,π^* excited state of imine. For simplicity, the diagram shows the component AOs that overlap to form the delocalised MOs. (c) A luminescent molecule with a very small singlet-triplet splitting. The carbazole R groups form the HOMO, while the dicyanobenzene forms the LUMO. The carbazole groups are twisted out of the plane of the benzene ring [2].

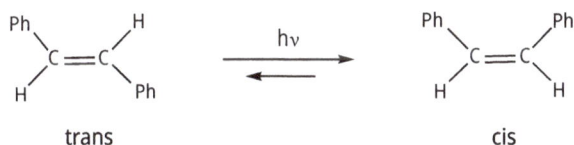

Figure 2.2: Trans-cis (*E-Z*) isomerisation of stilbene following excitation.

Until now, our diagrams have only shown electronic states and have neglected vibrational states. In general, excited vibrational levels lie in the mid-IR region (4000–100 cm^{-1}) at much lower energy than electronic excited states. In Chapter 1, we saw that an excited state may have a different geometry from the ground state at equilibrium,

---

1 The designation *n* refers to the non-bonding lone pair of electrons, in this case, on nitrogen.

but the Franck-Condon Principle (Chapter 1) states that transitions take place so fast that the nuclei do not move their position. The consequence of this apparent paradox is that excited vibrational levels are usually populated during an electronic transition. This concept is displayed in the *Jablonski diagram* which illustrates the primary photo-physical processes (Scheme 2.2). The diagram shows the electronic states arranged vertically according to their energy and horizontally according to their spin multiplicity. In most cases, the gap between the excited states decreases as their energies increase. Jablonski diagrams were originally designed for aromatic molecules.[2] For each electronic state, excited vibrational levels are also shown; the ground vibrational levels of each electronic states are shown with a thick line, and the vibrational levels above them with a thin line (Scheme 2.2). This scheme shows how absorption of a photon of energy $h\nu_1$ generates a vibrationally excited state of the electronic singlet S_2, which relaxes to the ground vibrational level of S_2 within ~ 10^{-13} s (labelled VR, vibrational relaxation). From there, relaxation to the lowest excited singlet, S_1, occurs – such relaxation between states of the same spin multiplicity is termed *internal conversion.* Additionally, it usually involves redistribution of energy between different modes of vibration. This is termed *intramolecular vibrational energy redistribution* (IVR). From here, S_1 can convert to the vibrational excited states of the ground electronic state S_0 either without emission of radiation (IC, radiationless) or by fluorescence (radiative emission of $h\nu_2$). The rate constant of a radiationless process (k_{nr}) between two states is inversely proportional to the energy gap between them ($k_{nr} \propto 1/\Delta E$, energy gap law, see below). In general, radiationless processes have a greater activation energy than the corresponding radiative processes. For this reason, radiationless processes are slowed down greatly on cooling, so weakly emitting states can be observed at low temperature. Measurements in frozen solutions that form rigid glasses at 77 K, the boiling point of nitrogen, often prove valuable.

For many organic molecules, the lifetime of the S_1 state is in the region of 10^{-9} s (1 ns). Since, for example, the frequency of a C–C stretching vibration is ca. 10^{13} s^{-1}, the bond undergoes ca. 10^4 vibrations during the lifetime of S_1. There is therefore plenty of time for conformational and geometric changes as well as changes to the arrangement of the solvent shell. For this reason, the S_1 state reaches equilibrium with its surroundings and is described as an *equilibrated excited state.* The same applies to the T_1 state that lives much longer.

As mentioned above, the energy gap between excited states shrinks with increasing n for neighbouring S_n states. Consequently, the lifetime of the higher states is very short, and light emission occurs (in appreciable yield) only from the lowest excited state of a given spin multiplicity (S_1 or T_1). This is Kasha's rule.[3] Azulene forms an ex-

---

2 These diagrams were developed by Aleksander Jablonski (1898–1980) and Jean-Baptiste Perrin (1870–1942, Nobel Prize in Physics 1926).

3 Michael Kasha (1920–2013) was a US spectroscopist who was the first to show that phosphorescence originates from triplet states that are paramagnetic, something scorned at the time by many organic chemists. He did his PhD with G.N. Lewis, one of the founders of photochemistry.

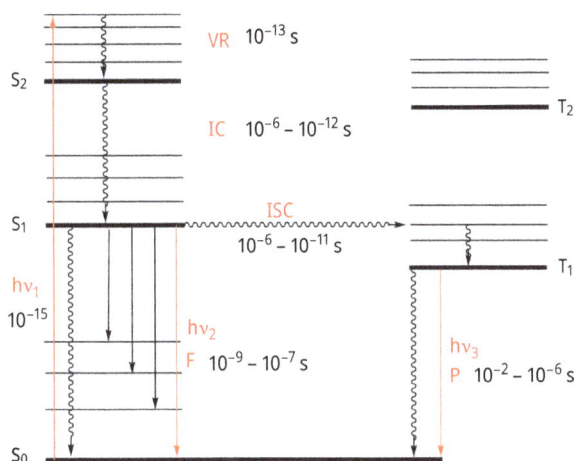

Scheme 2.2: Jablonski diagram enables the visualisation of primary photophysical processes for a molecule that emits from the states S_1 and T_1. Straight arrows indicate radiative processes, wavy arrows indicate radiationless processes that generate heat. The times given are appropriate to typical aromatic molecules. Very different times may be recorded for other systems. VR, vibrational relaxation; IC, internal conversion; ISC, intersystem crossing; F, fluorescence; P, phosphorescence.

ception in which $\Delta E(S_1\text{-}S_0)$ is smaller than $\Delta E(S_2\text{-}S_1)$. Consequently, the radiationless decay of $S_1 \rightarrow S_0$ is fast enough to prevent fluorescence. The radiationless $S_2 \rightarrow S_1$ conversion is slower, and therefore azulene, unlike isoelectronic naphthalene, fluoresces from S_2 and does not obey Kasha's rule.

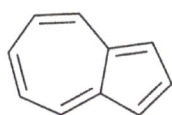

azulene

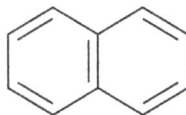

naphthalene

In addition to conversion between states of the same multiplicity, processes with a change of spin can occur, in general described as Intersystem Crossing (ISC). As mentioned in Chapter 1, they violate the selection rule $\Delta S = 0$ and are therefore forbidden, but weak coupling between spin and orbital angular momentum enables ISC to occur nonetheless. Since spin-orbit coupling grows with atomic number, ISC becomes more probable the higher the atomic number of the element concerned. The energy difference between S_1 and T_1 states (see Scheme 2.2) varies greatly: 20–40 kJ mol^{-1} for n,π^* excited states and about 80 kJ mol^{-1} for charge-transfer (CT) excited states (see below). The T_1 state can return to the ground state either non-radiatively (ISC) or radiatively (phosphorescence). The potential energy curves (Figure 2.3) plot the variation in potential energy (E_{pot}) vs bond length (r), in this case for a hypothetical molecule X–Y. The horizontal lines represent the vibrational levels as in Scheme 2.2. IC and ISC proceed with no change in energy (isoenergetically) at the crossing point of the singlet and triplet potential energy curves. Phosphorescence has become ever more important and there have been many recent advances in design of organic compounds and materials (see textbox) that exhibit long-lived phosphorescence [3].

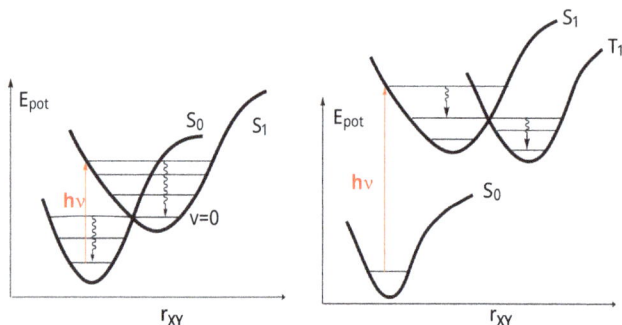

Figure 2.3: Internal conversion (left) and intersystem crossing (right) proceed at the vibrational levels where the potential curves of the relevant states cross.

Daylight makes cycle path light up at night

Phosphorescence occurs frequently in solids. It was first discovered by the cobbler and alchemist Vincenzo Cascariolo from Bologna in the early seventeenth century and marketed as a tourist attraction. He sold a modified mineral under the name "Lapis Solaris" as a collector's item to aristocratic tourists on their European educational "Grand Tour." Two hundred years later, Johann Wolfgang von Goethe briefly held one of these stones in the Sun and watched it light up yellow in a darkened room. Jacopo Beccari, a doctor and chemist also from Bologna, discovered a more expensive version of a luminescent stone (phosphor): when a certain patient came into his surgery from the sunshine into his dimly lit surgery, her diamond ring glowed. Nowadays, luminescent materials have numerous applications. High-purity crystalline materials such as the colourless zinc oxide or zinc sulfide luminesce on doping with silver (blue emission) or copper (green emission). In addition to excitation by daylight or artificial light, luminescence can occur through the action of radiation from radioactive sources (self-luminescent materials). The functional units for numerous applications, from tubes for fluorescent lights to light-emitting diodes (LEDs), from high-vis clothing to security marks on banknotes (Figure 1.8), depend on the principles of luminescence. A recent invention from Mexico is a phosphorescent concrete for roads that stores daylight and lights up in the night. In Holland, a half-kilometre-long cycle track between two watermills has been decorated with 50000 phosphorescent pebbles laid into the surface. They make a green shimmering pattern in the dark, a little like Vincent van Gogh's paint of "A starry night" – an almost mystical effect.

Reproduced with permission © Daan Rosegaarde, www.studioroosegaarde.net

2.1.2 Intensities and rates of photophysical processes

What determines the intensities of electronic transitions and the rates of the processes discussed in the previous section? We have already encountered selection rules (Chapter 1) and seen their influence on both intensities of absorption and rates of emission. We will now delve deeper into the principles. In classical physics, the absorption process is considered as the interaction between the electric field vector of the light wave and the dipole moment of the molecules. In contrast, quantum theory provides an expression for the probability that a molecule occupies an excited state (see ref. [4]). This probability is given by Equation 2.1 in which Ψ_G and Ψ_E represent the complete wavefunctions (core and valence shells) of the ground and excited states (G and E). The oscillator strength $f_{G \rightarrow E}$ corresponds to the effective number of electrons that contribute to a transition and is proportional to the square of the transition dipole moment **M** (written in Dirac notation as $<\Psi_G|\ \hat{\mu}|\ \Psi_E>$ which is a shorthand for an integral over all space) and takes the value of 1 for a fully allowed transition. The dipole moment operator $\hat{\mu}$ is given by $e \sum_i r_i$, where r_i is the distance of the electron e_i from the centre of charge:

$$f_{G \to A} \propto \left[\iiint \Psi_G \times \hat{\mu} \times \Psi_E \, dx \, dy \, dz \right]^2 \tag{2.1}$$

$$f_{G \to A} \propto\ <\Psi_G|\hat{\mu}|\Psi_E>^2$$

Since the mass of the electron is thousands of times smaller than the mass of nuclei, electrons move much faster than nuclei and one can separate the complete wavefunction into the nuclear (θ) and electronic (ψ) wavefunctions according to the Born-Oppenheimer approximation:

$$\Psi = \theta \times \psi \tag{2.2}$$

With the assumption that only one electron is excited in a transition, the full electronic wavefunction can be replaced by a one-electron function (in other words, an orbital), ϕ. This can be split in turn into a spatial wavefunction, φ, and a spin wavefunction, S:

$$\phi = \varphi \times S \tag{2.3}$$

The transition moment **M** can now be expressed as a product of three factors (Equation 2.4). The first two terms correspond to the overlap integral of the wavefunctions for the vibrations of the nuclei in the ground and excited states. The second part is the overlap integral for the spin wavefunctions and the third part is the electronic transition moment:

$$\mathbf{M} = \langle \theta_G | \theta_E \rangle \times \langle S_G | S_A \rangle \times \langle \varphi_G | \hat{\mu} | \varphi_E \rangle \tag{2.4}$$

In the following section, we will look briefly at the contributions of each of these factors and deduce when a transition is allowed (**M** $\neq$ 0) or forbidden (**M** = 0). Even forbidden transitions have some small intensity. For some transitions, the residual intensity arises because of the neglect of the interaction between electronic and nuclear wavefunction. As we have already seen, spin-forbidden transitions arise because of the neglect of spin-orbit coupling.

2.1.3 Qualitative estimates of the transition moment

2.1.3.1 Contribution of the molecular vibrations

The magnitude of the overlap integrals of the molecular vibrations can be visualised quantitatively in Figure 2.4, which shows the change in the potential energy of the molecule X–Y with respect to internuclear distance r_{XY} (Morse curve). The thin straight lines represent the wavefunctions of the lowest vibrational levels, and the thin waves represent the amplitude of the corresponding wavefunctions.[4] Positive amplitudes lie

---

4 The energy gap between the vibrational levels decreases as the vibrational quantum number increases.

above the line of the corresponding vibrational levels, and negative amplitudes lie below the line. According to the Franck-Condon Principle, light absorption can be represented as a vertical arrow since the positions of the nuclei cannot change during this instantaneous process. The resulting vibrational level of the electronic excited state is referred to as the Franck-Condon state. The initial point for the absorption process is the $v = 0$ vibrational level (the lower state is designated without a prime ' and the upper state with a prime '). If the equilibrium internuclear distance is the same in the ground state as in the excited state as shown in Figure 2.4, it can be seen that the overlap integral ($<v|v'>$) decreases as the vibrational quantum number of the upper level increases following the order 0→0, 0→1, 0→2. Correspondingly, the absorption spectrum should show the vibrational fine structure in the lowest absorption band with the most intense component being the 0→0 transition.

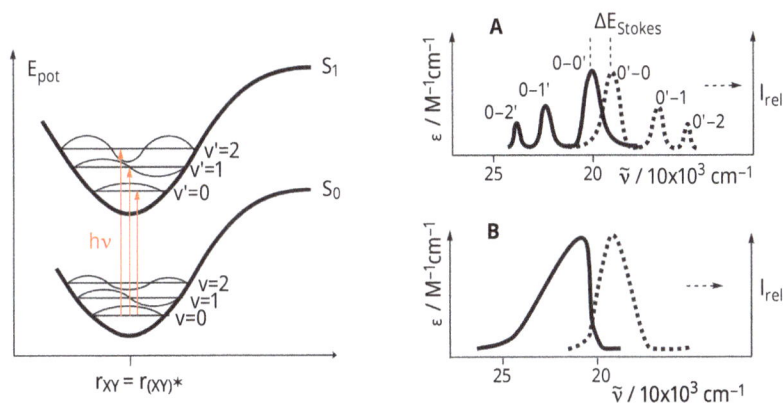

Figure 2.4: The origin of the vibrational fine structure of absorption and emission bands of a molecule X–Y in which the internuclear distance is the same in the ground and excited states. Left: potential energy diagrams; right: schematic spectra (solid lines: absorption spectra; broken lines: emission spectra). (**A**) Vibrational fine structure resolved, and (**B**) broadening obscures the vibrational fine structure. The numbers on the spectra represent the vibrational quantum numbers of the ground and excited states (v, v').

The emission spectrum of such a molecule is the mirror image of the absorption spectrum (Figure 2.4A). Rapid relaxation of the Franck-Condon state to the lowest vibrational level of S_1 underlies Vavilov's rule that emission takes place from $v' = 0$ level (Figure 2.5). Before emission occurs, the solvent shell also has sufficient time to reorganise so as to stabilise the S_1 state. As sketched in Scheme 2.3, the S_1 state is equilibrated with its surroundings. The result of the relaxation of the vibrational levels is that we see transitions in emission from $v' = 0$ to $v = 0$, $v' = 1$ to $v = 0$, $v' = 2$ to $v = 0$, etc. Because of the solvent relaxation, the 0→0 transition may be displaced slightly from its position in absorption – this is called the Stokes shift, ΔE_{Stokes}. In the absence of fine structure, the Stokes shift is taken as the separation of the absorption and emission maxima.

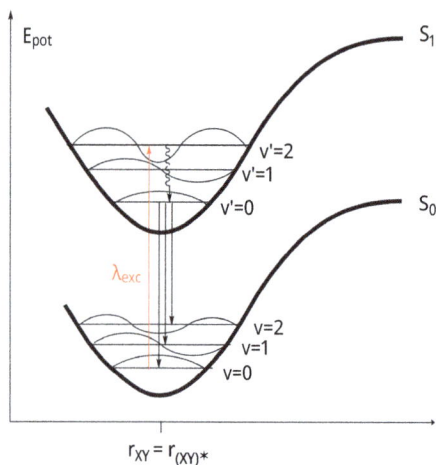

Figure 2.5: Primary processes of fluorescence leading to emission as the mirror image of absorption.

When there is no change in geometry between ground and excited electronic states as described here, the 0→0 transition is dominant and may be the only transition present in absorption and emission. Moreover, the bands are often very narrow. These features can be seen in the spin-forbidden transitions between quartet (ground) and doublet excited states of Cr(III) complexes such as $[Cr(urea)_6]^{3+}$ and $[Cr(N,N,N)]^{3+}$ (N,N,N is a chelating tris-pyridine derivative) [5]. In the case of more complex molecules, especially in polar solvents or cases of non-rigidity, a broader but asymmetric band may be observed (Figure 2.4B).

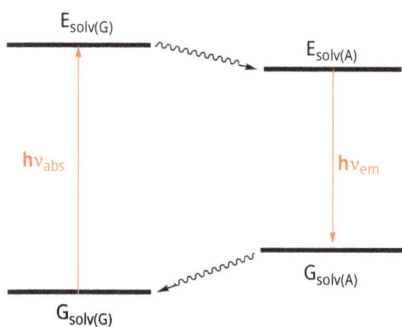

Scheme 2.3: The rearrangement of the solvent shell before the emission leads to a red shift (shift to lower energy) of the emission (Stokes shift). The magnitude of the effect is usually smaller than shown.

When the geometries of the ground and excited states are not the same, the S_1 state is displaced relative to the S_0 state (Figure 2.6A). We still may see a vibrational progression, but the most probable transition is not between $v = 0$ and $v' = 0$ but between $v = 0$ and $v' = n$ ($n > 0$) as shown in Figure 2.6B. The separation between the maxima corresponds to the wavenumbers of the vibration in the excited state that is most affected by the electronic transition. For acetone, the separation is ca. 1200 cm^{-1}, indicating that the electronic transition is located on the keto group. According to IR spectroscopy, the ground state vibration is found at ca. 1700 cm^{-1}, indicating that the C=O stretching vibration of acetone is substantially reduced in the S_1 excited state, as expected from the population of the π^* orbital. Thus, the absorption spectrum provides valuable information about the electronic structure of the *excited state*. Fine structure can clearly be seen in aqueous solution for the five-atom, tetrahedral permanganate ion (Figure 2.7). In this example, the fully allowed transition between S_0 and S_1 transfers an electron from the non-bonding O(2p) lone pairs to the 3d orbitals. The vibrational structure arises from excitation of the Mn–O symmetric stretching vibration that is reduced from 839 cm^{-1} in the ground state, as obtained from Raman spectroscopy, to 683 cm^{-1} in the excited state measured from the average separation of the vibrational peaks in this spectrum. For routine measurements on larger, low-symmetry molecules in which multiple vibrational states are excited, the vibrational fine structure is replaced by broad symmetric bands, indicating that there is a change in structure between ground and excited states. The same may apply in polar solvents or for non-rigid molecules (Figure 2.6C).

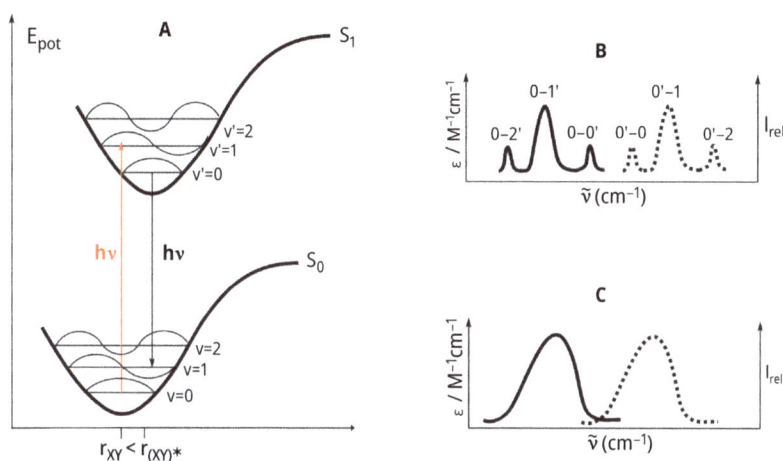

Figure 2.6: (**A**) Potential energy curves for a molecule X–Y where the internuclear distance is greater in the excited state than in the ground state. (**B**) Schematic of resulting absorption and emission spectra showing vibrational fine structure. (**C**) Schematic of absorption and emission spectra where the bands are broadened, so obscuring vibrational fine structure.

When a molecule is at room temperature, according to the Boltzmann distribution, many low-frequency vibrational levels of the ground electronic state may be populated since $kT \sim 200 \text{ cm}^{-1}$ at room temperature. Consequently, lowering the temperature reduces the population of these vibrational excited states and sharpens the electronic absorption bands. Further sharpening may be achieved in rigid media where motion of the surrounding molecules is minimised. This situation is illustrated by the laser-induced fluorescence of decamethylrhenocene ($Re(C_5Me_5)_2$) at 10 K in

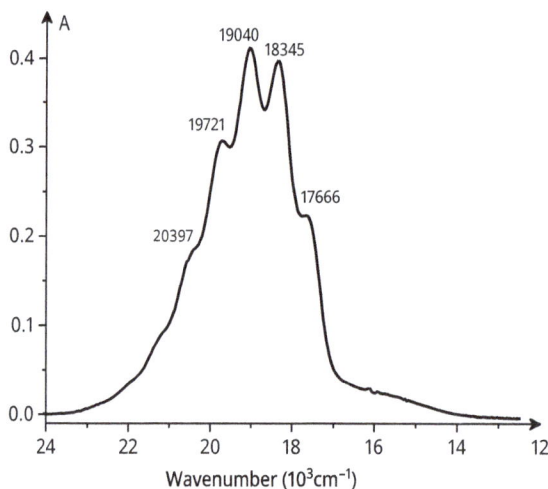

Figure 2.7: Absorption spectrum of potassium permanganate in water at 300 K showing vibrational fine structure.

solid argon that shows exceptionally sharp vibrational fine structure in this rigid, non-polar medium with almost no Stokes shift (≤ 2 cm^{-1}). The vibrations correspond to the symmetric ring–Re–ring stretching motion that is increased by 9 cm^{-1} in the excited state (Figure 2.8) [6, 7].

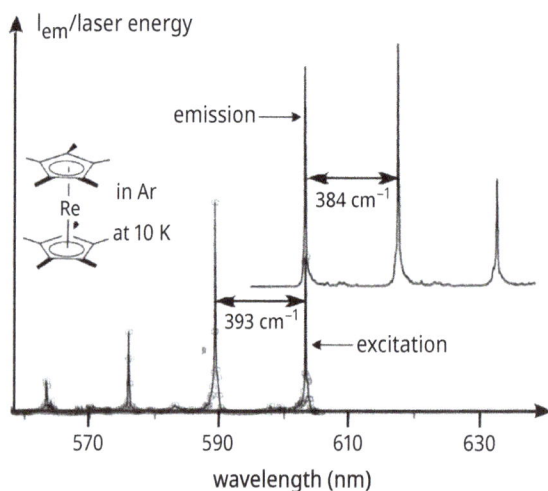

Figure 2.8: Laser-induced fluorescence of decamethylrhenocene (Re(C$_5$Me$_5$)$_2$) at 10 K in solid argon. Left: excitation spectrum measured at emission wavelength of 617.7 nm (16190 cm^{-1}); right: emission spectrum measured at excitation wavelength of 568.2 nm (17599 cm^{-1}). Adapted with permission from ref. [7]. Copyright 1996 American Chemical Society.

In summary, through their shape and vibrational fine structure, absorption bands provide indications of structural changes in excited molecules and in the nature of their chromophores.

2.1.3.2 Contribution of electron spin

According to quantum theory, the overlap integral is equal to 1 when the ground and excited states have the same spin: the transition is then spin allowed – this is one aspect of Wigner's Spin Conservation Rule for both radiative and non-radiative processes. It is spin forbidden when the states have different spins, for instance singlet-triplet transitions as in phosphorescence (Scheme 2.1). This is only true, however, when the interaction between spin and orbital angular momentum of the electrons, the spin-orbit coupling, is negligible. The spin-orbit coupling constant grows with the fourth power of atomic number.[5] As a result, the transition becomes less forbidden and its absorption coefficient increases. If the hydrogen atoms in the 9 and 10 positions of anthracene are replaced by halogen atoms, the absorption coefficient of the S_0-T_1 band doubles (*internal heavy atom effect*). Instead of placing a heavy atom within the molecules, if a solvent is used along with a heavy atom, then the spin-orbit interaction can also increase (*external heavy atom effect*). Even measurement under an inert gas such as xenon can enhance the band. A prominent heavy atom effect is thus a good indicator of an electronic transition with a change of spin.

2.1.3.3 Contribution of the electronic transition moment

The symmetry properties of the electronic wavefunction play a decisive role in determining whether a transition is allowed. Use of group theory allows us to determine if the integrals $\langle \psi_G \, |\hat{\mu}| \, \psi_E \rangle$ are non-zero without evaluating them.[6] We make use of the point group appropriate to the symmetry of the molecule under investigation in conjunction with the irreducible representations of the ground and excited *states* (not orbitals). The transition is allowed when the triple direct product $\psi_G \times \hat{\mu} \times \psi_E$ contains the totally symmetric representation. This can be simplified to determining if the direct product $\psi_G \times \psi_E$ contains the same irreducible representation as one of the three Cartesian coordinates (corresponding to those for $\hat{\mu}$). This result also reveals the direction of the transition within the molecule, something that can be exploited when studying substances with polarised light. The selection rule for the special case of a molecule with a centre of inversion[7] is known as the Laporte Rule: transitions are allowed between electronic states of different symmetry with respect to inversion, g→u

---

5 Values for neutral atoms of group 14: C 29 cm^{-1}, Si 210 cm^{-1}, Ge 1450 cm^{-1}, Sn 4100 cm^{-1}, Pb 13000 cm^{-1} [8].

6 For details of group-theoretical methods of analysing chemical problems, see ref. [9].

7 A molecule possesses the symmetry operation called a centre of inversion if it is indistinguishable after the coordinates of each atom, x, y, z, have been replaced by $-x$, $-y$, $-z$. Examples are the point

and u→g, but forbidden when they have the same symmetry. In addition to the symmetry criteria, the magnitude of absorption coefficients, ε, is affected by the degree of overlap of the orbitals in question. As shown in Figure 2.1, the value of ε is significantly higher for π,π^* transition than for n,π^* because of the greater separation of electron distributions for the latter.

Violation of these selection rules is enabled by coupling to vibrations that distort the molecule, thereby reducing its symmetry (vibronic transitions). The rules are obeyed less rigorously than the spin selection rule, meaning that absorption coefficients of electronically forbidden bands are larger than those of spin-forbidden bands.

2.1.4 Experimental determination of oscillator strength

We introduced the oscillator strength, $f_{G \to E}$, as the theoretical measure of the intensity of a transition. The quantum mechanical analysis of oscillator strength leads to Equation 2.5, according to which $f_{G \to E}$ is proportional to the area of the entire absorption band. If the band is symmetrical, its area is approximately equal to the product of its full width at half-height, $\Delta \tilde{v}_{1/2}$ (abbreviated *half-width*), times its absorption coefficient ε at the maximum (Equation 2.6, Figure 2.9):

$$f_{G \to E} = \text{const.} \int_{v_1}^{v_2} \varepsilon(v)dv \qquad (2.5)$$

$$f_{G \to E} \cong \text{const.} \times \varepsilon(\tilde{v}_{max}) \times \Delta \tilde{v}_{1/2} \qquad (2.6)$$

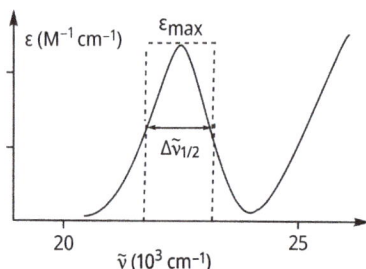

Figure 2.9: Definition of half-width of a symmetrical absorption band.

There is also a relationship between the rate of light emission and the oscillator strength. If it is assumed that the excited state decays by emission only, quantum theory leads to the result that the natural[8] rate constant for emission, $k_{em(nat)}$, is given by Equation 2.7.

---

groups D_{2h} and O_h. An irreducible representation of the corresponding point group may be symmetric or antisymmetric with respect to inversion, denoted as gerade (g) or ungerade (u).

8 The word "natural" means that the excited state only decays by light emission.

This leads to Equation 2.8 that shows that this rate can be derived from the oscillator strength and ε_{max}. We saw in Chapter 1 that the lifetime of the excited state, τ, is defined as the inverse of the rate constant for decay. Thus, the natural lifetime (Equation 2.9) is inversely proportional to the absorption coefficient which can be readily measured from the absorption spectrum. The smaller the absorption coefficient, the longer-lived is the excited state, and the higher the probability of interaction and bonding with a reaction partner:

$$k_{em(nat)} = \text{const.} \; \tilde{v}^2 \int v dv \cong \text{const.} \times \tilde{v}^2 \times \varepsilon_{max} \Delta \tilde{v}_{1/2} \tag{2.7}$$

$$k_{em(nat)} \cong f_{G \rightarrow E} \times \tilde{v}^2_{max} \tag{2.8}$$

$$\tau_{nat} = 1/k_{em(nat)} = 1/(f_{G \rightarrow E} \times \tilde{v}^2_{max}) \tag{2.9}$$

2.2 Primary photochemical processes and quenching of excited states

2.2.1 Prompt photochemical processes and equilibrated excited states

The four classes of events that can dispose of the excitation energy following light absorption were shown in Figure 1.4: heating (radiationless decay), luminescence, energy transfer (EnT) and chemical reaction. We describe the first three as photophysical and the fourth as *primary* photochemical processes – that is processes that occur directly from the excited state. We understand "chemical reactions" to mean cleavage and formation of chemical bonds, leading to reaction intermediates and products; electron transfer is also included. In Section 2.1.1, we introduced the concept of the *equilibrated excited state* when we considered the Jablonski diagram. The time required to reach the equilibrium geometry and for the solvent to rearrange around it is of the order 10^{-11} to 10^{-10} s. We now know that some chemical reactions occur in a much shorter time, often in less than 1 ps (10^{-12} s). These are *prompt* photochemical processes. Diffusion occurs on much slower timescales than this, meaning that there is no time for intermolecular reaction except with the solvent. Thus, we observe intramolecular photochemical events occurring by unimolecular mechanisms on the picosecond timescale. The main prompt photochemical reactions are dissociation, isomerisation and intramolecular charge-transfer, shown for a hypothetical molecule X–Y in Scheme 2.4. Famous examples are the photodissociation of ICN,[9] photodissociation of $Cr(CO)_6$ and ring-opening photo-

---

[9] Dissociation of ICN was the first photoreaction to be studied on a femtosecond timescale revealing coherent oscillations. It was studied by Egyptian-born Ahmed Zewail (1946–2016) who won the Nobel Prize in 1999 [10].

isomerisation of cyclohexadiene to hexatriene. Luminescence from non-equilibrated excited states occurs with very low quantum yields, if at all.

Prompt photoprocesses

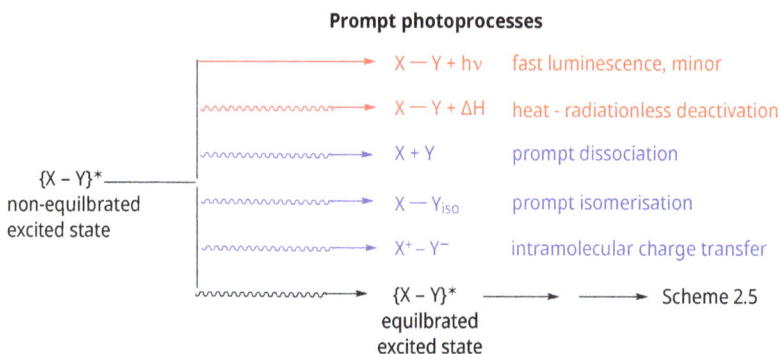

Scheme 2.4: Primary processes of an excited molecule {X–Y}* that has not equilibrated with its surroundings. Photophysical processes in red and photochemical processes there is a typo in radiationless in blue.

Associated with the concepts of prompt photoreaction and equilibrated excited states are the ideas of adiabatic and non-adiabatic photochemical reactions. In an adiabatic photoreaction, the reaction takes place on a single potential energy surface, whereas a non-adiabatic reaction requires a change of potential energy surface (a change in surface is shown in Figure 2.3). As we have seen, our descriptions of the potential energy surfaces in Figures 2.3–2.6 depend on the ability to separate the nuclear motion from the electronic motion (the Born-Oppenheimer approximation). Likewise, it is assumed that the Born-Oppenheimer approximation applies when distinguishing adiabatic and non-adiabatic photoreactions. Most prompt photoreactions are non-adiabatic – with the aid of the latest ultrafast spectroscopy, it is even possible to identify the sequence of changes in potential energy surface. In some of these reactions, the bond lengths start to change as soon as the photon is absorbed (within 10^{-14} to 10^{-13} s), as if the molecule had been hit by a bullet. The extraordinary feature is that this bullet is a photon that has energy but no mass. Such reactions involve a direct coupling of nuclear and electronic motion, in other words a breakdown of the Born-Oppenheimer approximation.

Next, we look more closely at the processes that occur from an *equilibrated* excited state for a hypothetical hydrogen-containing molecule {**G–H**}* in the presence of an acceptor **A** (Scheme 2.5). The latter can accept energy, electrons, protons or hydrogen atoms. The chemical reactions can now be both unimolecular (intramolecular) or bimolecular (intermolecular). Equilibrated excited states are typical for luminescence as we have described earlier.

Photoprocesses of equilibrated excited states

G — H + hν	luminescence
G — H + ΔH	heat - radiationless deactivation
G — H + A*	energy transfer
P_{intra}	intramolecular reaction (isomerisation)
P_{inter}	intermolecular reaction
[G–H]$^{•+}$+ A$^-$	oxidative electron transfer
[G–H]$^{•-}$+ A$^+$	reductive electron transfer
G$^-$ + [A–H]$^+$	proton transfer
G$^•$+ [A–H]$^•$	hydrogen atom transfer (HAT)

{G–H}* equilibrated

A Substrate A A A A

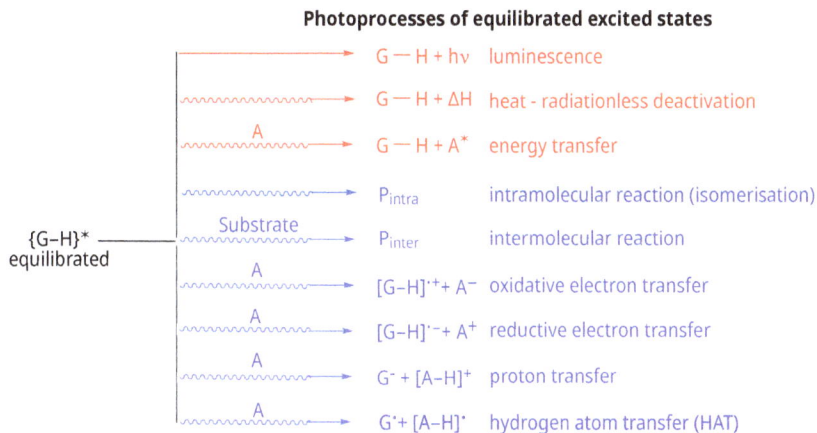

Scheme 2.5: Primary processes of an equilibrated excited state {G–H}*. Photophysical processes in red and photochemical processes in blue.

2.2.2 Quenching processes and their kinetics

In condensed phases, there are always interactions with unreactive compounds (*quenchers*) that reduce the lifetime of excited states, a process described as *quenching*. These radiationless decay processes almost always have a stronger temperature dependence, i.e. they have a higher activation energy, than radiative decay. Emission of light is often unobservable at room temperature but becomes observable at low temperature as a result of the slowing of radiationless processes (see textbox).

Lighting up cold hands

J. Beccari, an eighteenth-century doctor from Bologna, remarked on the difference between radiative and non-radiative processes when he observed the phosphorescence of human proteins. Only when the hands of his patients were washed with cold water and then briefly held in the Sun, did they light up on returning to a dark room.

One of the most efficient quenchers is dissolved molecular oxygen, especially of triplet states. Since oxygen has a triplet ground state (Scheme 1.4), it is capable of converting excited triplet state molecules to their ground states and forming singlet oxygen $^1O_2$ if the energetics are right. Not only does this quench the excited state but it supplies $^1O_2$, which is a powerful oxidising agent (see Section 2.2.4). For this reason, luminescence measurements should be made in oxygen-free solution.[10] Important deductions about

---

10 Saturation with N_2 is insufficient. Several freeze-pump-thaw cycles are needed for complete degassing: freezing with liquid nitrogen to 77 K, evacuation, thawing to allow degassing, then repetition of the whole process.

spin multiplicity and lifetime of reactive excited states can be made from the concentration dependence of quenching effects. For the simplest case, a unimolecular reaction **G** + hv → **P**, the following can be assumed:
(a) The quenching of **G*** proceeds by free diffusion.
(b) The reaction of **G*** to form product **P** proceeds only from the lowest singlet state, S_1.
(c) In addition to the formation of product, **G*** decays by radiative and non-radiative deactivation.

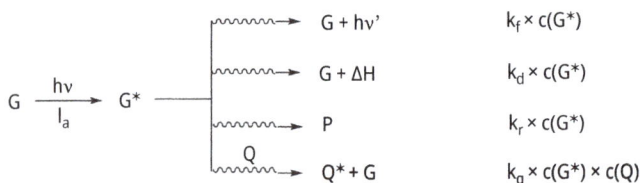

$$
G \xrightarrow[I_a]{hv} G^*
\begin{cases}
\longrightarrow G + hv' & k_f \times c(G^*) \\
\longrightarrow G + \Delta H & k_d \times c(G^*) \\
\longrightarrow P & k_r \times c(G^*) \\
\xrightarrow{Q} Q^* + G & k_q \times c(G^*) \times c(Q)
\end{cases}
$$

Scheme 2.6: Formation and deactivation of G* and their rates (c, concentration).

Scheme 2.6 summarises these processes and their rates. For unimolecular reactions, the rate of product formation is determined by the rate constants and the concentration of **G***; for the bimolecular quenching reaction, the rate is dependent on the concentrations of **G*** and quencher **Q**. In order to simplify the kinetic analysis of these multistep reactions, we assume that the concentration of a short-lived intermediate can be considered as a quasi-stationary state. That means that the rate of formation of **G*** (I_a) is equal to its rate of decay during continuous irradiation (Equation 2.10):

$$I_a = c(G^*)\left[k_f + k_d + k_r + k_q \times c(Q)\right] \tag{2.10}$$

$$\Phi_p = \frac{k_r}{k_f + k_d + k_r + k_q \times c(Q)} \tag{2.11}$$

$$\Phi_p{}^\circ = \frac{k_r}{k_f + k_d + k_r} \tag{2.12}$$

$$\frac{\Phi_p{}^\circ}{\Phi_p} = 1 + k_q \times \tau \times c(Q) \tag{2.13}$$

$$\frac{\Phi_p{}^\circ}{\Phi_p} = 1 + k_{SV} \times c(Q) \tag{2.14}$$

Since the quantum yield of the product, Φ_P, is given by the rate of reaction divided by I_a (Chapter 1), we obtain the expressions for quantum yields in Equations 2.11 and 2.12 for the reactions with and without quencher (note that I_a cancels out). If the excited state decays only by unimolecular processes, its lifetime τ is given by $1/(k_f + k_d + k_r)$. The ratio of the quantum yield with quencher to that without quencher is given by Equation

2.13; this can also be written as Equation 2.14. Both forms are described as the Stern-Volmer equation. The Stern-Volmer constant $K_{SV} = k_q \times \tau$ has dimensions M^{-1}.[11] If the quantum yield is measured without quencher and in the presence of increasing concentrations of quencher and the resulting values of Φ_p^0/Φ_p are plotted against the concentration of Q, a straight line is obtained with gradient K_{SV} (Figure 2.10). If, in addition, the lifetime of S_1 is known, the rate constant for quenching, k_q, can be obtained. Sometimes a curve is obtained instead of a straight line, indicating that conditions (a)–(c) are not fulfilled. The Stern-Volmer method can be used equally well to study the quenching of luminescence, whether of a singlet or a triplet state.

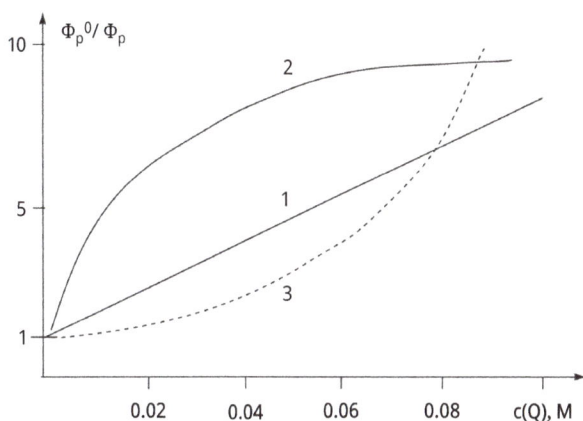

Figure 2.10: Stern-Volmer diagram: (1) reaction from S_1, (2) reaction from S_1 and T_1, (3) static quenching, i.e. the reactant and the quencher molecules form a complex before excitation.

Equation 2.14 can be modified for the special case that exactly half of the molecules G^* are quenched, i.e. $\Phi_p^0/\Phi_p = 2$, resulting in Equation 2.15. This version is useful for selecting the optimum conditions for experiments. Supposing that k_q takes the maximum value for a diffusion-controlled reaction ($k_q \approx 1 \times 10^{10}\ M^{-1}\ s^{-1}$ for hexane at water at ambient temperature, but depends on temperature and viscosity),[12] and the lifetime of S_1 is 10^{-9} s, the concentration required for half quenching is 0.1 M. In contrast, a triplet state with a lifetime of 10^{-5} s requires a concentration of 10^{-5} M:

$$c(Q)_{1/2} = 1/K_{SV} = (k_q \times \tau)^{-1} \qquad (2.15)$$

In many cases, the excited state that leads to photoreaction has not been identified and the primary process is unknown (see Scheme 2.5). Quenching experiments with

---

11 The dimensions of k_q, τ and $c(Q)$ are $M^{-1}\ s^{-1}$, s^{-1} and M, respectively.
12 The value of the diffusion-controlled rate constant k_{diff} can be estimated from the Stokes-Einstein equation as $8RT/3\eta$, where η is the viscosity in centipoise.

quenchers of a variety of properties designed following the principles below can deliver essential mechanistic information:

- Find a compound that quenches the reaction.
- Does this compound also quench the emission?
- Choose quenchers with different redox properties.
- Choose quenchers with different S_1 and T_1 energies.

2.2.3 Energy transfer mechanisms

As in chemistry without light, the transfer of energy or electrons plays a fundamental
role in photochemistry. We first tackle energy transfer that can be described by the
following equation, in which a donor **D** in an excited state transfers its energy directly
to an acceptor **A** so that **A** becomes excited:

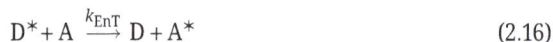

$$D^* + A \xrightarrow{k_{EnT}} D + A^* \tag{2.16}$$

There are four mechanisms for energy transfer:

1. Radiative mechanism (trivial mechanism). This is the trivial case in which the
donor **D** (the sun or a lamp) transfers its energy directly to the acceptor **A** (e.g. chlorophyll and rhodopsin). This requires the emission bands of the donor **D** to overlap with
absorption bands of the acceptor **A**.

2. Coulomb mechanism (also known as the resonance, dipole-dipole or *Förster*
***mechanism*, and in biology as *Förster resonance energy transfer (FRET))*.** Almost
every molecule possesses a dipole moment in its excited state that can alter the charge
distribution of an acceptor molecule. This dipole-induced mechanism is the basis of Förster energy transfer. In contrast to the trivial mechanism, the influence is strongly dependent on the distance between **D** and **A**, $r_{D\cdots A}$, and cannot be greater than about
10 nm. Like the trivial mechanism, however, the emission band of **D** must overlap with
absorption band of **A**. The rate constant for this energy transfer is proportional to the
square of the energy of the Coulomb (dipole-dipole) interaction[13] (Equation 2.17). This
rate includes the inverse sixth power of the distance $r_{D\cdots A}$, the absorption coefficient ε_A
of **A** and the overlap integral of the two bands, J, which is large for S→S, but small for
S→T transfer. Consequently, transfer is only "allowed" between states of the same multiplicity, but T→T transfer is unlikely because of the low value of ε_A. Förster energy transfer takes place between states of identical energy, hence, the term resonance. Like ISC in
Scheme 2.2, the energy is transferred horizontally between the donor **D** in its $v = 0$ level
of S_1 and a vibrationally excited level of the S_1 state of the acceptor, **A** (Scheme 2.7):

---

[13] The dipole-dipole interaction depends on the inverse cube of the distance between the dipoles.

$$k_{\text{EnT}} \propto (E_{\text{Coulomb}})^2 \propto [k_{\text{diff}} \times \varepsilon_{\text{A}} \times J / (r_{\text{D-A}})^6] \tag{2.17}$$

If energy transfer were to take place through collisions as in the quenching mechanism described above, the maximum value of the rate of energy transfer would be the diffusion-controlled rate constant, k_{diff}. Dipole-dipole energy transfer takes place over a longer range than the collisions. An experimental indication for the Förster mechanism is therefore a value of k_{EnT} that is greater than the diffusion-controlled rate constant, k_{diff} (see Section 2.1.5).

Scheme 2.7: Förster mechanism for dipole-induced dipole energy transfer.

3. **Exchange or Dexter mechanism**. This process can be considered as double electron transfer via the overlap of molecular orbitals (Scheme 2.8). In this case, the distance $r_{\text{D-A}}$ must be no more than 1 nm in order to achieve the overlap. The energy requirement is that the excitation energy of **D** exceeds that of **A**: $\Delta E(\mathbf{D}^*-\mathbf{D}) \geq \Delta E(\mathbf{A}^*-\mathbf{A})$. Like the Förster mechanism, transfer between states of the same multiplicity is allowed, but unlike Förster, k_{EnT} is now dependent on the solvent because diffusion of the partners is required for the close encounter. The Dexter mechanism provides the principal route for T→T energy transfer. This process is also described as triplet sensitisation and is of fundamental importance since an acceptor molecule **A** can be transferred to its triplet state without consideration of symmetry or spin selection rules (see Section 2.2.2).

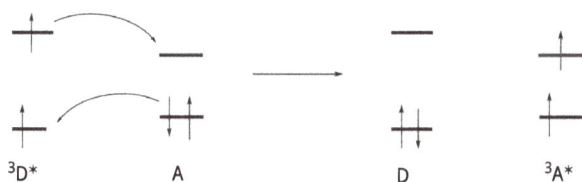

Scheme 2.8: Diagrammatic representation of energy transfer by the Dexter mechanism.

4. **Energy hopping or exciton transfer**. There is frequently a language gap between molecular and solid-state scientists. The world of solids refers to *excitons* instead of excited states. The excitons consist of an electron-hole pair just as a molecular excited state, but in solids the pair can be strongly or weakly coupled, depending on the solid con-

cerned. Excitons hop very rapidly through the solid until they are trapped at a defect where energy or electron transfer to substrate occurs.

In each of these four mechanisms, the Wigner Spin Conservation Rule applies to the *overall* process. This rule applies not only for photophysical processes but also for any elementary chemical reaction. That means the sum of spin states of reactants must be identical with that of all the products [1]. This rule has important consequences, for example, for the reactivity of oxygen, whether in its triplet or singlet state. Examples of allowed and forbidden processes involving singlet and triplet states are shown below:

$$S + T^* \rightarrow T^* + S \qquad \text{allowed}$$

$$S^* + S \rightarrow S + S^* \qquad \text{allowed}$$

$$S + T^* \rightarrow S^* + S \qquad \text{forbidden}$$

A further process $(T + T^*)$ is discussed in detail in Section 2.2.4.

2.2.3.1 Examples

A simple example of *Förster energy transfer* (FRET) can be found for the two dyes, coumarin (donor) and sodium fluorescein (acceptor), whose absorption bands overlap (Figure 2.11). The measured value of k_{EnT} in alkaline ethanol solution is ~2.5×10^{11} M^{-1} s^{-1}, far higher than the value of k_{diff} which is 6×10^9 M^{-1} s^{-1}, a clear indication of energy transfer. The average distance separating **D** and **A** for energy transfer is estimated as 5.3 nm [1]. Förster energy transfer has become a mainstay of cell and structural biology driven by the ability to install fluorescent tags. It is used for measuring distances between domains in proteins, for analysing signalling pathways, for investigating the binding of ions and small molecules to donor or acceptor and many others.

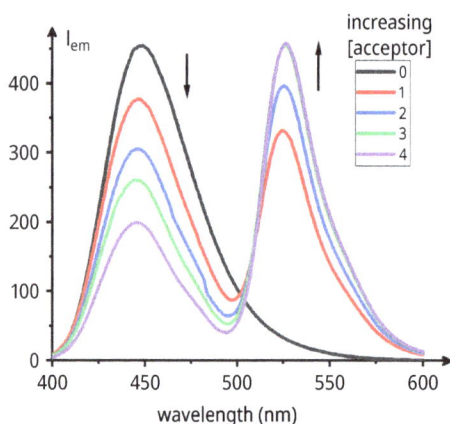

Figure 2.11: Fluorescence spectra following the energy transfer from excited coumarin (**D**) to sodium fluorescein (**A**) in ethanol containing NaOH. The concentration of **D** is identical in each run. The concentration of **A** is increased from zero in the black spectrum and increases in the order $1 < 2 < 3 < 4$.

Dexter energy transfer is the principal mechanism of triplet photosensitisation and is discussed further in Section 2.2.4. It is illustrated by the transfer from ^{3}(benzophenone) to naphthalene that occurs with $k_{EnT} = 1 \times 10^{10}$ M^{-1} s^{-1} in benzene solution, a similar rate to k_{diff} (Equation 2.18):

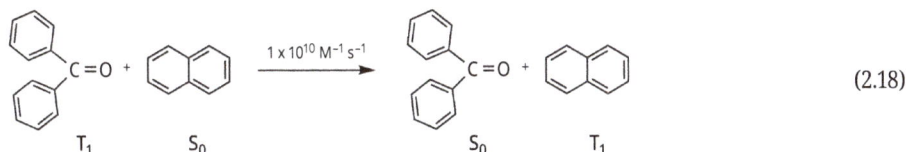

$$(2.18)$$

The most important example of *energy hopping* is found in photosynthesis: the transfer between antenna pigments and from them to the photosynthetic reaction centre in green plants (Section 6.1) [11]. The structure of one of the light-harvesting proteins is also shown in Figure 2.12.

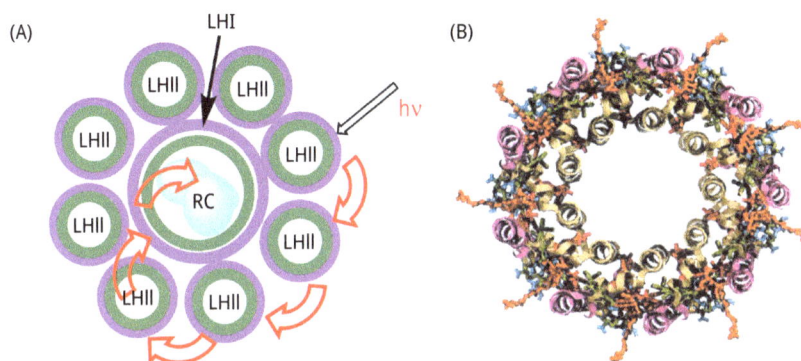

Figure 2.12: (A) Simplified model of energy hopping (red arrows) in purple bacteria between antenna pigments LHII, LHI and the reaction centre (RC) in photosynthesis. (B) Molecular structure of the light-harvesting protein LHII from purple bacteria showing ninefold symmetry with polypeptide chains in ochre and pink, bacteriochlorophyll in green, red and blue, and carotenoid in orange. There are two sets of bacteriochlorophyll, 18 of one type and 9 of another, arranged perpendicular to the first set. Reproduced with permission from ref. [12]. Copyright Federation of European Biochemical Societies and Wiley and Sons.

2.2.4 Triplet-triplet annihilation and singlet fission

In the above-mentioned examples, energy is transferred from an excited triplet to a singlet. An important alternative is for two triplets to interact to generate a ground state singlet and an excited singlet (Equation 2.19). This process is usually energetically favourable and obeys spin conservation (simplistically, consider one triplet as having two spin-up electrons and the other as having two spin-down electrons). It is called triplet-triplet annihilation (TTA) and is an important quenching mechanism, particularly for long-lived triplets:

$$T_1 + T_1 \rightarrow S_1 + S_0 \tag{2.19}$$

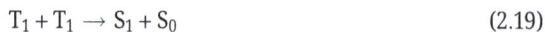

The resulting excited singlet may decay by fluorescence, so creating fluorescence at later times than the normal fluorescence lifetime, "delayed fluorescence." The conversion of ground state oxygen to singlet oxygen (Equation 2.20) represents a special case of triplet-triplet annihilation since the ground state of oxygen is a triplet and the excited state is a singlet (see Sections 4.2.5 and 6.2):

$$^3O_2 + T_1(\text{sensitiser}) \rightarrow {}^1O_2 + S_0(\text{sensitiser}) \tag{2.20}$$

In triplet-triplet annihilation, two photons generate one excited state. The reverse process in which one photon affords two excited states is called singlet fission (Equation 2.21 and Scheme 2.9) [13]. It requires that the excitation energy to form the singlet is at least twice the energy to form the triplet. Furthermore, competitive processes like intersystem crossing and chemical reactions should by highly disfavoured:

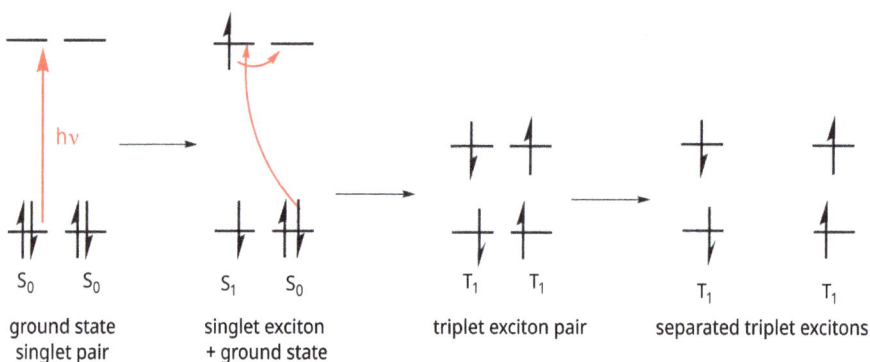

$$S_1 + S_0 \rightarrow {}^1\{T_1\ T_1\} \rightarrow T_1 + T_1 \tag{2.21}$$

Scheme 2.9: A simplified mechanism for singlet fission involving a pair of molecules. The red arrows indicate charge transfer between the molecules.

The molecular and solid-state structure of fused aromatics like tetracene and pentacene and their derivatives partly fulfil these criteria. Very many efficient singlet fission chromophores are of this type. Phosphorescence quantum yields exceeding 150% have been recorded (the ideal value would be 200%) [14]. This conversion of one short-wavelength photon into two long-wavelength photons, each of half the energy of the initial photon, could improve the energy utilisation of photovoltaic systems [13b].

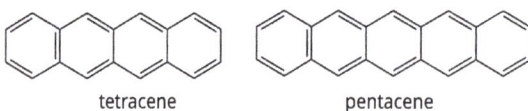

tetracene   pentacene

2.2.5 Thermally activated delayed fluorescence

In the previous section, we looked at singlet fission that depends on a very large singlet-triplet energy gap. Now we look at a situation where the gap is very small, and the lifetime of the triplet state is long enough to equilibrate with the singlet state. According to the Jablonski diagram (Scheme 2.2), ISC converts a short-lived fluorescent singlet to a phosphorescent triplet state. However, when the gap between the triplet and the singlet excited states, ΔE_{ST}, is very small (typically 0.2–8.0 cm^{-1}, 2–12 kJ mol^{-1}), the singlet excited state may be repopulated by thermal equilibration: this is reverse intersystem crossing (RISC) and leads to fluorescence at much later times than normal, thermally activated delayed fluorescence (TADF, Scheme 2.10). Three key characteristics for TADF are the long apparent lifetime, the temperature dependence of the lifetime and the redshift of the emission on lowering the temperature. The actual lifetime for singlet emission without RISC is usually very short. As the system is cooled, RISC becomes less probable and triplet emission takes over; the triplet emission maximum lies at longer wavelength and is much longer lived than the singlet emission.

The dye eosin provided the first example of TADF;[14] since then many more organic examples have been investigated. They are typically based on molecules containing a donor and an acceptor unit, where an electron is transferred from donor to acceptor in the excited state (Scheme 2.10). Steric hindrance forces the two units to be orthogonal to one another such that the overlap between the HOMO and LUMO is small and ΔE_{ST} is also small (see Section 2.3.1.2). Numerous sterically hindered Cu(I) complexes have also been shown to exhibit the phenomenon. The ideal geometry of Cu(I) complexes is tetrahedral, while Cu(II) complexes are ideally square planar. Steric hindrance forces an intermediate geometry allowing excitation to the metal-to-ligand charge-transfer state with little distortion (see Franck-Condon Principle and Section 2.3.2.2). The use of copper complexes ensures that the rate of ISC is fast because of moderate spin-orbit coupling. Compounds exhibiting TADF are advantageous for organic light-emitting diodes (OLEDs, see Section 3.2) and for light-emitting electrochemical cells (LECs) [15].

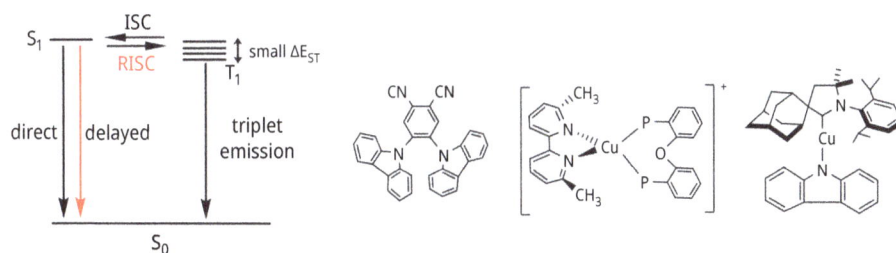

Scheme 2.10: Principle of thermally activated delayed fluorescence (TADF) with examples.

---

14 TADF was discovered by Parker and Hatchard in 1961, the same authors who invented the ferrioxalate actinometer.

2.2.6 Excited state electron transfer mechanisms

In addition to energy transfer, an excited state **GH*** can exchange electrons with an acceptor or donor. The direction of this process depends on the redox potentials of the partners (E_{ox} and E_{red}) since a redox potential is simply related to free energy via the following equation:[15]

$$\Delta G = -n \times F \times E_{redox} \tag{2.22}$$

where n is the number of electrons transferred and F is the Faraday (96500 C V^{-1}). If the potential of **GH*** is more negative than that of **A**, an electron will be given from **GH*** to **A** (**GH*** is oxidised) and if the potential of **D** is more negative than that of **GH*** an electron will be given to **GH*** (**GH*** is reduced, Scheme 2.5). Redox potentials of molecules in their ground state are relatively easily measured by electrochemical methods (in non-aqueous solution, typically by cyclic voltammetry) provided that they are reversible. The value for the excited state **GH*** depends on $E_{0,0}$ for the equilibrated excited state (i.e. the transition for the vibrational ground state $v = 0$ to the vibrational upper state $v' = 0$) via Equations 2.23 and 2.24, where $E_{0,0}$ is expressed in eV. A difficulty arises because $E_{0,0}$ can rarely be measured directly and must be estimated from the corresponding absorption and emission spectra. If ISC has occurred, as is the case commonly with transition metal complexes, it is $E_{0,0}$ for the triplet state that is required:

$$E°(G^+/G^*) = E°(G^+/G) - E_{0,0} \tag{2.23}$$

$$E°(G^*/G^-) = E°(G/G^-) + E_{0,0} \tag{2.24}$$

Applying these equations shows that **G*** is both a better reducing agent and a better oxidising agent than the ground state. This astonishing finding can be understood from the frontier orbitals of the excited state shown in Scheme 2.11. The HOMO-LUMO gap of the ground state is equated with $E_{0,0}$. The potential for oxidising **G** corresponds to the ionisation energy which is reduced in the excited state, and the potential for reducing **G** corresponds to the electron affinity which is increased in the excited state. The rates of electron transfer (ET) are linked to the free energy of electron transfer (the driving force, ΔG^0) in Marcus theory.[16] The Marcus equation (2.25) relates the free energy of activation $\Delta G^{\ddagger}$ for electron transfer to ΔG^0 and the rearrangement energy, λ [17]. The rearrangement energy includes both changes in the geometry of the reactants and

---

15 Redox potentials are defined for reduction, i.e. oxidised species + ne$^-$ → reduced species. Standard redox potentials E^0 are defined for standard conditions, i.e. 1 M concentrations, gas pressures of 1 atm and 298 K. They are altered under non-standard conditions according to $E = E^0 - (RT/nF)\ln Q$, where Q is the reaction quotient [16].
16 Rudy Marcus (born 1923) received the Nobel Prize in 1992 for the development of the theory of electron transfer. His first major paper on the theory was published in 1956 but it wasn't until 25 years later that experimental proof was obtained.

rearrangement of the solvent shell. This equation predicts that the activation energy will reach a minimum when $-\Delta G^0 = \lambda$ (Figure 2.13). Thereafter, the key counter-intuitive feature of Marcus theory is that the rate constants of ET decrease with increasing driving force after reaching a maximum – the inverted Marcus region. This reduction in rate may reduce the rate of electron-hole recombination in natural photosynthesis [18].

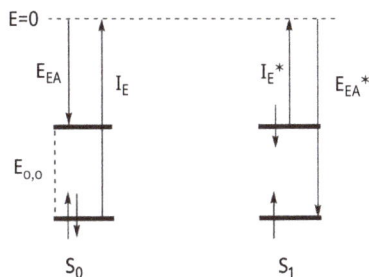

Scheme 2.11: Redox properties of the ground and excited singlet states in simplified form.

$$\Delta G^{\ddagger} = \frac{\lambda}{4}\left(1 + \frac{\Delta G^0}{\lambda}\right)^2 \qquad (2.25)$$

The classic example is $[Ru(bpy)_3]Cl_2$, bpy = 2,2'-bipyridyl, for which the potentials of ground and excited states are illustrated in Scheme 2.12.

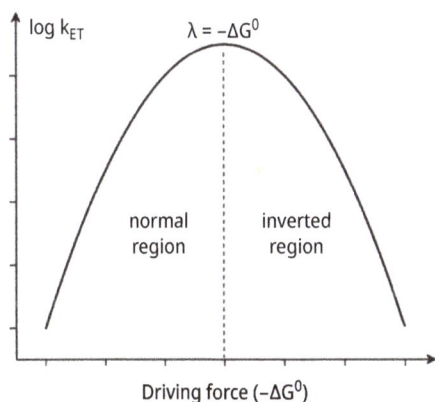

Figure 2.13: Plot of the rate of electron transfer log k_{ET} versus driving force, $-\Delta G^0$, showing normal and inverted Marcus region.

The redox potential $E^0(Ru^{3+}/Ru^{2+})$ is 1.26 V in the ground state meaning that Ru^{2+} is a poor reducing agent but $E^0(Ru^{3+}/Ru^{*2+})$ is –0.84 V, making Ru^{*2+} a good reducing agent. As an oxidising agent, Ru^{2+} is also poor since $E^0(Ru^{2+}/Ru^+)$ is –1.28 V, but this changes to +0.86 V in the excited state making it a good oxidising agent. The popularity of redox photocatalysis (Section 5.1) is driven by the availability of numerous

Redox potentials of $[Ru(bpy)_3]^{2+}$

$^3[Ru^{2+}]^*$ excited state

+0.86 V   −0.84 V

$E_{0,0}$ bpy =

$Ru^+ \leftarrow$ ——— $Ru^{2+} \leftarrow$ ——— Ru^{3+} ground state

−1.28 V   +1.26 V

Emission (at 300 K) 613 nm (2.0 eV), $E_{0,0}$ ~2.1 eV

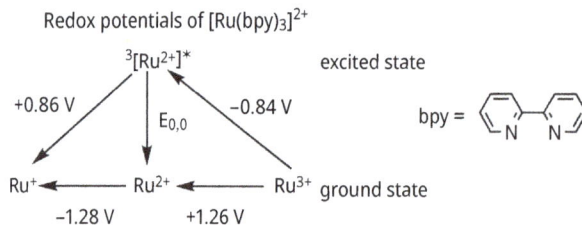

Scheme 2.12: Redox potentials of $[Ru(bpy)_3]^{2+}$ in ground and excited states. Note that the equilibrated excited state is a triplet.

transition metal complexes and organic dyes that behave in a similar manner to $[Ru(bpy)_3]^{2+}$ [19].

2.2.7 Proton-coupled electron transfer (PCET)

In addition to electron exchange, the excited state **GH*** may release a proton and an electron – *excited-state proton-coupled electron transfer* (PCET). First, we consider PCET in the ground state since many redox reactions only become possible when they are linked to uptake or release of protons. Suppose, for example, that a compound RCH_3 has a standard potential E^0 $(RCH_3{}^{+\bullet}/RCH_3)$ (Equation 2.26) of 2.6 V, an oxidising agent must have a potential at least as great as this to make the reaction favourable. However, if the oxidation is linked to deprotonation (Equation 2.27), an oxidation agent of potential 0.8 V will suffice:

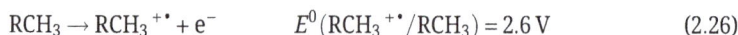

$$RCH_3 \to RCH_3{}^{+\bullet} + e^- \qquad E^0(RCH_3{}^{+\bullet}/RCH_3) = 2.6\,V \qquad (2.26)$$

$$RCH_3 \to RCH_2^{\bullet} + H^+ + e^- \qquad E^0(RCH_2^{\bullet}, H^+/RCH_3) = 0.8\,V \qquad (2.27)$$

An analogous situation applies to reduction reactions such as electron transfer to unsaturated compounds, O_2, or NO_2 (see Chapter 5). Such reactions play an important role in enzymic redox reactions, for example cytochrome oxidase (reduction of O_2 to water) or nitrite reductase (reduction of nitrite to ammonia). In the case of organic halogen compounds, the formation of radicals also takes place without participation of protons:

$$ArCH_2X + e^- \to ArCH_2^{\bullet} + X^- \qquad (2.28)$$

The fundamental reason for the change in redox potential when coupled to proton transfer can be seen from the thermodynamic cycle in Scheme 2.13 in which the redox potential $E^0(RCH_2^{\bullet}, H^+/RCH_3)$ is decomposed into the bond dissociation free energy (BDFE) of C–H of RCH_3 (3.2 eV) and the reduction potential of the proton to H atoms (−2.4 V in water). Alternatively, it may be equated to the sum of the redox potential $E^0(RCH_3{}^{+\bullet}/$

RCH$_3$) and the pK_a of RCH$_3^{+\bullet}$. The same question arises in excited state PCET (Scheme 2.5), but now we need to use excited state potentials as in the preceding section [20].

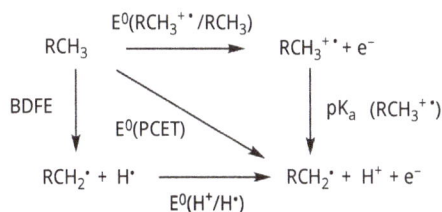

Scheme 2.13: Thermodynamic cycle for analysing PCET.

2.2.8 Excited state proton transfer, photo-acids and hydrogen atom transfer (HAT)

Transfers of protons or hydrogen atoms were represented among the primary reactions of **GH*** with an acceptor **A** (Scheme 2.5). Since the excited state usually has a quite different electronic structure from the ground state, its acidity can change drastically. As an example, 2-naphthol is a weak acid in its ground state (pK_a 9.5) but has a similar acidity to formic acid in its excited state (p$K_a^* = 3.7$). Substitution with two cyano groups yields an excited state super acid with p$K_a^* - 4.5$. Such compounds are described as *photo-acids*[17] [21]. It is equally possible to design photobases: usually they consist of pairs of ions. The anion is the light absorber that, in its excited state, abstracts a proton from the cation. For instance, the tetraphenylborate anion abstracts a proton from trialkylammonium ions on UV irradiation. Both photoacids and photobases are useful in light-induced polymerisation [22].

A famous example of *hydrogen atom transfer* (HAT, also referred to as hydrogen atom abstraction) is the hydrodimerisation of acetone in propan-2-ol (Equation 2.29) which proceeds in sunlight and was discovered as early as 1911.[18] In the initial step, the ^{3}(n,π*) state of acetone abstracts a hydrogen atom from the solvent (Equation 2.30); this is followed by C–C coupling of the resulting radicals to form pinacol. This class of reactions can be made to be catalytic with suitable metal salts or organic dyes that abstract hydrogen atoms on excitation [23].

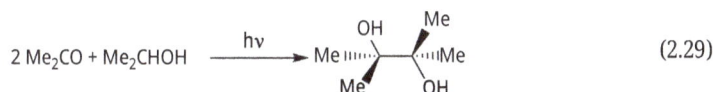

$$2\,Me_2CO + Me_2CHOH \xrightarrow{h\nu} \underset{Me}{\overset{OH}{\underset{}{Me\cdots}}}\overset{Me}{\underset{OH}{\cdots Me}} \tag{2.29}$$

$$^3(Me_2CO) + Me_2CHOH \longrightarrow Me_2\dot{C}(OH) + Me_2\dot{C}(OH) \tag{2.30}$$

---

17 The excited state pK_a^* is related to the ground state value by p$K_a^* = $ p$K_a - (E_{0,0}^{acid} - E_{0,0}^{base})/2.3RT$, where acid and base refer to the conjugate acid and conjugate base.

18 Discovered by Giacomo Ciamician (1857–1922), one of the fathers of organic photochemistry, who also foresaw the replacement of fossil fuels by fuels generated through solar energy.

There is a very close relation between HAT and proton-coupled electron transfer, since one mechanism for HAT is to transfer an electron and a proton. Moreover, the theory of HAT has been developed in an analogous way to Marcus theory for electron transfer [24].

2.2.9 Photosensitisation

Photosensitisation describes the process in which one molecule, the *photosensitiser* (**S**), absorbs light, but another molecule undergoes a change. This sensitisation can take place by one of the mechanisms described above, in most cases by energy transfer (EnT) or electron transfer (ET). Typically, **S** absorbs at longer wavelengths than the acceptor, **A**. The excited state formed first is the excited singlet, $^1\text{S}^*$, that is converted by ISC as efficiently as possible to the triplet state $^3\text{S}^*$ from which energy transfer occurs to form the excited triplet state $^3\text{A}^*$. Chemical reaction then takes place from this state to form the product **P**. The result is a photochemical reaction of **A**, although **A** has not absorbed the light (Equations 2.31–2.33). This triplet-triplet sensitisation represents an ideal way to enable selective reactions to triplet states without competing processes occurring from singlet states:

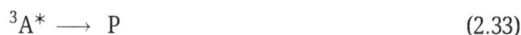

$$\text{S} \xrightarrow{h\nu} {}^1\text{S}^* \xrightarrow{\text{ISC}} {}^3\text{S}^* \tag{2.31}$$

$$^3\text{S}^* + \text{A} \longrightarrow \text{S} + {}^3\text{A}^* \tag{2.32}$$

$$^3\text{A}^* \longrightarrow \text{P} \tag{2.33}$$

The energetic requirements (Scheme **2.14**) demonstrate that least energy is lost when the singlet-triplet splitting of **S** is as small as possible and the two triplet states have energies as close as possible to one another.

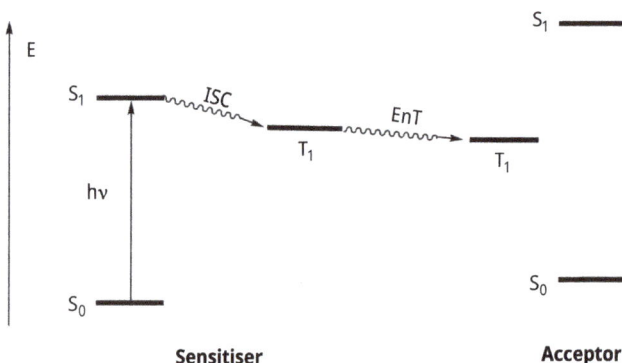

Scheme 2.14: Energetics for triplet-triplet photosensitisation.

The thermodynamic criterion for sensitisation through reductive electron transfer is that the potential $E^0(S^+/S^*)$ is more negative (or at least equal to) than the ground state potential of the acceptor $E^0(A/A^-)$. For oxidative sensitisation, the potential $E^0(S^*/S^-)$ must be more positive than the potential of the electron donor **D**, $E^0(D^+/D)$ (Scheme 2.15).

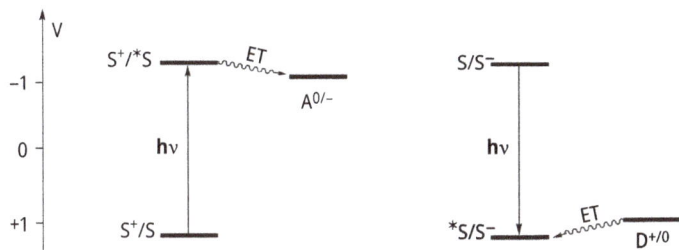

Scheme 2.15: Electrochemical potential energy diagram for reductive (left) and oxidative (right) sensitisation. Notice the arrows representing the excitation.

After the electron transfer has taken place, the sensitiser is in its oxidised form (reductive electron transfer) or reduced form (oxidative electron transfer). It must then be returned to its original state by a second substrate that acts as electron donor or acceptor. The steps of an ET photosensitisation reaction, $D + A \rightarrow D^+ + A^-$, are shown in Scheme 2.16.

Photosensitisation by these electron transfer routes forms the basis of countless reactions, including their use in artificial photosynthesis and photoredox catalysis including chiral syntheses (Section 5.1)[19] [19].

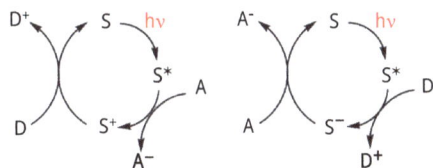

Scheme 2.16: Electron transfer photosensitisation of the reaction $D + A \rightarrow D^+ + A^-$. Left: reductive ET; right: oxidative ET.

2.2.10 Photon upconversion

Photon upconversion (UC) consists of the sequential absorption of two or more photons inducing emission of light with shorter wavelength as compared to the exciting light

19 David MacMillan (born 1968) is a Scottish chemist working in the USA. He won the Nobel Prize in 2021 for his work on asymmetric organocatalysis. He also pioneered the applications of photoredox catalysis in organic chemistry.

[25]. It is therefore described as anti-Stokes luminescence. Thus, shining green light onto an UC-capable compound generates an anti-Stokes emission in addition to the common Stokes emission (Figure 2.14 and Scheme 2.17 A,B). This differs from *two-photon absorption* where the photons are absorbed simultaneously, therefore requiring very intense lasers, which is a topic beyond this discussion.

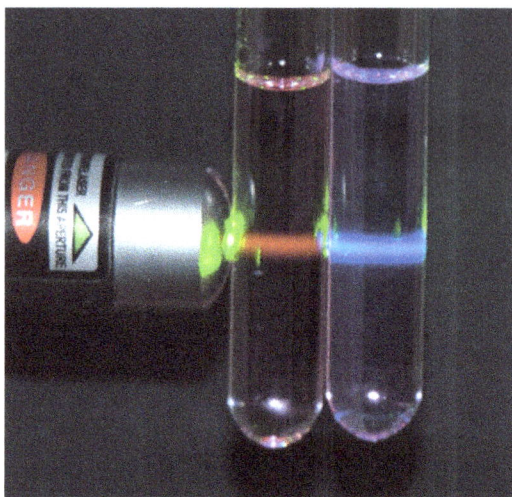

Figure 2.14: Green laser light generates normal-Stokes and anti-Stokes emissions [26].

Both organic and inorganic compounds exhibit UC but through different mechanisms as summarised for the most important ones in Scheme 2.17. In the case of Scheme 2.17B, the excited state of the emitting compound EM* absorbs a second photon generating EM** which returns to the ground state by emitting a photon of energy $h\nu_3$. In the case

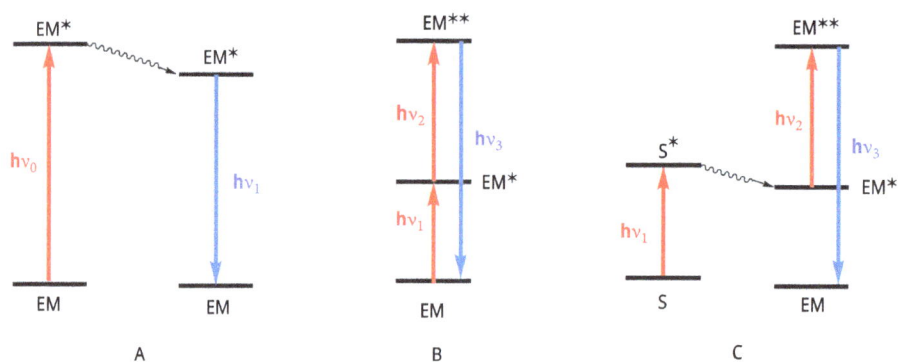

Scheme 2.17: Conventional luminescence (A), two-photon upconversion via excited state absorption (B) and sensitised upconversion (C).

of Scheme 2.17C, a sensitiser **S** generates the excited state EM* by energy transfer (EnT) which then absorbs a second photon generating EM**. Return to the ground state proceeds by emission of the photon $h\nu_3$. When the excited sensitiser is long-lived, like triplet states of polycyclic aromatics, triplet-triplet annihilation (TTA) may also generate EM** (Scheme 2.18). After formation of two singlet sensitiser excited states, intersystem crossing to the triplet states and triplet-triplet energy transfer generates two emitter triplet states. Their annihilation leads to the singlet state of one emitter molecule (^{1}EM**) which returns to the ground state through emission of a photon having almost the twice the singlet excitation energy of the sensitiser.

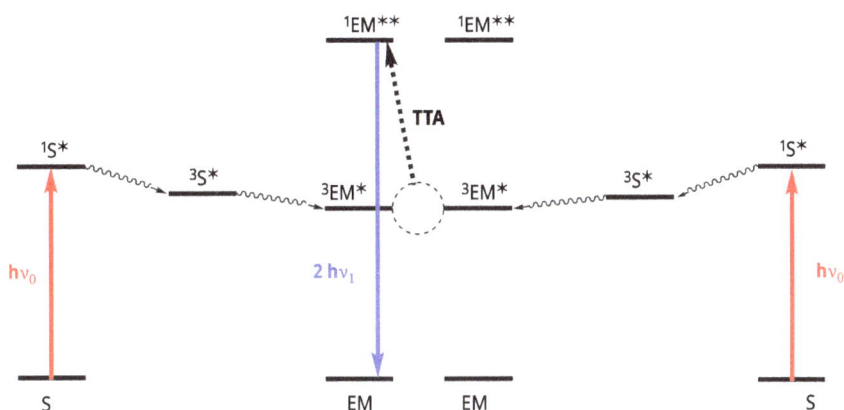

Scheme 2.18: Basic scheme of triplet-triplet annihilation (TTA) upconversion.

An efficient annihilator EM should (i) not absorb the exciting light, (ii) have a high quantum yield of fluorescence and (iii) have a long-lived triplet state located below that of the sensitiser but a little higher than the half of its excited singlet state ^{1}EM** energy. These criteria are met to a sufficient extent by polycyclic aromatics and inorganic materials containing ions of d-block or f-block elements like Ln^{3+}, Er^{3+} and Yb^{3+}. Their excited states have long lifetimes (ms), large anti-Stokes shifts of a few hundred nanometres and sharp emission lines. They are usually embedded in transparent nanocrystals of fluorides such as $NaYF_4$ and CaF_2 as host materials. These should possess low lattice-phonon energies since this minimises radiationless decay, which needs coupling to lattice vibrations. Fluorides have phonon energies in the range of 500 cm^{-1} and are chemically very stable. $NaYF_4$ is probably the most used host in the field of lanthanide-based upconversion nanoparticles. Guest upconverter ions are in most cases the pairs erbium-ytterbium (Er^{3+}-Yb^{3+}) or thulium-ytterbium (Tm^{3+}-Yb^{3+}). A typical example is the upconverter β-$NaYF_4$:18% Yb^{3+}, 2% Er^{3+} absorbing from 850 to 1050 nm. Ytterbium functions as a sensitiser absorbing near-infrared (NIR) light. Upon excitation at 980 nm, energy is transferred to erbium acting as an upconverter by generating characteristic green and red emissions. Embedding the upconverter lanthanide ions in core-shell host particles often suppresses undesired quenching processes.

In addition to lanthanide ions, semiconductor nanoparticles like CdSe, PbS and PbSe can also act as sensitisers. In combination with molecular emitters, they exhibit good upconversion yields through triplet-triplet annihilation. They have been used to upconvert near infrared light (NIR, 800–2500 nm) to visible light. These materials can therefore make the large IR part of solar radiation usable for photovoltaic solar cells as sketched in Figure 2.15. IR light not absorbable by the silicon p-n junction is absorbed by the upconversion and converted into light absorbable by silicon (see Chapter 3). With terrestrial sunlight, the photocurrent density was enhanced by about 6% [27].

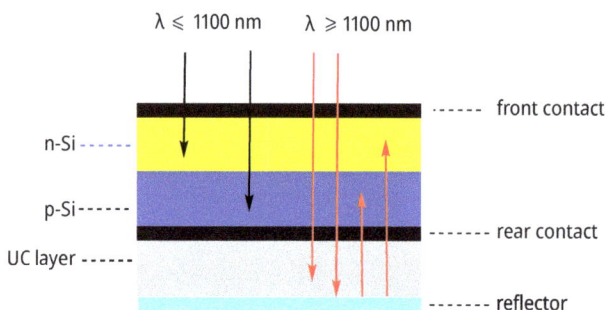

Figure 2.15: Layer structure of a silicon photovoltaic cell additionally utilising infrared light above 1100 nm for charge generation through upconversion (UC) [27].

2.3 Classification of electronic transitions

2.3.1 Organic molecules

The classification of electronic transitions anticipated for organic molecules follows from the nature of the HOMO and LUMO as Kasha observed. The absorption coefficient and the wavelength dependence on the solvent follow from the nature of the excited state (Table 2.1). The σ,σ^* transitions of single bonds occur at very short wavelengths (100–200 nm) and will not be discussed here.[20] They occur in saturated compounds such as ethers and dialkylsulfides and have small absorption coefficients (under 1000 M^{-1} cm^{-1}).

Even in simple molecules, identification of the type of transition can be ambiguous because of the presence of several functional groups or because transitions may have mixed character. Traditional methods of checking assignments rely on changes in substituents in a series of related molecules. Nowadays, both the wavelengths and oscilla-

---

20 For example, ethanol and *n*-hexane begin to absorb at about 200 nm. However, the σ,σ^* transitions of bonds between heavier main group elements such as Si-Si occur in the 200–300 nm region.

tor strengths of electronic spectra can be calculated by time-dependent density func-tional theory (TD-DFT) with considerable accuracy. These computational methods pro-vide important tests of assignments.

Table 2.1: Electronic transitions and the resulting excited states.

HOMO[a]	LUMO[b]	State
σ $\longrightarrow$	σ^*	σ,σ^*
π $\longrightarrow$	π^*	π,π^*
n $\longrightarrow$	π^*	n,π^*
n $\longrightarrow$	σ^*	n,σ^*

[a]Bonding and non-bonding.
[b]Antibonding.

2.3.1.1 π,π' and n,π' transitions

π,π^* transitions of unsaturated compounds range from ultraviolet to near infrared, since the HOMO-LUMO gap becomes smaller with the number of conjugated multiple bonds. Thus, benzene with its three double bonds absorbs at 260 nm, anthracene with seven double bonds at 380 nm and free-base porphyrins with nine C=C bonds at about 630 nm (Figure 2.16 and Table 2.2). If the energy differences of electronic states of these compounds are to be compared, wavenumbers must be used as they are di-rectly proportional to the energy (Equations 1.3 and 1.4).

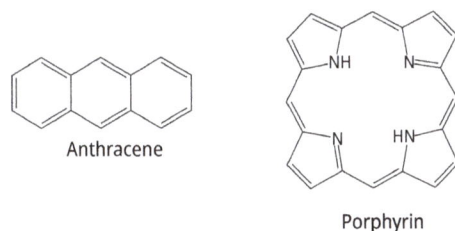

Anthracene

Porphyrin

Figure 2.16: Two classical aromatic multi-π-electron systems.

Table 2.2: Characteristic HOMO-LUMO transitions of chromophores.

Chromophore	λ_{max} (nm)	$\tilde{\nu}$ (cm^{-1})	ε_{max} (M^{-1} cm^{-1})	Transition
C=C–C=C	220	45000	2×10^4	π,π^*
Benzene	260	38000	200	π,π^*
Anthracene	380	26000	1×10^4	π,π^*
C=O	280	36000	20	n,π^*
N=N	350	29000	100	n,π^*

Further information about the nature of an absorption band can be obtained from its solvent dependence. The absorption of a photon effects a change in the electronic structure of the molecule with consequent changes in its dipole moment, and therefore, its interaction with the solvent. Comparisons of absorption spectra in polar and non-polar solvents allow estimates of how the dipole moment changes through electronic excitation. In a ketone such as acetone, the negative portion of the dipole is localised on oxygen. However, the electron density on oxygen is transferred to the π^* MO of the keto group through the n,π^* transition, resulting in a reduction in the dipole moment. Consequently, on changing from a non-polar to a polar solvent, the excited state is less well solvated than the ground state, and the absorption band is shifted to shorter wavelengths (hypsochromic shift, blue shift and negative solvatochromism). There is no similar clear relationship for the opposite effect (bathochromic shift, red shift and positive solvatochromism), although it may be observed for π,π^* transitions, especially when the molecule has a dipolar electronic structure like Reichardt's dye (Figure 2.17 and Scheme 2.19).

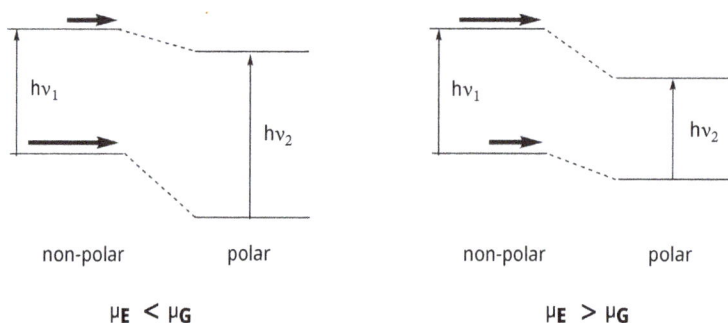

Scheme 2.19: Hypsochromic (left) and bathochromic (right) shifts of an absorption band on changing solvent from *n*-hexane to methanol. μ_G and μ_E are dipole moments of ground and excited states, respectively.

2.3.1.2 Charge-transfer (CT) transitions

2.3.1.2.1 Intramolecular CT

If a molecule includes both an electron donor and an electron acceptor group, a charge-transfer absorption band can occur. These bands often exhibit a marked solvent dependence. In the case of 4-nitrophenol, the intramolecular CT (ICT) band shifts from 312 nm in HCl (aqueous) to 392 nm in NaOH (aqueous) (Figure 2.17). Sometimes, the CT excitation results in a marked change in conformation, as observed for 4-cyano-dimethylaniline, where the NMe_2 group rotates from co-planar with the benzene ring to perpendicular to the benzene ring. These "twisted intramolecular charge-transfer" (TICT) states exhibit unusual absorption and emission properties with applications in materials [28]. Compounds that exhibit extreme solvatochromism are used in scales of polarity; their sensitivity often arises from the ICT character of their absorption bands. For example, the absorption maximum of Reichardt's dye (Figure 2.17) can be adjusted across the entire visible region by changing the solvent.

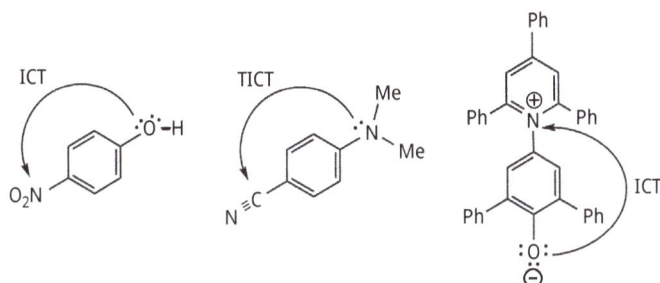

Figure 2.17: Intramolecular charge transfer between donor and acceptor groups in 4-nitrophenol (left), 4-cyano-dimethylaniline (centre) and Reichardt's dye (right) (Ph = C_6H_5).

2.3.1.2.2 Intermolecular CT

When measuring the absorption spectrum of benzoquinone, a new band appears at about 3×10^4 cm^{-1} if the solvent is changed from cyclohexane to 2,3-dimethylbutadiene. This band arises from a π,π^*-transition within an electron donor-acceptor (EDA) complex between benzoquinone and 2,3-dimethylbutadiene (Figure 2.18). For maximum orbital overlap, the two components should be aligned cofacially with an intermolecular separation of about 3.5 Å.

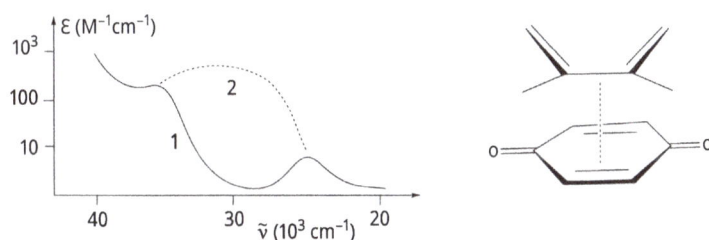

Figure 2.18: Absorption spectrum of benzoquinone in cyclohexane (1) and in 2,3-dimethylbutadiene (2) showing formation of the intermolecular CT band.

Figure 2.19 illustrates the formation of such an EDA complex through alteration in the potential energy when the two components approach one another. The extent of charge transfer is very slight in the ground state resulting in small potential well. On light absorption, the charge transfer and consequent Coulomb attraction of the CT complex are increased. This complex is stronger in the excited state than the ground state and is called an exciplex[21] – indeed it barely exists in the ground state. The energy of the CT state is lowered for a given acceptor if the electron density of the donor component is increased. A good example is the combination of the strong acceptor tetracyanoethene with enol ethers that are good donors. If the *cis*-hydrogen atom of ethoxyethene is substituted by a further ethoxy group, the energy of the CT state is lowered from about 280 to 220 kJ mol^{-1} [22] [29].

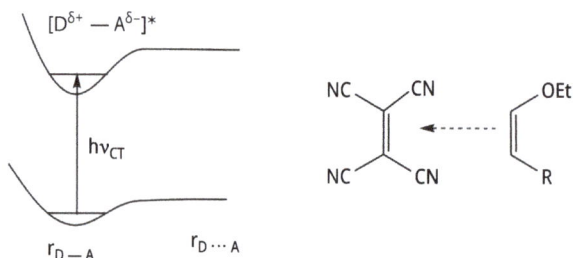

Figure 2.19: Change in the potential energy with the donor-acceptor separation during formation of a CT complex between tetracyanoethene and ethoxyalkenes (R=H, OEt, Et=CH_3CH_2).

In many cases of such two-component systems, no new absorption band is observed but a new, long-wavelength emission band appears (Figure 2.20). The fact that there is no new absorption indicates that there is a negligible donor-acceptor interaction in the ground state. In contrast to an EDA complex, there is no potential minimum in the ground state, but there is a minimum in the excited state (Figure 2.21). Exciplexes formed from aromatics such as pyrene together with amines represent classic examples.

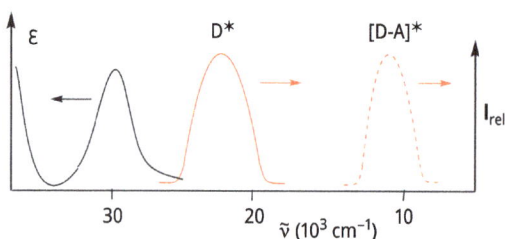

Figure 2.20: Schematic absorption and emission spectra of an exciplex with $\lambda_{exc} = 333$ nm (30.03×10^3 cm^{-1}).

---

21 An abbreviation of excited complex.
22 The energies correspond to the values of λ_{max} of 423 and 545 nm. See p. 248 of ref. [1].

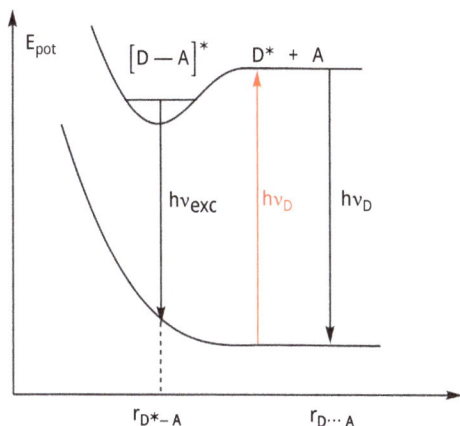

Figure 2.21: Change in the potential energy with donor-acceptor distance in an exciplex (exc).

An emission band shifted to long wavelengths also arises in a concentrated solution of strongly delocalised aromatic molecules such as pyrene. This band arises from electronically excited dimers or *excimers*. The interaction diagram in Scheme 2.20 illustrates the energetics of formation of weakly bonded exciplexes that are formed in the excited state. A similar diagram could be drawn for excimers, but the orbital energies of the two components would be identical.

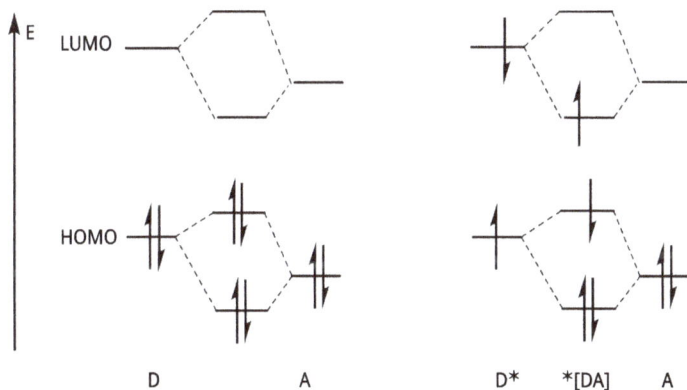

Scheme 2.20: Simplified interaction diagram of two molecules in the ground and excited states showing formation of an exciplex. In an excimer, D and A are identical.

According to MO theory, the electronic interaction of two orbitals leads to a bonding and an antibonding orbital. In the ground state of D + A, the orbitals resulting from interaction of the HOMO's contain two pairs of electrons. One pair is stabilised and the other destabilised, resulting in no net stabilisation.[23] However, if one component is excited (the donor D in Scheme 2.20), the HOMO and LUMO of that component each contain one electron. Now the interaction with the LUMO of A generates two MOs, with one electron in the lower-lying bonding MO while only one antibonding electron remains from interaction of the two HOMOs. Overall, there are now three bonding electrons and one antibonding electron, resulting in a net stabilisation.

2.3.2 Transition metal complexes

2.3.2.1 Overarching principles

The keys to understanding the spectra and photochemistry of transition metal complexes are their electron configurations, their geometries and the nature of the ligands.

Electron configurations. Two different characteristics are important in electron configuration: the d-electron configuration and the total valence electron count. The d-electron count n is given very simply by the following equation:

$$n = n_0 - \text{oxidation state} \ (0 \leq n \leq 10) \tag{2.34}$$

where n_0 is the group number of the metal in the Periodic Table. For instance, the group numbers for chromium, manganese and iron are 6, 7 and 8 respectively.

The phenomenon of high-spin complexes (see Scheme 1.4) results in two (or occasionally three) different possible electron configurations but is restricted to complexes of first-row metals with d-electron counts of 4–8. Notably all metal carbonyl and metal cyanide complexes adopt the low-spin configuration.

The total valence electron configuration, T, includes the contribution of the ligands and is given as follows:

$$T = n_0 + \sum_i l_i - z \tag{2.35}$$

where n_0 has the same meaning as above, $\sum_i l_i$ is the sum of the ligand contributions to the metal and z is the overall charge on the complex. Contributions of selected ligands are given in Table 2.3. The contributions of metal-metal bonds correspond to their order, one for single, two for double, etc. Very many complexes with metal-carbon bonds (organometallics) have $T = 18$ corresponding to a complete shell, just as main group compounds often have a valence electron count of eight (octet rule). This 18-electron rule is followed more strictly for second- and third-row transition metal complexes than for first-row complexes. Notable deviations are found for complexes

---

23 The stabilisation energy is the difference in energy between the stabilised and destabilised electrons. When both the bonding and corresponding antibonding orbitals are occupied as in the left-hand diagram, the net interaction is slightly repulsive – a so-called 4-electron repulsion.

with low d-electron counts ($n = 0$–3) and d^8 complexes with square-planar geometries that have 16-electron configurations.

Table 2.3: Selected ligand contributions to the total valence electron configuration.[a.]

No. of electrons contributed by the ligand	Ligand (R = H, alkyl, aryl)
0	$AlMe_3$, $ZnCl_2$, ZnR_2
1	H, alkyl, aryl, F, Cl, Br, I, OR, SR, NO_2, CN, NO (bent), η^1-allyl, η^1-cyclopentadienyl, carboxylate (monodentate)
2	CO, PR_3, C_2H_4, N_2, OH_2, NH_3, O,[b] S, carbene
3	NO (linear), N, carboxylate (bidentate), η^3-allyl
4	O,[b] η^4-diene, $R_2P(CH_2)_nPR_2$, $R_2N(CH_2)_nNR_2$ (n = 1, 2, 3)
5	η^5-Cyclopentadienyl
6	η^6-Arene

[a]η^x denotes the number of carbon atoms (x) within the bonding distance of the metal.
[b]O can be a 2e- or a 4e-donor.

Geometry. While the five 3d orbitals of a first-row transition metal are degenerate in the gas phase, they split into two or more levels in a complex, according to the geometry. In the second- and third-row transition metal complexes, the splitting of the corresponding 4d and 5d orbitals is generally greater than in the first-row complexes. The geometries of the complexes determine the d-orbital splitting diagrams (Scheme 2.21). According to the crystal field model, the low-lying orbitals in each case are less repulsive toward the ligands than the high-lying orbitals. In the MO model, appropriate for more covalent ligands, the low-lying orbitals are non-bonding or bonding while the high-lying orbitals are antibonding (see below). When the d-orbitals are partly occupied, electronic transitions between the orbitals are possible – these are termed metal-centred (MC) or d-d transitions.

Ligands. The nature of the ligands is the third factor determining spectra and photochemistry. The ligands may be divided into three classes:
   σ-only ligands: OH_2, NH_3, ethers, amines, H⁻
   σ-donor, π-donor ligands: halide, O^{2-}, OH⁻, OR⁻, NR_2^- (R = alkyl)
   σ-donor, π-acceptor ligands: CO, CN⁻, N_2, alkene, alkyne, benzene, cyclopentadienyl

Orbital overlap diagrams illustrate how these interactions arise. In σ-only ligands, the occupied ligand orbital donates to the empty metal orbital (Figure 2.22A), and no other orbitals are available for interaction or to accept electrons on excitation. In σ-donor, π-donor ligands, there are occupied π-orbitals (lone pairs) that contribute to the bonding (Figure 2.22A and C) in addition to the σ-donation and can transfer charge to the empty metal orbitals on excitation (ligand-to-metal charge transfer – LMCT). In σ-donor, π-acceptor ligands, there are empty antibonding π^* orbitals on the ligand

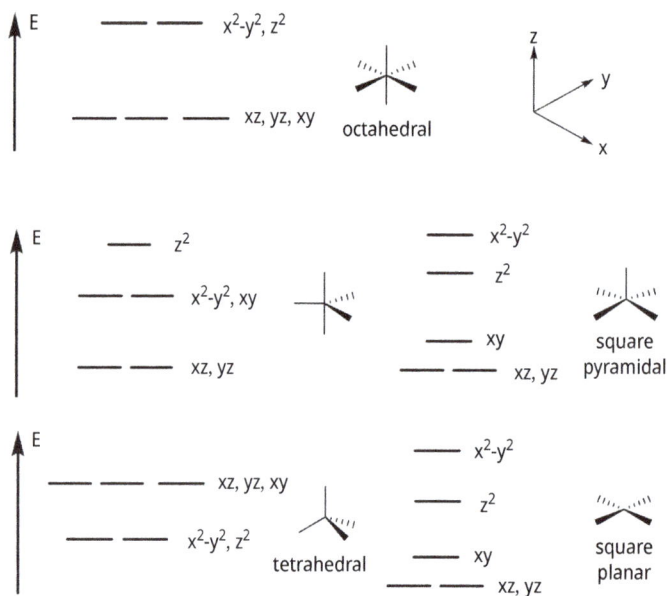

Scheme 2.21: d-Orbital splitting diagrams for common 4-, 5- and 6-coordinate geometries. The labels indicate the identity of each d-orbital. The d_{z^2} orbital is sometimes lower in square-planar complexes than shown here.

that contribute to the bonding (Figure 2.22A and B) and can receive charge from the metal on excitation (metal-to-ligand charge transfer – MLCT).

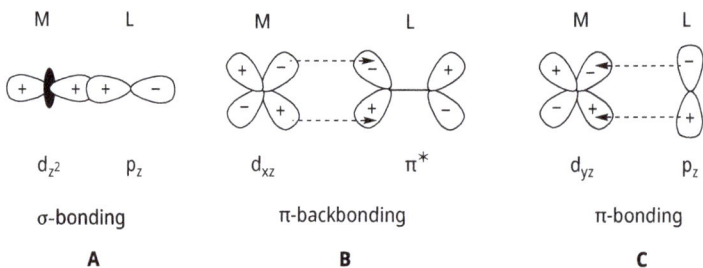

Figure 2.22: Schematic orbital interactions involved in forming metal-ligand bonds (M–L bonds).

Examples of transition metal complexes are shown in Figure 2.23 together with their electron configurations. Ligands such as ethene and benzene use their occupied π-orbitals as donors and their empty π^*-orbitals as acceptors with the consequence that they are oriented face-on to the metal.

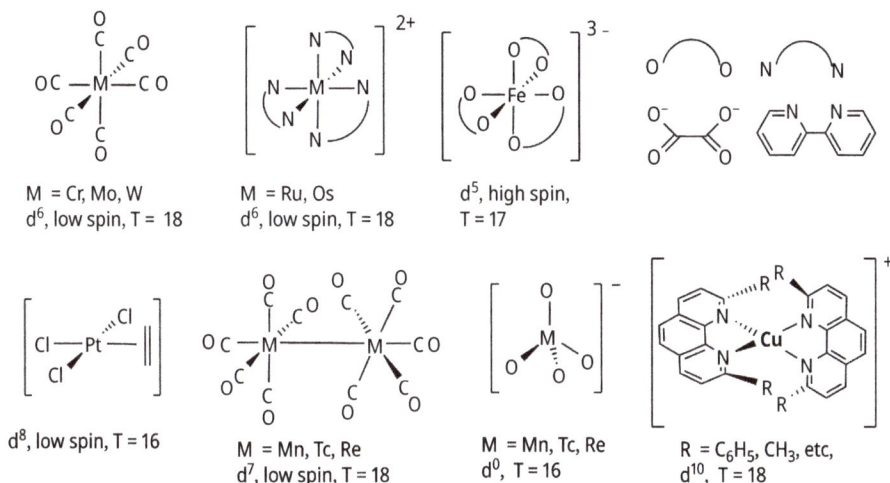

Figure 2.23: Examples of transition metal-ligand complexes and their electron configurations.

In the last few paragraphs, we have encountered metal-centred (MC) and charge-transfer transitions (LMCT and MLCT). These transitions are combined into one ideal-ised diagram for a d^6-low-spin octahedral complex in Figure 2.24. Additionally, the diagram includes a ligand-centred (LC) transition. In reality, we rarely observe more than two types of transitions in a single metal complex. Symmetry considerations allow a rigorous distinction between MC and CT transitions for an octahedral complex, but this is not possible in low symmetry complexes because they may mix together.

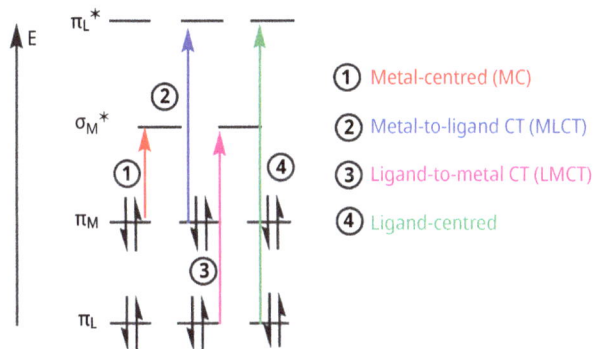

Figure 2.24: Simplified schematic MO diagram for a d^6 low-spin octahedral complex showing the possible electronic transitions.

The requirements for these different types of transitions are shown in the textbox below. Often these criteria and those mentioned below provide unambiguous assignments. If not, further information may be provided by variation of substituents and the charge or oxidation state of the metal and by use of TD-DFT.

Requirements for different electronic transition types
MC d^n, $n = 1$–9
MLCT d^n, $n = 1$–10, empty ligand π^*
LMCT d^n, $n = 0$–9, ligand lone pair not used for σ-bond to M
LC occupied ligand π orbitals and vacant ligand π^* orbitals

The types of charge-transfer discussed so far involve intramolecular excited states. We will also consider intermolecular excited states involving a solvent (charge transfer-to-solvent (CTTS) bands) or to a second substrate. If the complexes contain two or more transition metal centres, two more types of transitions become possible. Metal-to-metal charge transfer (MMCT) are observed when two metals are present in different oxidation states linked together by a bridging ligand. If, on the other hand, the two metals are linked by metal-metal bonds, excited states are available in which electrons are transferred from M–M bonding orbitals to their antibonding counterparts. In the next paragraphs, we describe each type in more detail with examples.

2.3.2.2 Metal-centred (MC) transitions

The overwhelming majority of MC transitions are of the d-d type and are therefore possible for electron configurations d^n, $n = 1$–9. The transitions can be understood on the basis of the splitting diagrams in Scheme 2.21, but a detailed analysis of the number of transitions and their wavelengths requires consideration of electronic states as well as electron configurations as described in textbooks of inorganic chemistry. Spin-allowed d-d transitions have absorption coefficients of 10–100 M^{-1} cm^{-1} for octahedral complexes, increasing to 100–1000 M^{-1} cm^{-1} for tetrahedral and other complexes without a centre of symmetry. The low value of ε is a consequence of violation of the Laporte rule forbidding transitions when the ground and excited states are both symmetric with respect to inversion (see Section 2.1.3.3). Spin-forbidden transitions have absorption coefficients that are about 100 times smaller than their spin-allowed counterparts and are therefore difficult to observe in absorption except in the case of d^5 high-spin configurations where no spin-allowed d-d transitions are possible. Figure 2.25A illustrates the spectrum of $[Rh(NH_3)_6]^{3+}$, a d^6 low-spin octahedral complex with two spin-allowed d-d absorption bands in the UV region and a spin-forbidden emission band in the visible region. In general, the wavelengths of d-d absorption bands with the same set of ligands and electron configuration decrease in the order 5d < 4d < 3d metals: λ_{max} $[Ir(NH_3)_6]^{3+}$ < $[Rh(NH_3)_6]^{3+}$ < $[Co(NH_3)_6]^{3+}$.

Figure 2.25B illustrates the absorption spectrum of d^3 $[Cr(H_2O)_6]^{3+}$ with two prominent features in the visible region. Emission spectra of Cr(III) complexes often show broad and narrow features. The spin-forbidden d-d emission bands of octahedral Cr(III) complexes are very narrow when the spin-flip keeps the electrons in the triply degenerate orbital so there is no change in Cr-ligand bond length (Franck-Condon principle). Ruby is a well-known example of Cr(III) in an oxide lattice showing spin-forbidden d-d emission bands (Section 1.7), but recent research has shown that such emission bands with high quantum yields may also be seen in some molecular complexes (Figure 2.25C) [30].

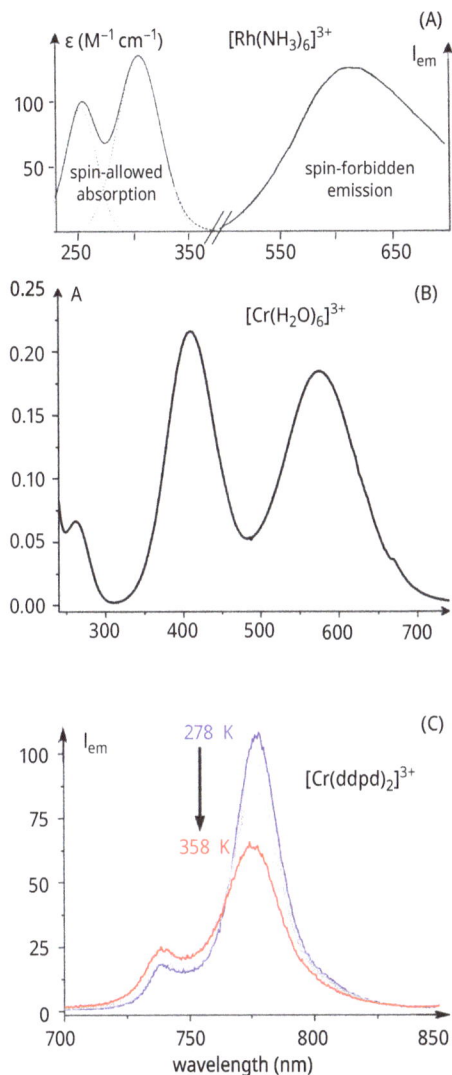

Figure 2.25: (A) Solution spectrum of $[Rh(NH_3)_6]^{3+}$. The two bands between 250 and 350 nm are due to spin-allowed transitions. The emission spectrum reveals the band at ca. 600 nm that is due to a spin-forbidden transition. Adapted from ref. [31]. (B) Absorption spectrum of $[Cr(H_2O)_6]^{3+}$ in water. (C) Near-IR spin-forbidden emission bands of $[Cr(ddpd)_2]^{3+}$ (ddpd is a tridentate pyridine derivative). Adapted from ref. [30]. Copyright 2020 Royal Society of Chemistry.

2.3.2.3 MLCT transitions

Metal-to-ligand charge-transfer transitions are characteristic of complexes with d^n, $n >$ 0, and low-lying π-acceptor orbitals on the ligand. MLCT transitions have absorption coefficients in the range 10^3–10^6 M^{-1} cm^{-1}. We address two examples, both with d^6 low-spin configurations, one leading to prompt dissociation and the other to an equilibrated excited state (see Section 2.2.1). Chromium hexacarbonyl, Cr(CO)$_6$, is a colourless complex with two prominent absorption bands in the UV region due to charge transfer from filled triply degenerate d orbitals to the CO-π^* orbitals. General trends in MLCT bands are summarised in the following textbox:

MLCT trends to *shorter* wavelengths

For series of isoelectronic (or closely related) molecules

As positive charge on metal increases: λ_{max} [Mn(CO)$_6$]$^+$ < Cr(CO)$_6$ < [V(CO)$_6$]$^-$

With change to more electron donating substituents on e-accepting ligand

---

Irradiation into these bands of Cr(CO)$_6$ leads to formation of singlet MLCT excited states and prompt dissociation of CO. This behaviour is unexpected because this transition does not populate M–CO antibonding orbitals directly. However, excitation causes the molecule to distort and the MLCT excited state crosses over into an MC excited state that is σ-antibonding and CO is expelled. Since the photo-reaction is complete within 10^{-12} s or less, there is no time for equilibration and there is no fluorescence. The resultant Cr(CO)$_5$ is a short-lived, intermediate with a square-pyramidal structure that picks up a solvent molecule as a sixth ligand on the picosecond timescale. Displacement of the solvent molecule leads to more stable products (Equations 2.36–2.38). The mechanism is described in more detail together with examples of applications in synthesis and catalysis in ref. [32]:

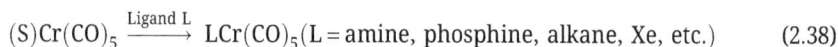

$$\text{Cr(CO)}_6 \xrightarrow{h\nu, \text{ MLCT}} \text{Cr(CO)}_5 + \text{CO} \tag{2.36}$$

$$\text{Cr(CO)}_5 \xrightarrow{\text{Solvent S}} \text{(S)Cr(CO)}_5 \, (\text{S = alkane, arene, liquified Xe, etc.}) \tag{2.37}$$

$$\text{(S)Cr(CO)}_5 \xrightarrow{\text{Ligand L}} \text{LCr(CO)}_5 (\text{L = amine, phosphine, alkane, Xe, etc.}) \tag{2.38}$$

A second paradigm for an MLCT excited state is [Ru(bpy)$_3$]$^{2+}$ (bpy = 2,2'-bipyridyl). In this case, the singlet MLCT excited state converts to the corresponding triplet state on an ultrafast timescale. The ^{3}MLCT state is sufficiently long-lived to equilibrate with its surroundings and is luminescent. There is a large Stokes shift because of the change in spin states (Figure 2.26). The triplet state is demonstrated by time-resolved Raman spectroscopy to be best described as oxidised at Ru and reduced at a single bipyridine ligand [Ru^{III}(bpy)$_2$(bpy$^-$)]$^{2+}$. This structure leads to its extraordinary ability as an excited state oxidant and excited state reductant described in Section 2.2.6. This complex and thousands of its derivatives have found innumerable applications in photocatalysis, luminescence sensing and even molecular logic. Related phenomena extend to first-row transition metals such as copper with sterically hindered phenanthroline ligands (Figure 2.23) that form complexes with high emission quantum yields in solution at room temperature.

Like organic CT transitions, MLCT transitions can be highly sensitive to the surroundings. For example, [Ru(bpy)(CN)$_4$]$^{2-}$ shows extreme sensitivity to the solvent both in absorption to the ^{1}MLCT state and in emission from the ^{3}MLCT state. The excited states are also sensitive to formation of hydrogen bonds and halogen bonds with the cyanide ligands, and to the presence of metal cations from groups 1 and 2.

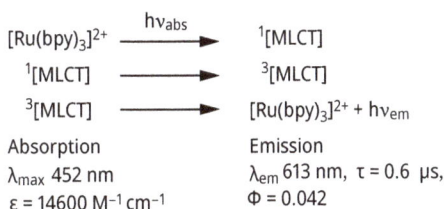

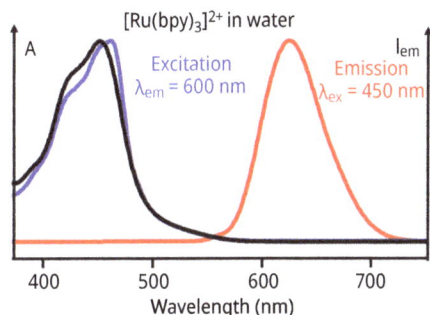

$$[Ru(bpy)_3]^{2+} \xrightarrow{h\nu_{abs}} {}^1[MLCT]$$

$${}^1[MLCT] \longrightarrow {}^3[MLCT]$$

$${}^3[MLCT] \longrightarrow [Ru(bpy)_3]^{2+} + h\nu_{em}$$

Absorption

λ_{max} 452 nm

$\varepsilon = 14600$ M^{-1} cm^{-1}

Emission

λ_{em} 613 nm, $\tau = 0.6$ µs,

$\Phi = 0.042$

Figure 2.26: Above: Excited state properties of [Ru(bpy)$_3$]$^{2+}$ in aqueous solution. Below: absorption, excitation and emission spectra.

2.3.2.4 LMCT transitions

Ligand-to-metal charge-transfer transitions require a lone pair on the ligand in a π-orbital (i.e. a lone pair that is not used to form the coordinate σ-bond to the metal). They also require a vacancy in the d-orbitals. A simple example is permanganate, [MnO$_4$]$^-$, a tetrahedral complex with a d^0 configuration and lone pairs on the oxo ligands. The LMCT transition from the full triply degenerate ligand orbitals to the doubly degenerate d-orbitals is fully allowed and gives rise to the familiar purple colour (spectrum in Figure 2.7). LMCT transitions have absorption coefficients in the same range as MLCT transitions. Trends in LMCT bands are shown in the following textbox:

LMCT trends to *shorter* wavelengths

For series of isoelectronic (or closely related) molecules

As negative charge on metal increases:	λ_{max} MnO$_4^-$ > CrO$_4^{2-}$ > VO$_4^{3-}$
From 3d to 4d to 5d metals:	λ_{max} MnO$_4^-$ > TcO$_4^-$ > ReO$_4^-$
With change to more electronegative e-donating ligands:	λ_{max} VSe$_4^{3-}$ > VS$_4^{3-}$ > VO$_4^{3-}$

With change to more electron withdrawing substituent on e-donating ligand

---

The tris(oxalato)iron(III) chelate complex forms another example that is important because of its photo-induced intramolecular redox reaction. Not only is it remarkable for having a quantum yield exceeding 1, but it finds use as a chemical actinometer to measure light intensities and hence to measure quantum yields of other reactions.[24] It is also important for understanding the flux of CO_2 from iron carboxylate complexes to the atmosphere. On excitation to the LMCT state, one of the oxalate ligands undergoes prompt photodissociation to form CO_2 and an Fe(II) intermediate with a coordinated CO_2 radical anion (step a). The latter dissociates rapidly to release the CO_2 radical ion (step b). Finally, this ion reduces another molecule of the starting material in a dark reaction (step c, Figure 2.27) [33]. Consequently, the quantum yield for product formation could reach 2, although only one photon excites one molecule in keeping with the Stark-Einstein law (Section 1.6). In reality, the maximum observed quantum yield is 1.2. The transfer of two electrons following absorption of a single photon has been named *photo-redox amplification* [34].

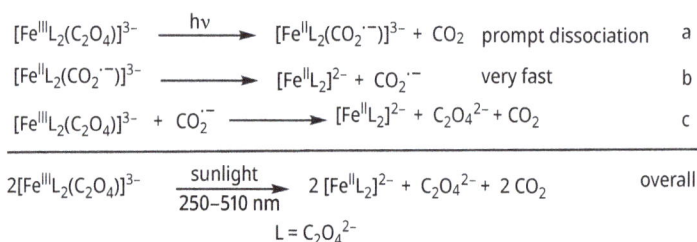

$$[Fe^{III}L_2(C_2O_4)]^{3-} \xrightarrow{h\nu} [Fe^{II}L_2(CO_2^{\cdot-})]^{3-} + CO_2 \quad \text{prompt dissociation} \qquad a$$

$$[Fe^{II}L_2(CO_2^{\cdot-})]^{3-} \longrightarrow [Fe^{II}L_2]^{2-} + CO_2^{\cdot-} \qquad \text{very fast} \qquad b$$

$$[Fe^{III}L_2(C_2O_4)]^{3-} + CO_2^{\cdot-} \longrightarrow [Fe^{II}L_2]^{2-} + C_2O_4^{2-} + CO_2 \qquad c$$

$$2[Fe^{III}L_2(C_2O_4)]^{3-} \xrightarrow[250–510\ nm]{\text{sunlight}} 2[Fe^{II}L_2]^{2-} + C_2O_4^{2-} + 2CO_2 \qquad \text{overall}$$

$$L = C_2O_4^{2-}$$

Figure 2.27: Mechanism of photoreaction of tris(oxalate)iron(III) following LMCT excitation, leading to a quantum yield exceeding 1.

It has proved difficult to design iron(III) complexes exhibiting strong luminescence but there has been recent success by using tridentate *N*-heterocyclic carbene ligands that are very strong σ-donors. These ligands enforce low-spin configurations and raise the energy of MC excited states allowing observation of spin-allowed LMCT absorption and emission (Figure 2.28) [35].

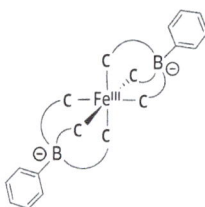

Figure 2.28: Schematic diagram illustrating the structure of [Fe(tris-carbene)$_3$]$^+$ complex exhibiting spin-allowed LMCT absorption and emission. From ref. [35].

----

24 Also known as the Hatchard-Parker actinometer after its inventors. The quantum yields are measured between 250 and 520 nm under conditions defined by an international agreement.

One more example is relevant to the self-cleaning bodies of water by sunlight. Iron(III) complexes are typically present in aqueous solution at pH 7 as hydroxide complexes such as high-spin $[Fe(H_2O)_5(OH)]^{2+}$. The LMCT band corresponding to charge transfer from OH to Fe overlaps sufficiently with the sun's UV radiation for the photoreaction in Equation 2.39 to occur. The resultant OH radical has a redox potential at pH 7 of +1.9 V and can therefore decompose most organic impurities by oxidation. Dimerisation of OH⁻ leads to hydrogen peroxide, also a powerful oxidant:

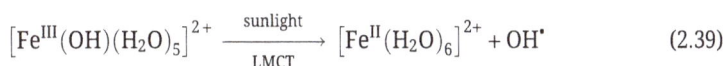

$$\left[Fe^{III}(OH)(H_2O)_5\right]^{2+} \xrightarrow[\text{LMCT}]{\text{sunlight}} \left[Fe^{II}(H_2O)_6\right]^{2+} + OH^{\bullet} \qquad (2.39)$$

This photo-Fenton reaction is also used in the laboratory to generate OH⁻ radicals.[25]

2.3.2.5 Ligand-centred (LC) and ligand-to-ligand charge transfer (LLCT)

Ligand-centred transitions are characteristic of extended aromatic systems. Metallo-porphyrins form classic examples in which the visible absorption and emission spectra are dominated by LC transitions, and some of which have absorption coefficients as high as 10^5–10^6 M^{-1} cm^{-1}. Thus, the red colour of blood depends on LC transitions since haemoglobin contains four iron centres, each bound to protoporphyrin IX. Transitions are also possible from one ligand to another, ligand-to-ligand charge transfer (LLCT).

2.3.2.6 Charge transfer to solvent (CTTS)

Simple inorganic ions with low ionisation energies, both cations and anions, often exhibit broad absorption bands in water and other polar solvents, assigned to charge-transfer to solvent (CTTS). For example, the CTTS bands of metal sulfides are observed at ca. 360 nm in water. Solvated electrons are formed on irradiation into these transitions with a lifetime of ~ 0.6 ms and an absorption band at ~ 720 nm in water. Such hydrated electrons are very strong reducing agents with a standard potential of –2.5 V. They may be detected by their reaction with nitrous oxide (Equation 2.40) which generates an OH radical that is itself a strong oxidising agent as we have seen above:

$$e_{aq}^- + N_2O + H_2O \rightarrow N_2 + OH^- + OH^{\bullet} \qquad (2.40)$$

2.3.2.7 Metal-to-metal charge transfer (MMCT)

Dinuclear or polynuclear metal complexes with at least two metal centres in different oxidation states often exhibit additional absorption bands at long wavelengths that are not present when the metals have the same oxidation state. These bands are

---

[25] The concentration of H_2O_2 in a Bavarian lake has been recorded to reach 1 μM on a sunny September day.

known as metal-to-metal charge-transfer (MMCT) bands or as intervalence charge-transfer bands (IVCT or IT). The classic example is the Creutz-Taube ion[26] in which the band appears in the near-IR region at 1570 nm. The position of the band, its absorption coefficient and band width vary substantially according to the bridging ligand (Scheme 2.22). Such "mixed-valence" complexes are distinguished according to the Robin-Day classification by the degree of interaction between the metal centres: no ground state interaction (type I), mild ground state interaction (type II) or fully delocalised (type III). Those of type II are of special interest because they act as intramolecular models for inner-sphere[27] electron transfer reactions (Marcus-Hush theory). Such MMCT bands are also observed in the solid state, famously in Prussian blue and its analogues (see textbox for application to photography) and in metalloenzymes such as hemerythrin. The MMCT bands can also result in electrochromism since they are linked to redox properties [36].

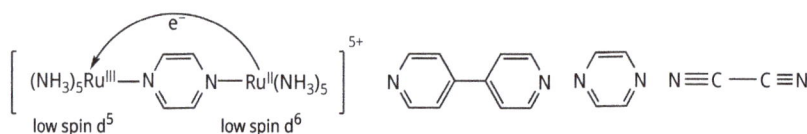

Scheme 2.22: The Creutz-Taube mixed-valence ion showing the MMCT transition and three alternative bridging ligands.

We saw the Marcus relationship between the free energy or driving force for electron transfer and the activation energy for electron transfer in Equation 2.25. Hush related the activation energy for thermal electron transfer $\Delta G^{\ddagger}$ to the energy for optical electron transfer E_{CT}, i.e. the MMCT band. In the limit that the interaction energy between the metal centres is much less than the rearrangement energy λ, the energy of thermal electron transfer band corresponds to $(\lambda + \Delta G^0)/4$, where ΔG^0 is the standard free energy for transfer of an electron between the components (Equation 2.41). In the case of a symmetrical mixed-valence ion like the Creutz-Taube ion, ΔG^0 is zero:

$$4\,\Delta G^{\ddagger} = \lambda + \Delta G^0 = E_{MMCT} \tag{2.41}$$

---

26 Carol Creutz (1945–2013) was the first to synthesise and study a soluble mixed-valence complex when working as a research student under Henry Taube (1915–2005). The latter received the Nobel Prize in 1983 for his studies of electron transfer reactions in metal complexes. See chapter 5 for another contribution from Creutz.

27 The inner sphere of a metal complex refers to ligands directly bound to the metal. The outer sphere refers to counterions, solvent molecules and other solute molecules that are not directly bound or interact very weakly with the metal centre.

Cyanotype photography and blue prints: John Herschel and Anna Atkins

In 1842, shortly after the invention of silver bromide-based photography, John Herschel invented an alternative photographic process, known as cyanotype. Solutions of potassium hexacyanoferrate(III) and ammonium iron(III) citrate (or recently oxalate) are applied to a surface such as paper or cloth. The required image is laid on the surface blocking light from reaching that part of the material. Irradiation with sunlight reduces the ammonium iron(III) citrate to $Fe(H_2O)_6^{2+}$ releasing CO_2 and the reduction product $CO(CH_2CO_2H)_2$. Fe^{2+} reacts with $[Fe(CN)_6]^{3-}$ to form the mixed-valence complex, Prussian blue $KFe^{III}Fe^{II}(CN)_6$, staining the material blue. After washing off the unreacted chemicals, a white image is left on the material with a blue background. The same process was used on large sheets of paper to form "blueprints" for technical drawings until the advent of computer-aided design. Cyanotypes have been revived in recent years by many artists.

Anna Atkins (born 1799) published *Photographs of British Algae: Cyanotype Impressions* in 1843. This was the first book to be printed and illustrated with photos (public domain).

2.3.2.8 Ion-pair charge transfer (IPCT)

When a pair of ions possesses donor-acceptor properties and the ions are present in solution in close proximity, an absorption band may be observed representing charge transfer between the ions. This band is described as an ion-pair charge-transfer (IPCT) band or an outer-sphere charge-transfer band.[27] An example is the 1:1 ion pair formed from the planar metal dithiolene anion with the planar N,N'-dimethyl-4,4'-dipyridinium cation (MV^{2+}, Figure 2.29). As in the Marcus-Hush equation (Equation

(2.41)), there is a linear relationship between the energy of the IPCT band and the free energy for electron transfer between the dianion and the dication $\Delta G^0{}_{IPCT}$. This driving force can be estimated independently from the redox potentials of the ions. The equation can be used for a series of ion pairs to calculate the rearrangement energy λ for electron transfer as 59 kJ mol^{-1}. Remarkably, it is possible to predict and direct the electrical conductivity of pressed powders from the IPCT band. The conductivity varies over a range from 10^{-11} to 10^{-3} Ω^{-1} cm^1 and correlates with the free energy of activation $\Delta G^{\ddagger}{}_{IPCT}$ derived from the frequency of the IPCT band between 0 and 0.7 eV [37].

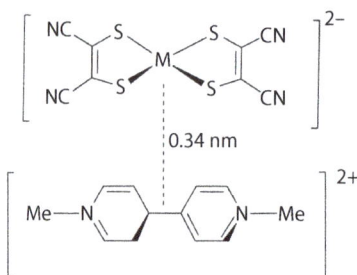

Figure 2.29: Cofacial arrangement of ion pairs (M = Ni, Pt) leading to IPCT bands.

2.3.2.9 Metal-metal bonds

Whereas main group compounds are limited to bond orders of 3, transition metal complexes may contain metal-metal bonds with a bond order up to 5 (or perhaps, even 6). The shortest CrCr quintuple bond known at present has a length of 1.706(1) Å. A main group compound such as ethyne has one σ and two mutually perpendicular π components to its $C\equiv C$ bond. A pair of transition metals can use their d_{z^2} orbitals to form a σ bond, while their d_{xz} and d_{yz} pairs form π bonds. Overlap of pairs of d_{xy} orbitals yields a further component of δ symmetry (two nodal planes containing the metal-metal bond) resulting in a quadruple bond; quintuple bonds use pairs of d_{xy} orbitals and pairs of $d_{x^2-y^2}$ orbitals to generate two components of δ symmetry (Scheme 2.23). If we look at the d-electron counts of each metal, a dinuclear complex with a d^3:d^3 configuration can form a bond of maximum order 3, a d^4:d^4 complex a quadruple bond and a d^5:d^5 complex a quintuple bond. Just as the bond order decreases with main group compounds through population of antibonding orbitals, the bond order of transition metal dimers goes down as further electrons are added. Fully allowed transitions are possible between occupied bonding and corresponding unoccupied antibonding orbitals, $\sigma-\sigma^*$, $\pi-\pi^*$, $\delta-\delta^*$. Irradiation into these transitions weakens the metal-metal bond or even causes dissociation.

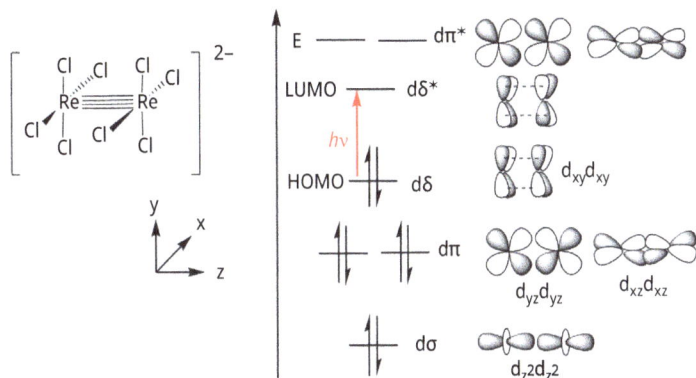

Scheme 2.23: $[Re_2Cl_8]^{2-}$, a quadruply bonded dinuclear complex with d-orbital overlap scheme.

d^4:d^4 quadruple bonds

Examples of complexes with quadruple M–M bonds include $[Re^{III}_2Cl_8]^{2-}$, $[Mo^{II}_2(CH_3)_8]^{4-}$ and numerous analogues with bridging O or N donor ligands. As shown in Scheme 2.23, the HOMO-LUMO transition corresponds to δ–δ^* excitation that can be observed in the visible region. This transition to the $^1\delta\delta^*$ state is allowed ($\varepsilon \sim 1000$ M^{-1} cm^{-1}), and the vibrational fine structure corresponding to the M–M stretching vibration can be observed at low temperature. Emission from the triplet state $^3\delta\delta^*$ has been observed for numerous quadruply bonded complexes with emission lifetimes in the 1–100 μs range [38].

d^7:d^7 single bonds

A classic example of a metal carbonyl with a metal-metal single bond is $Mn_2(CO)_{10}$ with an electron configuration $n = 7$, $T = 18$ at each metal. This complex has been studied extensively by matrix isolation and time-resolved spectroscopy (Figure 2.23). The absorption spectrum of $Mn_2(CO)_{10}$ exhibits an extra band in the near UV when compared with $Cr(CO)_6$. Irradiation of $Mn_2(CO)_{10}$ results in two competing photochemical processes: Mn–Mn bond dissociation to form two square-pyramidal $Mn(CO)_5$ radicals and Mn–CO dissociation to yield $Mn_2(CO)_9$ with a structure containing a semibridging CO group. Both $Mn(CO)_5$ and $Mn_2(CO)_9$ are transient intermediates in solution or in the gas phase and proceed to react further to form the final products. The quantum yields for the two processes are dependent on the wavelength of irradiation indicating that dissociation occurs before the excited state has a chance to equilibrate. On irradiation into the near-UV σ–σ^* transition, Mn–Mn fission dominates, while at shorter wavelengths, Mn–CO dissociation is dominant [32]. Formation of the $Mn(CO)_5$ radicals proves effective as an initiating method for photopolymerisation [39].

d^8:d^8 dimers

If you have followed the electron counts above, you will see that a pair of square-planar complexes with d^8 configurations should not have a metal-metal bond. Indeed, such dimers held together by bridging ligands have long M–M distances and low M–M stretching frequencies because both M–M $d\sigma$ and $d\sigma^*$ orbitals are occupied. However, they have remarkable excited-state properties because another M–M σ-bonding orbital formed from metal p-orbitals lies above the $d\sigma^*$ orbital. Consequently, the excited states have much shorter M–M bonds and higher M–M stretching frequencies than the ground states. Two classic examples are $[Rh_2(CN(CH_2)_3NC)_4]^{4+}$ and $[Pt_2(P_2O_5(BF_2)_2)_4]^{4-}$ (abbreviated $[Pt(pop-BF_2)]^{4-}$; Scheme 2.24 and Figure 2.30A). These complexes exhibit prominent $d\sigma^*$-$p\sigma$ absorption bands in the visible or near-UV regions. They form singlet and triplet excited states, both of which luminesce (lifetimes in MeCN at room temperature, 1.6 ns and 8 μs, respectively, for $[Pt(pop-BF_2)]^{4-}$). Thus they bear a resemblance to organic excimers. The striking increase in Pt–Pt stretching frequency in $[Pt(pop-BF_2)]^{4-}$ can be seen through vibrational progressions in the emission and excitation spectra (Figure 2.30B). The decrease in bond length in the excited state has been measured by a variety of time-resolved X-ray methods. For $[Pt(pop)]^{4-}$ (the parent complex in which the O–BF_2–O groups are replaced by O···H···O bridges), the triplet excited state Pt–Pt bond length is 2.92 ± 0.08 Å according to the time-resolved X-ray scattering, an astonishing 0.24 Å shorter than in the ground state. Another striking demonstration of the compression comes from femtosecond spectroscopy that reveals coherent sinusoidal oscillations, which on Fourier transformation correspond to the ground and excited state Pt–Pt frequencies. These complexes also show interesting photochemical properties, including excited state electron transfer [40].

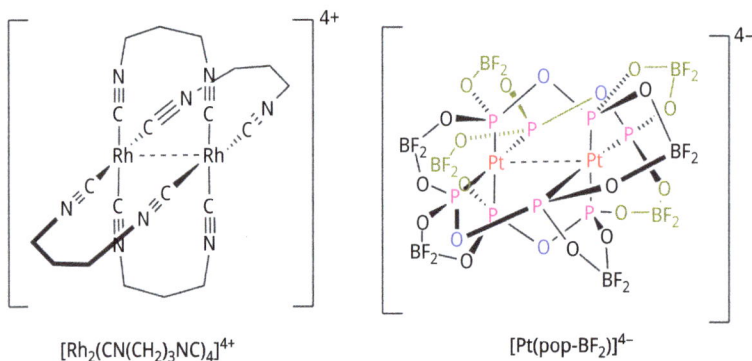

$[Rh_2(CN(CH_2)_3NC)_4]^{4+}$

$[Pt(pop-BF_2)]^{4-}$

Scheme 2.24: Two d^8:d^8 complexes which are more strongly bonded in their excited states.

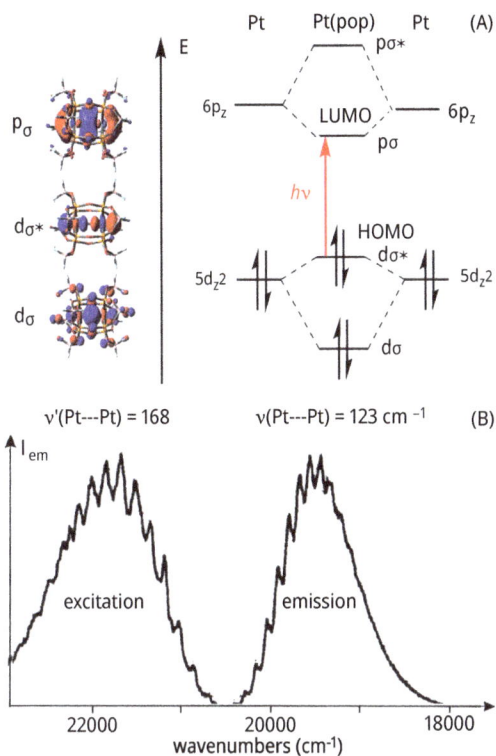

Figure 2.30: (A) MO diagram for $[Pt(pop-BF_2)]^{4-}$. (B) Low-temperature excitation and emission spectra of $(Bu_4N)_4[Pt(pop-BF_2)]$ showing an increase in vibrational spacing in the excited state. Adapted from ref. [40], *Coordination Chemistry Reviews*, *345*, 297, H. B. Gray et al. Copyright 2017 Elsevier.

2.3.3 Lanthanide complexes

Complexes of the lanthanide metals have absorption and emission spectra that are distinct from those of the transition metal complexes. Lanthanide complexes in their typical +3 oxidation state, Ln(III), have electron configurations between $4f^0$ and $4f^{14}$ and these electrons interact extremely weakly with the surrounding ligands. The coupling between the electrons through their orbital and spin angular momentum (Russell-Saunders coupling) is more significant than the energy of interaction with the ligands. We will consider two types of transition MC and CT transitions.

MC transitions

For each configuration in which the 4f shell is partially filled, a variety of electronic states are possible because of Russell-Saunders coupling. As a consequence of the Laporte rule and the lack of interaction with the ligands, absorption spectra show a series of very weak but very sharp bands corresponding to f-f transitions. Photoluminescence involving f-f excited states is also observed with lifetimes in the millisecond range. There is one oxidation state available to all lanthanides: +3. Among Ln(III) ions, only for Ce^{3+} with an f^1 configuration is an excited state also accessible in which the electron is transferred to a d-orbital. In that case, the $4f^1$-$5d^1$ absorption band in the UV region is broad and has a much higher absorption coefficient. However, emission from d-f states may be observed in other lanthanides, notably Tb^{3+}.

When we turn to lanthanides with a stable Ln(II) oxidation state, Eu^{2+} ($4f^7$), Sm^{2+} ($4f^6$) and Yb^{2+} ($4f^{14}$), f-d absorption bands as well as the corresponding d-f emission bands are observed in each. Recently, it has proved possible to synthesise salts of anionic complexes of the type $[Cp'_3Ln(II)]^-$ (Cp' = substituted cyclopentadienyl) for all the lanthanide elements that display absorption bands in the visible region with high absorption coefficients. Those of several of the earlier lanthanide elements (Ln = Pr, Nd, Gd, Tb, Dy, Ho and Er) have $4f^n5d^1$ ground states, rather than the usual $4f^{n+1}$ states.

The f-d absorption band still has enormous technological importance in fluorescent lamps and scintillation devices, even if many lamps have been displaced by LEDs. In these familiar lamps, an electric discharge excites mercury vapour which emits at 254 nm. The inside of the tube is coated with a solid colourless "phosphor" that includes Ce^{3+}. The f-d absorption band of Ce^{3+} overlaps with the emission of mercury resulting in excitation of Ce^{3+}; this energy is transferred in turn to other lanthanide ions that emit in the visible region through allowed transitions, notably Tb^{3+}, Eu^{3+} and Eu^{2+}.

In Section 1.4, we encountered Nd:YAG lasers that function by populating an f-f excited state of Nd^{3+} that undergoes stimulated emission at 1064 nm to a lower f-f excited state (Scheme 1.2). We also saw the red emission of Eu^{3+} embedded in Euro banknotes as a security measure (Figure 1.8). The long luminescence lifetime of f-f transitions of lanthanide ions can be put to advantage for sensing analytes and for imaging if shorter-lived luminescence is eliminated by time-gating. How do you overcome the small f-f absorption coefficients? The answer is to build in a strong absorber as an *antenna* and a sensitiser for the f-f excited states. This concept has been specially productive for biological systems in which short-lived luminescence would otherwise interfere. A multidentate ligand that binds to the lanthanide ions sufficiently strongly for the complex to be kinetically and thermodynamically inert is employed. A pendant chromophore acts as an antenna – its triplet excited state is quenched by energy transfer (either Förster or Dexter transfer) or electron transfer to the lanthanide ions. In the case of Eu^{3+}, all these mechanisms may occur. The excited lanthanide ion may be observed directly through its own emission or may undergo Förster energy transfer to another acceptor whose luminescence is imaged at very long wavelengths. The presence of analytes may be sensed through changes in (a) the emission intensity, (b) the ratio of emission from two different lanthanides, for instance Eu^{3+} and Tb^{3+} (Figure 2.31), (c) the emission lifetime or (d) circularly polarised emission when the lanthanide lies in an enantiopure environment [41].

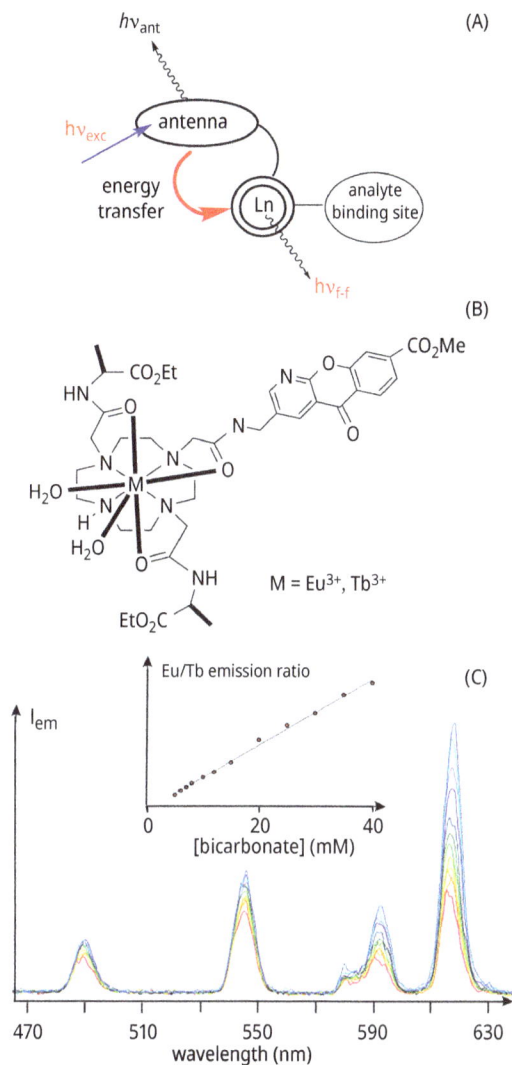

Figure 2.31: (A) Principle of antenna complexes; (B) bicarbonate binding complexes of Eu^{3+} (red emission) and Tb^{3+} (green emission); (C) selective determination of bicarbonate concentration from the ratio of Eu^{3+} to Tb^{3+} emission. Spectra measured in buffer solutions of human serum albumin. The inset shows a plot of the ratio of emission intensities against $[HCO_3^-]$. Adapted with permission from ref. [42], *Chemistry – a European Journal*, *18*, 11604 Parker et al. Copyright 2012 Wiley-VCH Verlag GmbH, Weinheim.

Questions

1. (a) Draw a simple Jablonski diagram, including vibrational energy levels, for a molecule exhibiting fluorescence and phosphorescence. (b) Explain the terms *vibronic relaxation, internal conversion, fluorescence, intersystem crossing and phosphorescence.* (c) Give approximate lifetimes of these processes for an organic molecule. (d) Formulate the rules of *Vavilov and of Kasha.* (e) Compare the activation energies of radiative and non-radiative processes and explain why this is relevant in experimental emission spectroscopy in a condensed phase.

2. (a) Give the energy-level schemes for the electron configurations of the frontier molecular orbitals for the ground and excited states of ethyne. (b) Draw the corresponding diagram of the three resulting energy states.

3. According to quantum theory, the probability that a molecule is in its excited state is given by the oscillator strength. It corresponds to the effective number of electrons transferred from the ground state to the excited state by light absorption and is proportional to the square of the transition dipole moment consisting of the product between the overlap integral of the vibronic and spin (electronic) wavefunctions and the dipole moment operator. (a) What approximation is necessary to justify that separation into independent contributions of nuclei and electrons? (b) What are the values of the oscillator strength for an allowed and a forbidden transition?

4. (a) Draw the potential energy curves including the three lowest vibronic energy levels for a diatomic molecule X–Y having *identical* bond lengths in the ground and excited states. (b) Also assign the vibrational quantum numbers. (c) What principle enables electron movement to be represented as a vertical arrow? (d) Mark and explain the Stokes shift in the above potential energy scheme. (b) What factors determine the value of the Stokes shift?

5. (a) Draw the potential energy curves including the three lowest vibrational energy levels for a diatomic molecule X–Y having a *larger* X–Y bond length in the ground state than in the excited state. (b) How can the vibrational energy of the ground and excited states be determined from the electronic spectra?

6. XeF has a much shorter bond length in its excited state than in its ground state. What applications depend on this property and why?

7. (a) Give the two most important mechanisms of excited state quenching and discuss the relevant properties of the corresponding quenchers. (b) Give the thermodynamic requirements for the two mechanistic types. (c) For which quenching type is molecular oxygen an extremely efficient quencher; explain the underlying principle? (d) What does the Wigner spin conservation rule mean? (e) Give an example with molecular oxygen.

8. (a) What is a Stern-Volmer diagram? (b) What information can be gained therefrom?

9. Discuss the four basic mechanisms of energy transfer.

10. (a) Give a definition for photosensitisation and discuss the two most important mechanisms. (b) How can O_2 be activated by visible light?

11. Discuss the most important excited states of organic molecules.
12. (a) Under what circumstances do electronic absorption bands show a strong dependence on the nature of the solvent? (b) Explain the difference between hypsochromic and bathochromic shift.
13. (a) Describe the difference between charge-transfer complex, exciplex and excimer. (b) Draw the potential energy diagram for exciplex formation.
14. What information is required to distinguish between a d-d transition and a charge-transfer transition of a transition metal complex containing a single metal? (b) What types of transition are possible in transition metal complexes containing two metals that are impossible in mononuclear complexes? (c) How do spectra of lanthanide complexes differ from those of transition metal complexes?
15. Use the principles of d-electron counting and the requirements for different types of electronic transition to complete the following table. The first entry provides a worked example:

	d^n	d-d	MLCT	LMCT
TiI_4	$n = 0$	No	No	Yes
$Cu(NCMe)_4^+$				
$Co(NH_3)_6^{3+}$				
$Co(NH_3)_5I^{2+}$				
$Fe(CO)_5$				
MnO_4^{2-}				

16. The MLCT absorption maxima of the isoelectronic series follow the trend λ_{max} $[Mn(CO)_6]^+ < Cr(CO)_6 < [V(CO)_6]^-$. Further isoelectronic complexes are now known for Ti and Fe. (a) Predict the wavelength trends of the complete series of five isoelectronic complexes. (b) What are their d-electron counts and total valence electron counts at the metals?
17. Predict the wavelength trends of the LMCT series: $Co(NH_3)_5X^{2+}$ (X = F, Cl, Br, I).
18. $Mo_2(NMe_2)_6$ contains a MoMo triple bond. What UV/VIS transitions are possible because of this bond?
19. What data are required to estimate an excited state redox potential? Give an example.
20. Why is the product quantum yield of LMCT excitation of tris(oxalate)iron(III) larger than 1.0?

References

[1]   N. J. Turro, V. Ramamurthy, J. C. Scaiano, *Modern Molecular Photochemistry of Organic Molecules*, vol. 188, University Science Books, Sausalito, CA, **2010**.

[2]   H. Uoyama, K. Goushi, K. Shizu, H. Nomura, C. Adachi, *Nature* **2012**, *492*, 234–238.

[3]   a) Z. Wu, J. Nitsch, T. B. Marder, *Advanced Optical Materials* **2021**, *9*, 2100411; b) A. S. Marfunin, *Spectroscopy, Luminescence and Radiation Centers in Minerals*, Springer, Berlin, **1979**; c) G. Blasse, P. H. M. de Korte, A. Mackor, *J. Inorg. Nuclear Chemistry* **1981**, *43*, 1499–1503; d) H. Shi, W. Yao, W. Ye, H. Ma, W. Huang, Z. An, *Accounts of Chemical Research* **2022**, *55*, 3445–3459.

[4]   P. Klan, J. Wirz, *Photochemistry of Organic Compounds*, John Wiley, United Kingdom, **2009**.

[5]   F. Reichenauer, C. Wang, C. Forster, P. Boden, N. Ugur, R. Báez-Cruz, J. Kalmbach, L. M. Carrella, E. Rentschler, C. Ramanan, *Journal of the American Chemical Society* **2021**, *143*, 11843–11855.

[6]   S. E. Bell, J. N. Hill, A. McCamley, R. N. Perutz, *Journal of Physical Chemistry* **1990**, *94*, 3876–3878.

[7]   J. N. Hill, R. N. Perutz, S. M. Tavender, *The Journal of Physical Chemistry* **1996**, *100*, 934–940.

[8]   W. Martin, *Journal of Research of the National Bureau of Standards. Section A, Physics and Chemistry* **1971**, *75*, 109.

[9]   D. C. Harris, M. D. Bertolucci, *Symmetry and Spectroscopy*, Dover Books, **1990**.

[10] A. H. Zewail, *Angewandte Chemie International Edition* **2000**, *39*, 2586–2631.

[11] T. Mirkovic, E. E. Ostroumov, J. M. Anna, R. Van Grondelle, Govindjee, G. D. Scholes, *Chemical Reviews* **2017**, *117*, 249–293.

[12] R. J. Cogdell, N. W. Isaacs, A. A. Freer, T. D. Howard, A. T. Gardiner, S. M. Prince, M. Z. Papiz, *FEBS Letters* **2003**, *555*, 35–39.

[13] a) J. C. Johnson, A. J. Nozik, J. Michl, *Accounts of Chemical Research* **2013**, *46*, 1290–1299; b) O. P. Dimitriev, *Chemical Reviews* **2022**.

[14] C. Hetzer, D. M. Guldi, R. R. Tykwinski, *Chemistry–A European Journal* **2018**, *24*, 8245–8257.

[15] a) Z. Yang, Z. Mao, Z. Xie, Y. Zhang, S. Liu, J. Zhao, J. Xu, Z. Chi, M. P. Aldred, *Chemical Society Reviews* **2017**, *46*, 915–1016; b) C. E. Housecroft, E. C. Constable, *Journal of Materials Chemistry C* **2022**, *10*, 4456–4482.

[16] P. W. Atkins, J. de Paula, J. Keeler, *Physical Chemistry*, 12th ed. Oxford, **2022**.

[17] R. A. Marcus, *Angewandte Chemie International Edition in English* **1993**, *32*, 1111–1222.

[18] H. Makita, G. Hastings, *Proceedings of the National Academy of Sciences* **2017**, *114*, 9267–9272.

[19] C. K. Prier, D. A. Rankic, D. W. C. MacMillan, *Chemical Reviews* **2013**, *113*, 5322–5363.

[20] a) E. C. Gentry, R. R. Knowles, *Accounts of Chemical Research* **2016**, *49*, 1546–1556; b) W. B. Swords, G. J. Meyer, L. Hammarström, *Chemical Science* **2020**, *11*, 3460–3473.

[21] L. M. Tolbert, K. M. Solntsev, *Accounts of Chemical Research* **2002**, *35*, 19–27.

[22] N. Zivic, P. K. Kuroishi, F. Dumur, D. Gigmes, A. P. Dove, H. Sardon, *Angewandte Chemie International Edition* **2019**, *58*, 10410–10422.

[23] L. Capaldo, D. Merli, M. Fagnoni, D. Ravelli, *ACS Catalysis* **2019**, *9*, 3054–3058.

[24] J. M. Mayer, *Accounts of Chemical Research* **2011**, *44*, 36–46.

[25] J. Zhou, Q. Liu, W. Feng, Y. Sun, F. Li, *Chemical Reviews* **2015**, *115*, 395–465.

[26] https://en.wikipedia.org/wiki/Photon_upconversion#/media/File:Stokes_and_Anti-Stokes_emission.jpg. Accessed on April 15th, 2004.

[27] B. S. Richards, D. Hudry, D. Busko, A. Turshatov, I. A. Howard, *Chemical Reviews* **2021**, *121*, 9165–9195.

[28] S. Sasaki, G. P. Drummen, G.-I. Konishi, *Journal of Materials Chemistry C* **2016**, *4*, 2731–2743.

[29] S. V. Rosokha, J. K. Kochi, *Accounts of Chemical Research* **2008**, *41*, 641–653.

[30] C. Förster, K. Heinze, *Chemical Society Reviews* **2020**, *49*, 1057–1070.

[31] M. Brorson, M. R. Dyxenberg, F. Galsbøl, *Acta Chemica Scandinavica* **1996**, *50*, 289–293.

[32] J. J. Turner, M. W. George, M. Poliakoff, R. N. Perutz, *Chemical Society Reviews* **2022**, *51*, 5300–5329.

[33] F. H. Pilz, J. Lindner, P. Vöhringer, *Physical Chemistry Chemical Physics* **2019**, *21*, 23803–23807.

[34]   H. Kisch, *Semiconductor Photocatalysis: Principles and Applications*, Wiley-VCH Weinheim, **2015**.

[35]   K. S. Kjær, N. Kaul, O. Prakash, P. Chábera, N. W. Rosemann, A. Honarfar, O. Gordivska, L. A. Fredin, K.-E. Bergquist, L. Häggström, *Science* **2019**, *363*, 249–253.

[36]   a) J. Ribas Gispert, *Coordination Chemistry*, Wiley-VCH Weinheim, **2008**; b) K. D. Demadis, C. M. Hartshorn, T. J. Meyer, *Chemical Reviews* **2001**, *101*, 2655–2686.

[37]   H. Kisch, *Coordination Chemistry Reviews* **1997**, *159*, 385–396.

[38]   L. A. Wilkinson, *Organometallic Chemistry* **2020**, *43*, 111–143.

[39]   C. P. Simpson, O. L. Adebolu, J.-S. Kim, V. Vasu, A. D. Asandei, *Macromolecules*, **2015**, *48*, 6404–6420.

[40]   H. B. Gray, S. Záliš, A. Vlček, *Coordination Chemistry Reviews* **2017**, *345*, 297–317.

[41]   D. Parker, J. D. Fradgley, K.-L. Wong, *Chemical Society Reviews* **2021**, *50*, 8193–8213.

[42]   D. G. Smith, R. Pal, D. Parker, *Chemistry–A European Journal* **2012**, *18*, 11604–11613.

3 Solids

3.1 Spectra of solids

3.1.1 Absorption spectra and quantum size effect

A crystal consists of an infinite three-dimensional arrangement of atoms undergoing electronic interactions. While the interaction generates sharp energy levels up to a few hundred atoms, broad bands, extending through the whole crystal, are obtained above that number. Like the HOMO and LUMO of a molecule, all electronic states are fully occupied in the valence band whereas the conduction band contains only empty states. In a metal both bands overlap, and the electrons can move freely in the conduction band inducing electrical conductivity. In contrast, there is no overlap between the two bands in a semiconductor since they are separated by an energy gap (E_g) in the range of 1–6 eV (ca. 1250–250 nm).

An electronic absorption spectrum offers a way of probing the energy gap. We discussed in Chapter 1 how the spectrum of molecules in a liquid or gaseous state can be measured. Essentially, the absorbed light intensity is obtained as the difference in intensities before and after passing through the sample cuvette. For an insoluble solid this is possible only when it can be prepared as a thin transparent layer

$$I_{tr} = I_o \times \exp(-\alpha d) \tag{3.1}$$

exhibiting no scattering and reflection. The Beer-Lambert law can now be written as Equation 3.1 where d is given in centimetres and the absorption coefficient α now has the dimension of cm^{-1}. It is analogous to the molar absorption coefficient of a dissolved molecule. The distance at which the intensity of the light wave is lowered to its $1/e$ value (about 63%) is called the *absorption length* or *penetration depth* and the plot of α as function of wavelength or energy represents the absorption spectrum of the solid. Figure 3.1 displays the spectra of silicon and gallium arsenide as examples for an *indirect* and a *direct* semiconductor, respectively. The notation *direct* corresponds to an allowed electronic transition and *indirect* to a forbidden transition. As expected from theory, and unlike the absorption spectrum of a molecule, no maxima are observable. Accordingly, absorption of a photon with an energy equal or larger than the bandgap (E_g) results in the transfer of an electron from the occupied valence to the unoccupied conduction band. From the onset of a plot of α vs energy of exciting light the value of E_g becomes available (Scheme 3.1). The much steeper decay of the absorption coefficient in the case of GaAs is characteristic for a direct semiconductor.

The absence of an absorption maximum can be rationalised by considering that in MO theory the number of orbitals increases with the number of atoms (N) present in a compound. Up to about 200 atoms, distinct energy levels are present, but above they collapse to broad *energy bands* (Scheme 3.1) [1]. This is the case for TiO_2 crystals of a

https://doi.org/10.1515/9783111029375-003

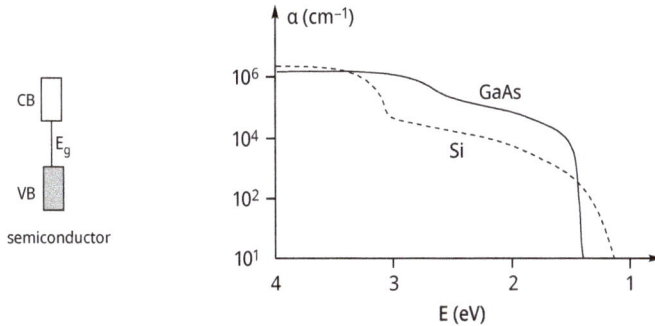

Figure 3.1: Schematically drawn absorption spectra of Si (indirect bandgap of 1.12 eV) and GaAs (direct bandgap of 1.40 eV). 1 eV = 8065 cm^{-1}.

size larger than about 5 nm having a bandgap of 3.2 eV. Decreasing the size to 3 nm increases the bandgap by 0.25 eV. This type of bandgap widening upon decreasing the particle size is called the *quantum size effect* and such nanoparticles are named

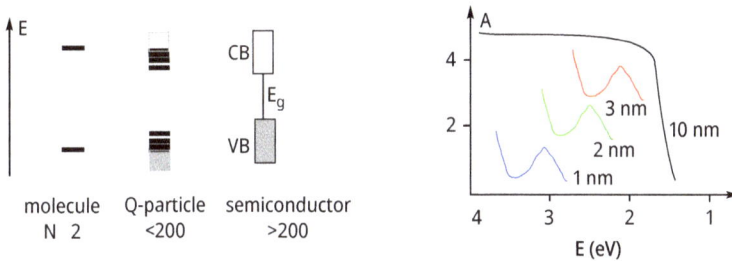

Scheme 3.1: Left: Decrease of HOMO-LUMO energy separation upon increasing number of atoms (N). VB, CB, and E$_g$ correspond to valence band, conduction band and bandgap, respectively. Right: Blueshift of absorption maximum with decreasing particle size for a direct semiconductor.

Q-particles, quantum dots (QD) or nanocrystals.[1] Absorption of a photon of energy larger or equal to the bandgap generates a loosely bound electron-hole pair (Mott-Wannier exciton).[2] Theory predicts that, when the crystal diameter approaches the exciton diameter (Bohr radius), a blueshift of the bandgap should be observable. For cubic CdS the lattice constant is 0.58 nm and the Bohr radius is 2.70 nm. An exciton is usually delocalised over a sphere of several lattice constants.[3]

---

1 Moungi Bawendi, Louis Brus and Aleksey Yekimov won the 2023 Nobel Prize for chemistry for the discovery and synthesis of quantum dots.
2 In the language of solids, an exciton is equivalent to an excited state for a molecule.
3 Lattice constants describe the dimensions of a unit cell, which is the smallest unit of a crystal lattice from which the complete structure can be built by translation operations.

Quantum dots usually suffer from fast agglomeration which is prevented by addition of capping agents (e.g. $HSCH_2COOH$). Classical examples are sulfides and selenides of cadmium and other metals. For instance, the bandgap of CdS can be tuned from 2.40 eV to 3.07 eV upon diminishing the crystal size from 10 nm to 3 nm, respectively, corresponding to a colour change from yellow to almost colourless. These quantum dots are utilised in a great variety of light-based technologies [2].

In general, solid compounds cannot easily be obtained as transparent films, but are rather present as powder or opaque layers. Therefore, the measurement of a is not possible by standard absorption techniques. Methods of choice are *Diffuse Reflectance Spectroscopy*[4] (DRS) and *Photoacoustic Spectroscopy* (PAS).

3.1.2 Diffuse reflectance spectroscopy

In diffuse reflectance spectroscopy a conventional absorption spectrophotometer must be equipped with an integrating sphere measuring the diffuse reflectance of the powder relative to a white standard such as alumina or barium sulfate [3]. The Kubelka-Munk function $F(R_\infty)$ relates the diffuse reflectivity R_∞ of an infinitely thick sample layer to the absorbance according to Equation 3.2 where a and S are the absorption and scattering coefficients, respectively. Equation 3.3 relates R_∞ to the ratio of the reflectivity of the sample and standard.

$$F(R_\infty) = (1 - R_\infty)^2 / 2\ R_\infty = a/S \qquad (3.2)$$

$$R_\infty = R_{sample} / R_{standard} \qquad (3.3)$$

Equation 3.2 applies only for *monochromatic irradiation, infinitely thick sample layers (for most powders reached at about 5 mm), low sample concentrations, uniform distribution, and absence of fluorescence.* Dilution with a white standard considerably improves results affording spectra of good resolution. Unfortunately, this method of sample preparation is only rarely reported [3a].

Assuming the scattering coefficient is independent of wavelength, the Kubelka-Munk function becomes proportional to the absorption coefficient. A plot of $F(R_\infty)$ as a function of wavelength (Figure 3.2) of the exciting light therefore affords an "absorption spectrum" or more correctly, a diffuse reflectance spectrum of the powder.[5]

---

4 Linguistically the name is not correct, since the method is not a "diffuse" spectroscopy, but the spectroscopy of light reflected by a diffusively scattering material.

5 When comparing DRS spectra of various powders, it is important to measure diluted samples of identical concentrations. The Kubelka-Munk function is available in standard menus of spectrometers.

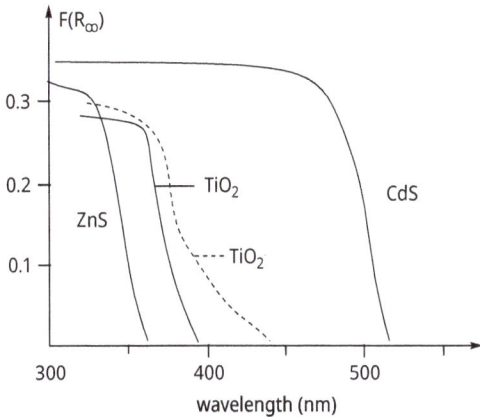

Figure 3.2: Schematically drawn diffuse reflectance spectra of the three classical ultrapure semiconductors titania (anatase modification), zinc and cadmium sulfide (both cubic crystal structure). Corresponding bandgaps are 3.6 eV (347 nm), 3.2 eV (390 nm), and 2.4 eV (520 nm). Commercial titania contains small amounts of impurities as indicated by low-energy shoulder (dashed line).

3.1.3 Measurement of bandgap

Since semiconductors are very seldom used as transparent materials in chemistry, but rather as opaque layers and powders, the bandgap cannot be obtained directly from the absorption spectrum (Figure 3.1). However, this is possible by electrochemical and diffuse reflectance measurements [4]. For the latter, the absorption coefficient is related to the bandgap according to Equation 3.4 [5]. Values of j are ½ and 2 for direct (i.e. allowed) and indirect (i.e. forbidden) transitions, respectively.

$$[\alpha\,(h\nu\,)]^{1/j} = A(h\nu - E_g) \qquad (3.4)$$

According to Equation 3.2 the absorption coefficient α is given by the product $F(R_\infty) \times S$. With this value and combining S with A to give a new constant B, we obtain Equation 3.5 [5a, 6].

$$[F(R_\infty)h\nu]^{1/j} = B(h\nu - E_g) \qquad (3.5)$$

Thus, from a plot of $[F(R_\infty)h\nu]^{1/j}$ vs $h\nu$ and extrapolation of the linear part to $[F(R_\infty)h\nu]^{1/j} = 0$ as exemplified in Figure 3.3, the bandgap is obtained. A good linear fit is found only for $1/j = 0.5$, as expected for the indirect band-to-band transition of TiO_2 at 3.16 eV. Often the linear part is too short to enable reliable extrapolation (Figure 3.3).

This deviation at energies smaller than the bandgap suggest the presence of localised energy levels, so-called *surface states* since they are not delocalised through the whole crystal like the valence and conduction bands. Often the data do not allow a

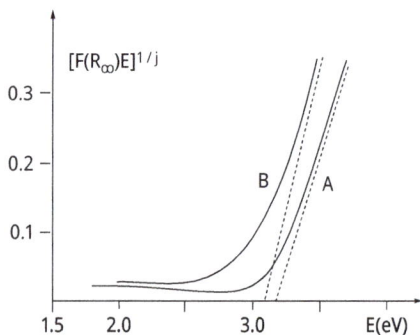

Figure 3.3: Plot of the modified Kubelka-Munk function vs. energy of exciting light leads to bandgaps of 3.16 eV for genuine anatase (A) and 3.11 eV for titania containing impurities (B).

precise linear extrapolation and a decision between direct and indirect character of the bandgap transition is not possible [4b, 7].

Figure 3.2 illustrates how the presence of surface states influences the shape of the diffuse reflectance spectrum. Whereas both sulfides and one titania sample exhibit an ideal shape, the other titania (dashed line) shows a distinct shoulder at long wavelengths. This suggests that localised, i.e. sharp, energy levels are present within the bandgap as depicted in Figure 3.4. The reason that no transitions are drawn between surface states is that the probability and therefore the intensity of an electronic transition is proportional to the density of states involved. Transitions between surface states are therefore "forbidden". Surface states can be introduced in a controlled way by doping or surface modification, i.e. introducing donor or acceptor ions in the bulk or only at the surface.

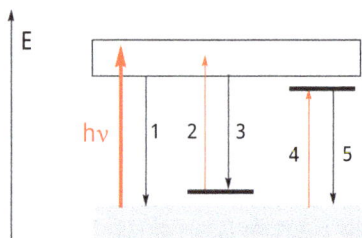

Figure 3.4: Simplified scheme of some basic electronic transitions in the absence (1) and presence (2–5) of surface states.

3.1.4 Emission spectra of solids

The principles of emission spectroscopy described in Chapter 1.2 and 1.3 apply equally to solids. Excitation of ZnS powder at 300 nm affords an emission spectrum schematically drawn in Figure 3.5. Three maxima are found at 360 nm, 400 nm and 690 nm, whereas a rather broad band appears at 490–530 nm [8]. As the excitation spectrum recorded at 420 nm exhibits peaks at 345–360 nm, the first two maxima of the emission spectrum originate from band-to-band transitions and emissions involving surface states (Figure 3.4, processes 1 and 3, 5). By recording excitation spectra at 550 nm or 720 nm, it was shown that the maximum at 370 nm is probably due to traces of zinc oxide. This is confirmed by the disappearance of these broad bands upon removing ZnO through washing with acetic acid. Recording at 420 nm corresponds to the band-to-band absorption.

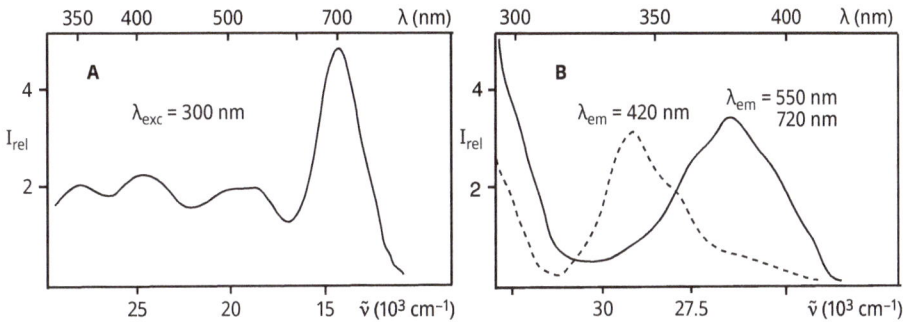

Figure 3.5: Schematically drawn emission (A) and excitation (B) spectra of *dry* ZnS powder at room temperature. Adapted from ref. [8].

The spectrum of an *aqueous* suspension of zinc sulfide is different from the emission of dry zinc sulfide (Figure 3.6): it contains the band-to-band emission at 345 nm and only one further but intense band at 430 nm (2.90 eV). The latter is known in the literature as *self-activated* emission [9] originating from excess adsorbed zinc ions. This is supported by its disappearance upon addition of sulfide ions. As expected, the emission is increased upon addition of zinc ions due to formation of zinc sulfide.

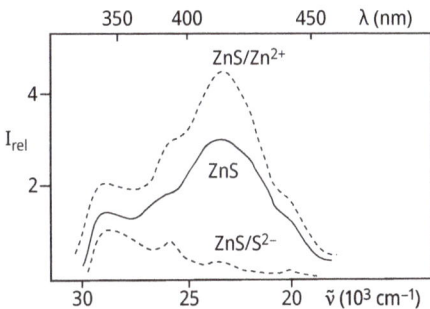

Figure 3.6: Schematically drawn emission spectra of an aqueous suspension of ZnS in the absence and presence of zinc or sulfide ions. Adapted from ref. [8].

Not only are many nanoparticles highly emissive with long emission lifetimes, but their emission wavelengths can be tuned by their size thanks to the quantum size effect. Quantum dots (QDs) based on semiconductors such as MS, MSe, MTe (M = Zn, Cd, Pb) as well as carbon and gold QDs are photostable and often have broad absorption bands.

By suitable choice of semiconductor and size, QDs are available with emission wavelengths from 300–1300 nm. They can be surface-modified for binding to biological targets (termed bioconjugation) or can be synthesised within a live cell (yeast or *E. coli*) to make them biocompatible. When synthesised in organic solvents, ligand-exchange methods can be employed to render them water-soluble. Numerous applications have been reported in imaging and sensing that depend on the ability to introduce specific labels into a biological system. They have also been used for measurement of temperature in vivo and for drug delivery. Near-IR emitting QDs are particularly useful for imaging deep tissue. When acting as sensors, QDs can be stimulated by all the mechanisms highlighted in Chapter 1.1 (photoluminescence, electroluminescence, chemiluminescence, etc.). Fluorescence resonance energy transfer (FRET, Chapter 2.2.3) can be used to measure distance. Luminescence imaging can also be combined with other imaging modes such as positron emission tomography [10].

3.1.5 The photonic bandgap

As discussed above on the quantum size effect, the crystal size may have a significant influence on the (optical) bandgap. Additionally, the *lattice structure* may also influence light absorption, but without changing the bandgap. This happens when the refractive index varies periodically in space and when its dimension is in the range of the wavelength. Now, propagation of certain waves along a specific crystallographic axis becomes forbidden. In simple terms, the *photonic bandgap* can be considered as a band of wavelengths that are Bragg-diffracted.[6] In that case the material is called a *photonic crystal* [11]. As a result a new band, the so-called *stop-band*, appears in the specular reflectance spectrum of the material.[7]

A classic example is the mineral opal which consists of a cubic or hexagonal close-packed lattice of about 200-nm sized silica spheres. Their stacked planes generate a diffraction grating and their orientation relative to the incident light determines the colour of the material. Consequently, the opal exhibits a beautiful play of colour. Examples are the opalescent (or iridescent) colours of fish scales, mother-of-pearl and some minerals. Artificial materials with opal structures are obtainable from colloidal systems which may spontaneously form close-packed spheres of various size. To observe a photonic effect, the refractive index of the spheres must be very high, whereas it should be very

---

6 Diffraction of X-rays in crystals is described as a set of reflections off regularly spaced planes leading to peaks in the diffraction pattern. The spacing of the planes corresponds to the wavelength of the X-rays. A parallel phenomenon can be observed in the visible region if there are periodically spaced features separated by distances corresponding to the wavelength of visible light.

7 The specular reflectance is the mirrorlike reflectance of light at a planar surface. In this case the angles to the surface normal are identical for incident and reflected rays, but they are different for diffuse reflectance.

low for the macropores between them. This condition is fulfilled for titania as light absorber when it contains air-filled pores. It is easily synthesised by first spreading a slurry of nanocrystalline titania and monodisperse polystyrene spheres on a glass substrate [12]. Subsequent calcination at 500 °C burns off the organic spheres and leaves a titania matrix filled with close-packed air spheres. This type of arrangement is known as an inverse opal structure. As predicted by theory, the photonic bandgap increases in wavelength (decreases in energy) with increasing sphere size. For titania it changes from 520 nm to 1120 nm upon varying the diameter from 395 nm to 770 nm. A titania inverse opal exhibited a 4-fold higher photocatalytic activity in aerial dye degradation [13].

3.1.6 Photoacoustic spectroscopy

In Figure 1.4B we summarised the major processes of excited state deactivation. Whereas energy transfer, luminescence, and chemical reaction were already mentioned briefly, this was not the case for one of the major deactivation paths: the conversion of light into thermal energy through radiationless conversion to the ground state. This process results in a temperature rise of the absorbing material. When a *pulsed* light source is employed, the periodic temperature pulsations generated in a solid or gas give rise to thermal waves which can be directly sensed by a piezoelectric sensor. In indirect sensing, the thermal waves in the solid sample generate acoustic waves in the gaseous or liquid phase in contact with the solid, in some cases even audible by the human ear.[8] Thus, absorbed photons are converted into a sound wave, the intensity of which depends on the amount of light absorbed and on the absorption coefficient of the sample. This is the basis of photoacoustic spectroscopy (PAS). Therefore, changing the wavelength of exciting light, changes the intensity of the photoacoustic signal and a plot of sound intensity vs wavelength affords an absorption spectrum of the material.

The basic parts of a photoacoustic spectrometer are depicted in Figure 3.7 for the case of a suspended solid. After leaving the excitation monochromator, light passes through a chopper and arrives in the sample cell, generating thermal waves at particles of different distances from the surface. Their thermal diffusion length μ_s decides if they reach the surface and generate a sound wave detectable by microphone M. Obviously, waves generated at particles 1 have a lesser chance of reaching the sensing gas phase than particles 2.

To a first approximation, the intensity of the PA signal of a solid sample depends on the penetration depth of light (see Equation 3.1) and on the thermal diffusion length. The latter is given by Equation 3.6, where κ is the thermal

---

8 This was first observed in the nineteenth century by Alexander G. Bell, the inventor of the telephone. He used chopped sunlight and a microphone as detector.

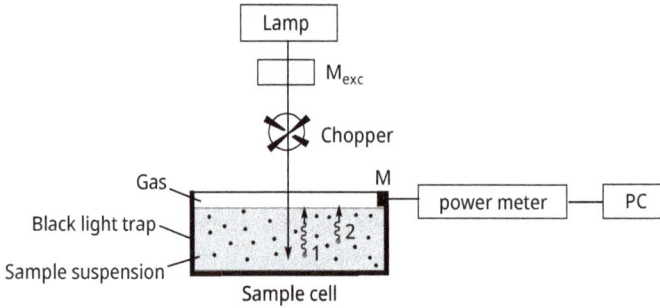

Figure 3.7: Basic components of a photoacoustic spectrometer for measuring the absorption spectrum of a suspended solid. Thickness of the sensing gas layer (air, N_2) is smaller than 1 mm. Inner walls of the sample cell are highly polished to prevent light absorption. A slightly modified sample cell is used for the measurement of gases and solids. LEDs or pulsed lasers are used as light sources.

$$\mu_s = (2\kappa / C\rho\,\omega)^{1/2} \qquad (3.6)$$

conductivity (W m^{-1} · K^{-1}), C the specific heat (J · kg^{-1} · K^{-1}), ρ the density (kg · m^{-3}), and ω the angular frequency $\omega = 2\pi f$ with f being the modulation frequency of the chopper (or the laser pulse repetition rate in the case of laser irradiation). The latter is usually in range of 1–1000 Hz. Typical thermal diffusion lengths vary from 4×10^{-6} to 170×10^{-6} m.

According to Equation 3.6 the diffusion length increases with a smaller modulation frequency. This relationship enables measurement of the absorption of a solid material for different penetration depths as schematically drawn in Figure 3.8 for a waxed red apple. At f = 220 Hz only the spectrum of the thin surface wax layer is obtained, but at 33 Hz the red skin beneath is observed through the strong absorption in the green at about 510 nm [14].

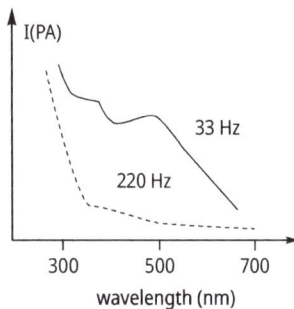

Figure 3.8: Photoacoustic spectrum of a waxed red apple at two modulation frequencies. Schematically redrawn from ref. [14a].

This unique depth dependence and the relatively simple instrumentation are the basis for the impressive range of practical applications in technology and science, including biology and medicine [15]. Examples include the remote sensing of explosive gases and drugs [16], photoacoustic tomography in medicine [17], herbicide monitoring [18], analysis of human breath [14b], and a photoacoustic fibre for ultrasound generation [19].

Depending on details of the photophysical processes occurring after light absorption, a phase shift between optical excitation and emitted acoustic wave may be observed. This shift often allows detection and lifetime measurements of excited states unobservable by other methods. Photoacoustic methods can also be used to measure quantum yields (see Textbox Chapter 1.6)

3.2 Solar cells, light emitting diodes and diode lasers

Semiconductors are critical to numerous applications: here we consider solar cells, light-emitting diodes and diode lasers in which light absorption or emission occurs across the bandgap, E_g. We note that E_g is defined as the energy difference between the upper and lower edges of the valence (E_v) and conduction (E_c) bands (Figure 3.9). Obviously, absorption of light having the energy $h\nu \geq E_g$ generates an electron in the conduction band and therefore also electrical conductivity. Materials with bandgaps above 6 eV, corresponding to the far ultraviolet, are electrical insulators.

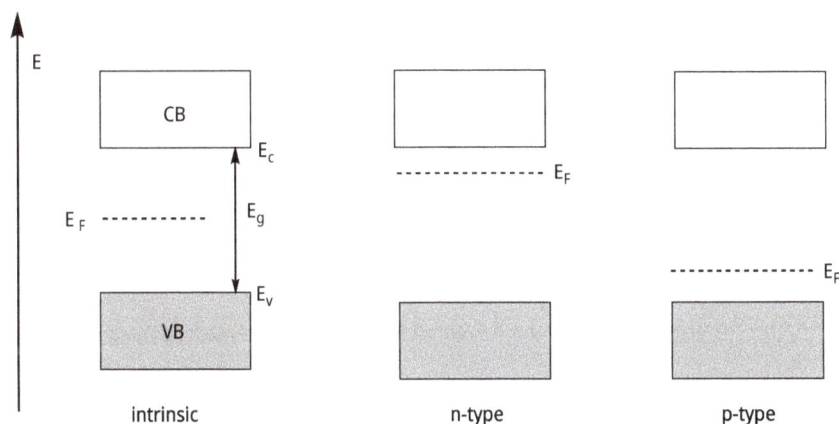

Figure 3.9: Schematic sketch of basic semiconductor electronic band structures.

For the understanding of the photoredox properties of semiconductors, the concept of the *Fermi level* (E_F) is of basic importance. Its value is defined by the energy necessary to add one electron to a solid. According to band structure theory, it corresponds for an ideal semiconductor to the hypothetical energy level of an electron located in the middle of the bandgap. Such a material is an intrinsic semiconductor. However, a real material always contains some impurities generating localised energy states within the bandgap (i.e. surface states, see Figure 3.4). They can be introduced in a controlled way through doping with electron accepting or electron donating atoms generating Fermi levels close to the valence and conduction band, respectively. Typical distances to the band edges are 0.1–0.3 eV. The resulting materials are named p-type and n-type semiconductors since the majority charge carriers are holes and electrons, respectively. Adding very small amounts of an acceptor allows an electron from the filled valence band to populate the energy level in the gap generating a positive charge (named electron hole, or just 'hole') in the valence band. Therefore, the resulting p-material now becomes electrically conductive. Doping with electron donors generates electrons in the conduction band leading to n-type conductivity. The classical example is doping of silicon with boron or phosphorus atoms.[9] When phosphorus replaces a lattice silicon atom the excess electron is placed in the conduction band, allowing electron migration in the resultant n-Si. The opposite occurs when silicon is replaced by boron. It abstracts one electron from the valence band and generates p-Si. When the two silicon semiconductors get in contact, a p-n junction is formed, and the system tries to reach charge equilibrium through migration of electrons from n-Si to p-Si recombining there with the holes (Figure 3.10A). Simultaneously holes migrate from p-Si to n-Si and recombine there with electrons. As a result, charges are missing on both parts and an electric field is generated at the p-n interface at which only the ionised doping atoms (P^+ and B^-) remain in fixed lattice positions. As a consequence, a stationary electric field is generated in this space-charge region (Figure 3.10B).

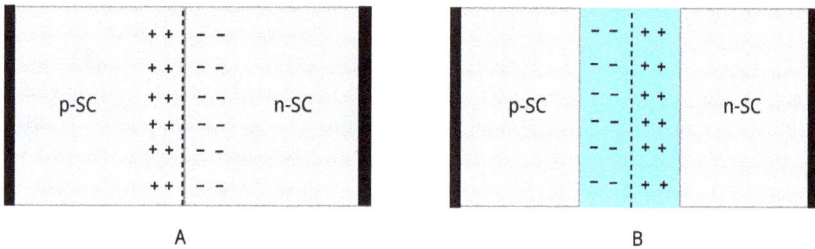

Figure 3.10: Schematic charge distribution at a p-n contact before (A) and after reaching charge equilibrium (B).

When light of energy larger than E_g hits such a p-n contact, the system is shifted away from equilibrium so that electrons and holes are formed in the conduction and valence band, respectively. The resulting electron-hole pairs (excitons) can recombine or be separated by the electric field of the space-charge region. Electrons migrate to n-Si and holes to p-Si since they are stabilised in these directions. As result, an electrical voltage is generated (U_{ph}). Upon adding electrical contacts, the p-n contact becomes a silicon solar cell as schematically depicted in Figure 3.11.

---

9 The atomic ratio Si/P is about 10^7.

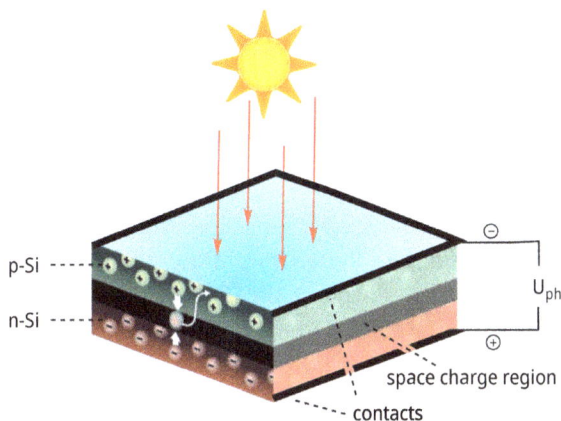

Figure 3.11: Principle of a silicon solar cell. Red and green spheres symbolise electrons and holes, respectively [20].

Since solar radiation is available for free almost anywhere on earth, photovoltaic generation of electricity has become increasingly popular. In China, Europe, and USA the share of total electricity has reached about 6, 6, and 3% [21]. However, there is a theoretical limit to the efficiency of conversion of light into electricity in a solar cell of about 33% (the Queisser-Shockley limit). In recent developments, silicon solar cells are coated with a layer of $MPbX_3$ (M = Cs or alkylammonium, X = halide with a perovskite structure) that has a lower bandgap than silicon increasing efficiency. The efficiencies of these new solar cells have exceeded the maximum observed values for pure silicon. An alternative method of broadening the range of light absorption through photon upconversion was discussed in Section 2.2.10.

It is also possible to create a solar cell with organic components or metal complexes as light absorbers though they haven't reached such high efficiencies (dye-sensitised solar cells). One of the most successful metal complexes is $Ru(bpy')_2(NCS)_2$ (in bpy' the 4,4' positions are substituted with CO_2H) that absorbs across a very wide range and injects electrons into titania (see below). In the organic variety, the light-absorbing layer consists of acceptor-donor-acceptor compounds that contain organo-sulfur groups as donor and carbonyl derivatives as acceptors. These cells also require a hole-transporting layer that is usually the organo-sulfur polymer PEDOT (Scheme 1.1).

The opposite of the solar cell principle is that of a light emitting diode (LED). In this case a voltage is applied to the p-n junction in such a way that the p-layer becomes the positive terminal. The charges generated within the space-charge region recombine by the emission of light of the wavelength as determined by the semiconductor bandgap. Since this is spontaneous emission, incoherent light is emitted requiring the construction of ultrathin p-n junctions (OLEDs) [22].

Instead of using a semiconductor, the luminescent layer often consists of electro-luminescent complexes of precious metals based on iridium (Figure 3.12) that are promoted to their MLCT (or mixed MLCT/LC) excited state by the potential. The complexes with phenylpyridine (ppy) and related ligands are chosen for their high quantum yields, tunability and the large spin-orbit coupling constant of iridium. The applied potential generates a statistical mix of singlet and triplet excited states, but it is desirable to use light emission from all the states that are generated. In the iridium complexes, the high spin-orbit coupling enables the singlets to undergo rapid ISC to the triplet states that are highly emissive. The complexes can be tuned to emit different wavelengths of light by variation of the ligands [23]. An alternative way of overcoming the statistics of triplets and singlets is to use thermally activated delayed fluorescence (Chapter 2.2.5). Here the principle is to convert the triplets back into singlets and use the resulting singlet emission. For this purpose, organic molecules or copper complexes are favourable, so avoiding the use of precious metals [24]. Many of the smaller displays used in mobile phones, sensors and even distance warning signals for cars are based on OLEDs.

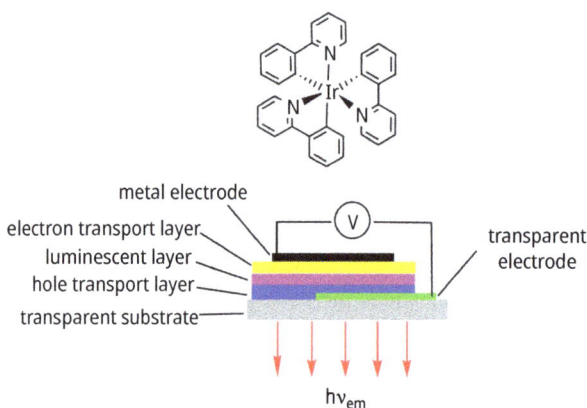

Figure 3.12: Principle of an organic light emitting diode (OLED) with structure of an iridium phenylpyridine complex (Ir(ppy)$_3$).

A semiconductor laser is essentially an LED with light-emitting parts that enable stimulated instead of the usual spontaneous emission (Figure 3.13). For laser action the electron population must be higher in the conduction than in the valence band. This inversion of electron population is achieved with a high current density. A fully reflective backside and semi-transparent front side induce formation of a standing wave. When the latter reaches a certain minimal intensity, it may stimulate the emission of coherent light. The active region consists of such materials as AlGaAs on GaAs. Diode lasers are to be found in laser pointers, CD players and laser printers and are used in fibre-optic communication.

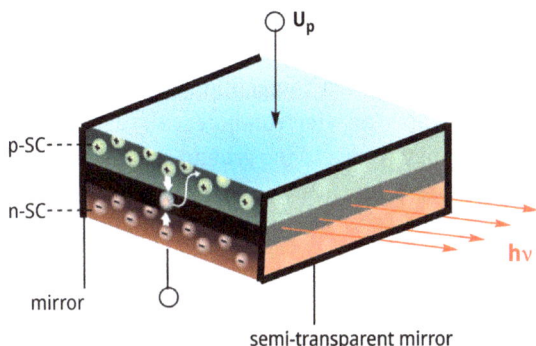

Figure 3.13: Principle of a diode laser. U_p = pump voltage.

3.3 Charge separation at the semiconductor-liquid interface

In the section above we discussed how the space-charge region in a semiconductor p-n junction prevents charge recombination within the light-generated electron-hole pair. We also included several devices that depend on charge-separation at a semiconductor-liquid interface, often using semiconductors in nanoparticle form. The resulting photovoltage is physically used to drive various machines. A chemical application is also possible when the photovoltage is utilised for an electrochemical process like the splitting of water into hydrogen and oxygen. This photovoltaic + electrolysis two-step process represents about 4% of the world wide hydrogen production and other methods are under development for commercial solar hydrogen production [25]. However, a one-step "wireless" process is also possible when light absorption on a semiconductor surface generates reducing and oxidising centres capable of splitting water or driving other redox reactions. In photoelectrochemistry, a light absorbing semiconductor electrode is wired to an inert counter electrode whereas in semiconductor photocatalysis only a single material is required to drive the reductive and oxidative steps of a redox reaction.

In the following we summarise the basic processes relevant for the photocatalytic action of semiconductor materials. The primary processes occurring after light absorption resemble those of a molecule as discussed with help of the Jablonski scheme (Scheme 2.2).

We first discuss as an example the primary processes occurring after light absorption by n-ZnS (Scheme 3.2). This white solid has a bandgap of 3.60 eV corresponding to a threshold wavelength of 347 nm. Quantum theoretical calculations afford effective charges of +1 and −1 for zinc and sulfur suggesting the presence of a remarkable degree of covalent bond character. Therefore, absorption of a photon $h\nu$ generates an exciton of the electronic configuration 'Zn(0)S(0)' [26] having a binding energy of 0.029 eV. The electron-hole distance, $i.e.$ the Bohr radius of the exciton, is in the

range of a few nanometres.[10] According to its electronic structure the exciton may decompose into elemental zinc and sulfur. This highly undesirable process known as
photocorrosion leads to photocatalyst deactivation.[11] In competition, vibronic relaxation to the lowest energy level (E_c), generates an electron-hole pair delocalised
throughout the entire crystal. From there it may return to the ground state directly or
be *trapped* at localised unreactive (e_{tr}^-, h_{tr}^+) and reactive (e_r^-, h_r^+) surface sites. As a
result, some major radiative and nonradiative processes may be observable as de-
picted in Scheme 3.2. It should be noted that in this and all the following schemes an
electron or hole is energetically stabilised when moving down and up, respectively.
Before discussing the interfacial electron exchange between reactive surface charges
and dissolved redox partners, we briefly describe the photoelectrochemical model of
this key step for the efficiency of charge separation.

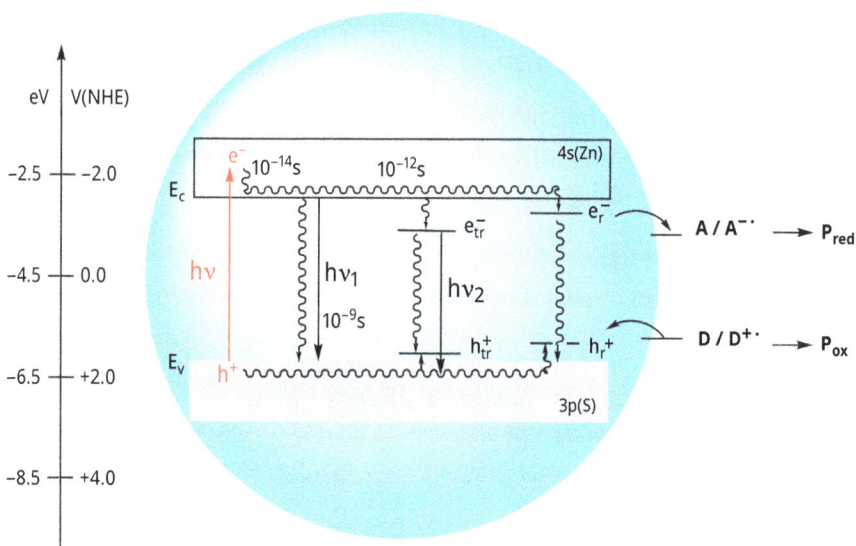

Scheme 3.2: Semiconductor photocatalysis of the redox reaction $A + D = P_{red} + P_{ox}$ in the presence of zinc
sulfide powder. The sphere symbolises a 10–100 μm particle composed of nanocrystals. Wavy arrows
correspond to nonradiative charge recombination and charge migration, straight arrows to radiative
processes. Photocorrosion processes are omitted. The band edge positions are given as electrochemical
potentials and apply for ZnS in contact with water of pH 7. For the relation between energy (eV) and redox
potential (V/NHE) see Equation 2.22.

---

10 Binding energies of excitons are in the range of 0.001 to 1.0 eV. A 1.5 nm radius was reported for
TiO_2 [27].
11 In the presence of air, the final product is zinc sulfate. This leads to "chalking" of colours containing ZnS and TiO_2 as pigments.

3.3.1 Interfacial electron transfer at a semiconductor electrode

Electron transfer in solution and at the metal/liquid and semiconductor/liquid interface (IFET) was investigated in the 1950s and 1960s by Marcus and by Gerischer [28]. We briefly discuss the basic results important for the understanding of charge-separation at the semiconductor-liquid interface. Since most semiconductors (SC) used in photocatalysis are of the n-type, we concentrate on these materials although p-SC follow the same principles [29].

Electrons prefer to flow in the direction of decreasing electrochemical potential, i.e. the gradient of this potential is the driving force for a directed move of electrons through a single phase or across an interface.[12] The opposite applies for holes which move towards increasing potential. Therefore the IFET is thermodynamically allowed when the reduction potentials of A and D are located within the bandgap (see Scheme 3.2). For the rate of the process the following apply:

- the Franck-Condon principle is valid
- the IFET occurs iso-energetically
- the rate of IFET is proportional to the density of states in the solid and in the dissolved substrates.

The Gerischer model describes the effect of solvent fluctuation on the potential energy of the redox species. As consequence of the Franck-Condon principle the initially formed redox products still have the original solvent shell and therefore must rearrange it to the product shell. As a consequence the density of electronic states of a redox system deviates from the standard redox potential $E^0{}_{red,ox}$ (rel. to NHE) by the solvent reorganisation energy λ. Finally, a partition function for the *density of states* was obtained (Figure 3.14A,B). Gerischer deduced that $E^0{}_{red,ox}$ can be expressed in the dimensions of electron volts according to Equation 3.7 (*n* is the number of electrons transferred). It was named "Fermi energy of the redox electrolyte" ($E^0{}_{F,red/ox}$).

$$E^0{}_{F,red/ox} = -4.5 - nE^0{}_{red,ox} \qquad (3.7)$$

Reorganisation energies may reach at most 1.0 eV and depend on the strength of electronic interactions between the two redox species. The model assumes only a weak interaction. When it is increased, the reorganisation energy is lowered, resulting in a lower energy of activation and therefore faster reaction (Figure 3.14C).

Figure 3.15 describes how the electronic structure of the solid/liquid contact changes as compared to the separated components. Assuming that $E^0{}_{F,red/ox}$ lies below

---

12 We recall that electrons and holes *drift* when they move under the influence of an electric field, but *diffuse* under the influence of a concentration gradient. Since experimental evidence for this differentiation is very rarely available in photocatalysis, we prefer to use more general formulations.

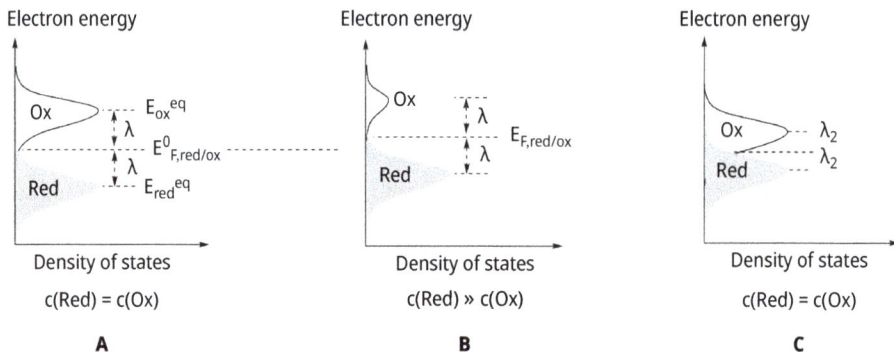

Figure 3.14: Density of states and influence of concentration (A, B) and increased electronic interactions (C) on the solvent reorganisation energy.

the Fermi level, electrons from the inner part of the semiconductor will move to the solution until the Fermi levels in both phases are the same, i.e. thermodynamic equilibrium is established. Since the density of free electrons (10^{16}–10^{19} cm^{-3}) in an n-SC is very low,[13] electrons will originate essentially from the bulk and only a little from the surface. As a consequence the band edges at the surface remain almost unchanged but become increasingly more anodic towards the bulk resulting in band-bending (Figure 3.15A,B).[14] The thickness of the resulting space-charge region increases to about 100–1000 nm. Charge neutralisation occurs through a negative solution layer. Since the position of the Fermi level of the SC/contact depends on the solution redox potential, it is no longer an intrinsic material property.[15] However, it is an intrinsic property in the absence of a dissolved redox component and this energy is named the flat-band potential E_{fb}. This differentiation is relevant in photoelectrochemistry, but not in photocatalysis as discussed in this book.

Figure 3.15C depicts very schematically how the n-SC/liquid interface acts as photocatalyst for the redox reaction $A + D \rightarrow A^{-\bullet} + D^{+\bullet}$. The formation of a space-charge layer resembles the silicon p/n-contact and is capable of preventing recombination of light-generated charges. Analogously, irradiation of the SC/contact with light of energy equal or larger than the bandgap generates surface charges capable of interfacial electron transfer (IFET) with donor and acceptor molecules. This mechanism applies only when the crystal size is above about 1000 nm, the width of the space-charge region. Charge-separation at smaller particles very likely occurs through kinetic competition between recombination and IFET (see Scheme 3.3).

---

13 In a metal the density of free electrons is about six orders of magnitude higher.
14 In the case of a p-semiconductor a cathodic band shift and negative space-charge region is observed since the Fermi level is close to the valence band edge. By definition, oxidation occurs at the anode so anodic means more positive potential, and cathodic means more negative potential.
15 In a photoelectrochemical cell a corresponding shift of the Fermi level can be induced through application of an anodic potential.

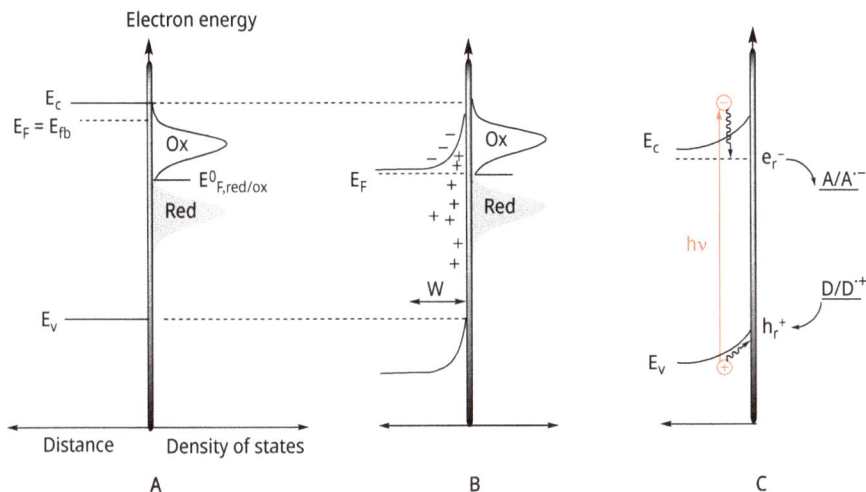

Figure 3.15: Electronic structure of the interface of a large n-semiconductor crystal and a dissolved redox system before (A), after reaching equilibrium (B), and under irradiation in the presence of donor and acceptor molecules (C). Energies of band edges E_c and E_v are shown as a function of the distance from the surface to the bulk of the solid. *Red* and *Ox* have the same concentration. *W* is the width of the depletion layer. D = donor, A = acceptor.

3.3.2 Charge separation at nanoparticles

Photoelectrochemical experiments with a titania powder electrode reveal that the current density does not increase when the Fermi level is cathodically shifted by applying a corresponding voltage. According to Figure 3.15B this would increase the band-bending and therefore the efficiency of charge separation as observed at a large crystal SC electrode. However, such an increase was observed when reducing agents of increasing reduction potential were present in solution. This strongly suggests that charge-separation is not due to the presence of a space-charge region but rather to a very fast rate of IFET from the reducing agent to the reactive hole. In addition to this kinetic competition between recombination and IFET there is some evidence that the light-generated electron may undergo an intercrystallite electron transfer (ICET) as indicated by arrow 1 in Scheme 3.3. The resulting spatial separation of the two charges should decrease their recombination. ICET between quantum size particles of ZnO was recently evidenced. Upon mixing small with larger particles electron transfer from small to large particles was observed since the conduction band edge of the smaller particle is located more cathodically (see Scheme 3.1) [30]. Spectrochemical experiments with nanocrystalline titania films point to a mutual ICET during anaerobic oxidation of formic acid [31].

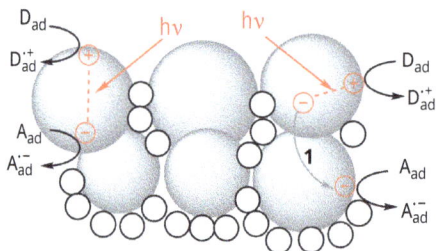

Scheme 3.3: Schematic view of the photogeneration of reactive electron-hole pairs in an aggregate of n-type semiconductor nanocrystals (spheres) without (left part) and with intercrystallite electron transfer (process 1, right part). Empty circles symbolise the solid/solute. D = donor, A = acceptor, ad = adsorbed.

3.3.3 Interfacial electron transfer

The thermodynamic criterion for interfacial electron transfer is that the reduction potentials of the acceptor and donor molecules A and D must be located within the semiconductor bandgap as drawn in Scheme 3.2 and Figure 3.15C.[16] While the molecular reduction potentials are easily obtainable, this is more difficult for the Fermi level (E_F) of a solid. If not already known, it can be measured by photoelectrochemical methods. Since for an n-SC it is very close to the conduction band edge, we assume in the following that they are about the same. With that approximation the valence band edge is obtained by subtraction of the bandgap energy from E_c (see Scheme 3.2). The reactive charges trapped at the surface must have weaker redox properties than electron-hole pairs trapped at the band edges.[17] Additionally, the interfacial electron transfer reactions (IFET) must be much faster than charge recombination (Equations 3.8–3.10).

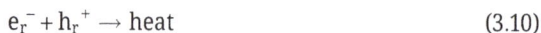

$$A + e_r^- \rightarrow A^{-\bullet} \tag{3.8}$$

$$D + h_r^+ \rightarrow D^{+\bullet} \tag{3.9}$$

$$e_r^- + h_r^+ \rightarrow heat \tag{3.10}$$

This can be achieved by surface deposition of nanometre-sized oxidation and reduction catalysts at distinct surface sites (see Chapter 5.3). The reactive charges are often localised at structural defects like under-coordinated metal ions and partially oxidised anions. This may explain the experimental fact that for instance high purity zinc sulfide is an excellent luminophore but not a photocatalyst unless it contains a slight ex-

---

16 The Fermi level and therefore band edge positions depend on solvent pH value. For oxidic semiconductors a cathodic shift of about 59 mV per pH unit is observed in the direction of more basic solutions.

17 We note that in this estimation the solvent reorganisation energies of the redox pairs are neglected. See also Chapter 5.3.2 on water splitting.

cess of sulfide ions [32]. Quite often traces of water strongly enhance the reaction rate, probably by enabling fast proton-coupled IFET reactions. As an example, we take the oxidation of alkanes (RH). Assuming the reactive hole is located at a potential of 2.0 V and the RH oxidation potential is 2.6 V, oxidation to the radical cation is *endergonic* by 0.6 eV. However, when the IFET is coupled with C-H cleavage the reaction is *exergonic* by 0.8 eV (Equations 3.11–3.12, see also Scheme 2.13).

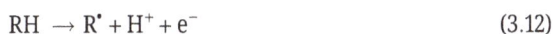

$$RH \rightarrow RH^{+\bullet} + e^- \tag{3.11}$$

$$RH \rightarrow R^\bullet + H^+ + e^- \tag{3.12}$$

3.4 Comparing photocatalytic activities

As discussed in Chapter 1.6 the quantum yield[18] can be formulated as the rate of product formation divided by the rate of monochromatic photon absorption (Equation 3.13). It is the only reliable parameter for comparing the degree of light utilisation of various photoreactions.

$$\Phi_p(\lambda) = r(P)/I_a(\lambda) \tag{3.13}$$

Both, product formation and photon absorption rates are obtainable for homogeneous reactions, but in heterogeneous systems like semiconductor photocatalysis it is extremely tedious to measure quantitatively the photon absorption rate: imagine a semiconductor suspension giving rise to irregular reflectance depending on stirring speed etc.

In the literature the comparison of photocatalytic activities is very often made by using turnover numbers (TON) or turnover frequency (TOF)[19] by analogy with thermal reactions. Since these numbers require the knowledge of number of active sites, a number in general unknown for the semiconductor photocatalyst, it cannot be applied in heterogeneous photocatalysis. Irrespective of these facts, TON in particular is commonly used to compare the efficiency of a photocatalyst claiming even differences in the range of 10% as significant [34]. A practical solution of this serious problem is the use of a "black body" photoreactor. In this, complete light absorption is achieved by placing the light source through a wave guide in the centre of the reaction vessel. With sufficient catalyst powder concentration all the incoming light is absorbed. Therefore, I_0 becomes equal to I_a and the as obtained quantum yields (monochromatic) or apparent quantum yields (polychromatic) are now comparable [35].[20] Even

---

18 Note that a reaction may have a high quantum yield but a very low *product yield*.

19 TOF is defined as number of molecules reacting per second and per active site. In biochemistry TON has the same meaning as TOF. However, in general TON is defined as the total number of product molecules per molecule of catalyst. See ref. [33].

20 Quite often polychromatic light is used and instead of quantum yield expressions like quantum efficiency or apparent quantum yield are employed [36].

in very careful experiments, the quantum yield measurements have an error bar of about 10%.

Within a research group a more practicable procedure is recommendable: that is to use one standard photoreactor containing the suspension and measure the reaction rate as function of increasing photocatalyst concentration. As in a homogeneous system the rate increases linearly with photocatalyst concentration due to the increase of absorbed light (photon flux), and then ends in a plateau. It was proposed to name that constant rate an optimal rate (v_{op}) and the photocatalyst concentration (e.g. in g/L) at the plateau onset an optimal concentration [37]. Obviously, both parameters depend on absorbed light intensity and photoreactor geometry. Since within one research group in general only one unique reactor type is employed, semi-quantitative comparisons become feasible.[21] In general only a qualitative comparison of "photocatalytic activities" is meaningful since many reactions have not been conducted in the plateau region. Quite often, rates are referenced to the weight of catalyst powder as practised in dark heterogeneous catalysis. However, this is not reliable for semiconductor photocatalysis where the amount of light absorbed depends on additional reaction parameters (see above). As a rule, quantitative comparisons of photocatalytic activities should be discussed with great care [29, 36].

Imagine you have tested a particular reaction in the presence of various semiconductor photocatalysts and try to interpret the observed changes of the quantum yield. The most practicable method is to separate the product quantum yield into a product of three efficiencies[22] (Equation 3.14) [38].

$$\Phi_p = \eta_r \times \eta_{ifet} \times \eta_p \tag{3.14}$$

where $\boldsymbol{\eta_r}$ is the efficiency of formation of the reactive electron-hole pair (e_r^-, h_r^+, see Scheme 3.2) and depends on intrinsic semiconductor bulk and surface properties. To the bulk belong the crystal phase, exciton dissociation energy, charge diffusion constant, crystal and aggregate size; to the surface belong the surface defects, surface charge and nature of the solvent-solute surface layer [39]. In general, the photogenerated charges are localised at these defects and they may be identified as reactive or just trapped through emission quenching and product inhibition experiments. A rare example for such basic studies is the dehydrodimerisation of 2,5-dihydrofuran (2,5-DHF) photocatalysed by zinc sulfide in an aqueous suspension (Equation 3.15, R = 2,5-dihydrofuryl radical).

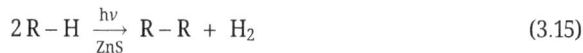

$$2\,R - H \xrightarrow[\text{ZnS}]{h\nu} R - R + H_2 \tag{3.15}$$

Upon addition of zinc or cadmium sulfate to a ZnS suspension the two emission bands of ZnS at 366 nm and 430 nm (see Figure 3.5), corresponding to band-to-band emission

---

21 In such a case the reproducibility of optimal rates is in the range of ±10%.
22 The term "efficiency" does not contain any aspect of light absorption. The expression *quantum efficiency*, which often is used in the literature instead of quantum yield, should therefore be abandoned.

and emission from trapped charges, are not significantly influenced. There is also no remarkable effect when the substrate 2,5-DHF is added as quencher. Unlike emission, product formation is strongly inhibited when the cadmium or zinc salts are added. This indicates that emitting and reacting surface sites are different (see Scheme 3.2). Importantly, the emission and inhibition experiments have to be conducted under the same experimental conditions in order to obtain reliable mechanistic information. This is often overlooked in the literature and false conclusions have frequently been made.

η_{ifet}, the efficiency of the two interfacial electron transfer (IFET) processes (Equations 3.16, 3.17) depends primarily on the detailed nature of the surface-solution interface and on redox and adsorption properties of the substrates [40]. Minor alterations in the synthesis of the photocatalyst powder and in the adsorption of reaction components (i.e. solvent and substrates) can strongly affect the lifetime of the reactive electron-hole pair. In general, the donor/acceptor substrates must adsorb onto the semiconductor surface to enable efficient IFET. Adsorption at the solid/liquid interface is described in the literature in various degrees of complexity. Hiemenz's approach provides relatively convenient basic information on the structure of the solid-solvent-solute interface [41]. This is illustrated by the adsorption of 2,5-DHF onto cubic ZnS powder from an aqueous solution [42]. The adsorption isotherm[23] consists of two plateau regions. The first one corresponds to a 2,5-DHF monolayer, the second to a multi-adsorption layer. One can estimate that within the first ZnS/2,5-DHF/water layer almost every zinc site is occupied by the ether positioned perpendicular to the surface.[24] The small downfield shift of 1.5 ppm measured for the $C(sp^3)$ atoms by NMR indicates that the oxygen atom of 2,5-DHF interacts with zinc sites through hydrogen bonding via coordinated water (Figure 3.16).

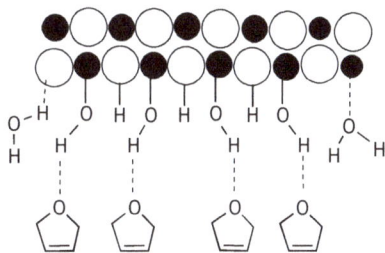

Figure 3.16: Proposed structure of the ZnS/water/2,5-dihydrofuran interface.

η_p, the efficiency of product formation from the primary IFET products often requires multiple reaction steps. In the case of the reaction system ZnS/2,5-DF/H$_2$O the primary product dimerises to the final dehydrodimer (Equation 3.18, R = 2,5-dihydrofuryl radical).

$$2\,e_r^- + 2\,H_2O \rightarrow H_2 + 2\,OH^- \tag{3.16}$$

---

23 The adsorption isotherm is a plot of amount of compound adsorbed from solution onto a solid.
24 From the specific surface area of 170 m^2 g^{-1} and the density of zinc sites in cubic ZnS (11.4 × 10^{-6} mol m^{-2}).

$$2\,h_r^+ + 2\,R - H \rightarrow 2\,R^\bullet + 2\,H^+ \tag{3.17}$$

$$2\,R^\bullet \rightarrow R - R \tag{3.18}$$

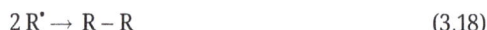

C–C coupling of the 2,5-dihydrofuryl radicals most probably occurs within the water-2,5-DHF surface layer as suggested by competition experiments with THF. Although the unsaturated ether 2,5-DHF reacts only ten times faster than THF, no THF dehydro-dimers or cross-products are detected when the THF concentration is ten times larger, but they become detectable at a 500-fold excess.

3.5 Pristine and surface modified semiconductor powders

The absorption spectrum and bandgap of semiconductor powders can be modified in a wide variety of ways with major consequences on the electronic properties and photo-catalytic activity:

– Phase change
– Change of particle size (quantum size effect)
– Surface modification
– Solid solutions with an isostructural compound
– Doping (p- or n-type) by substitution with an element of different valence
– Defects (oxygen vacancies, interstitial metal ions)

The physical effects of some of these changes may be familiar such as Al_2O_3 doped with Cr_2O_3 (ruby) and topics such as the particle size effect have already been mentioned. In the following we give a brief summary of the consequences emphasising surface changes. Crystal effects can have an enormous effect on the efficiency of formation of the reactive electron-hole pairs as mentioned in Chapter 3.3.

3.5.1 Pristine semiconductor powders

Figure 3.17 summarises the band edge positions of some inorganic semiconductor materials with good photocatalytic properties. In all these examples, the valence and conduction bands have predominantly anion and cation character, respectively. Accordingly, in Ta_2O_5, complete replacement of the oxide ions by nitride shifts the bandgap of 3.9 eV to 2.1 eV and valence band edge from 3.6 V to 1.6 V. The much smaller shift of –0.3 to –0.5 V observed for the conduction band edge agrees with its cationic character [43]. The two dashed horizontal lines indicate the potentials –0.41 V and +0.82 V, which are required for the splitting of neutral water into hydrogen and oxygen. Accordingly, oxygen formation is thermodynamically possible for all materials shown in Figure 3.17, but hydrogen generation only for titania, strontium titanate, tantalum oxide, tantalum nitride, cadmium sulfide and zinc sulfide.[25] Importantly, the energy differences between the water splitting potentials and the band edge positions are lost as heat. Only when this difference is very small may a high quantum yield be obtained. We further point to the fact that the light-generated charges may have extreme redox properties. In the case of titania, the hole generated by visible light has an oxidation potential of about +2.7 V![26]

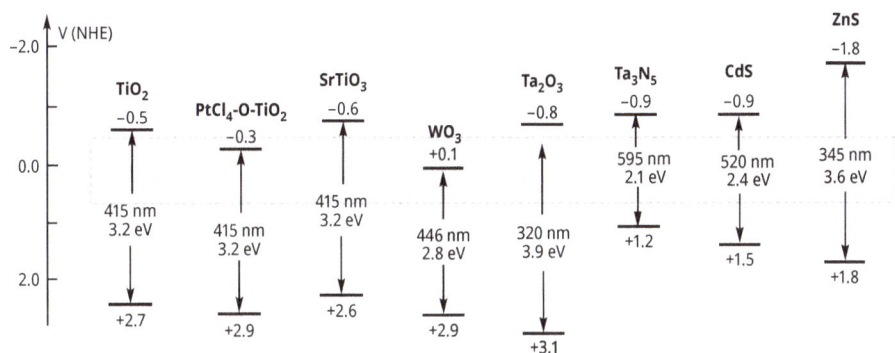

Figure 3.17: Bandgaps and band edges of some semiconductor powders in contact with neutral water. E_v and E_c values of silicon are +0.6 V and −0.5 V, respectively. The dashed lines indicate water splitting potentials [44].

In general semiconductor powders are prepared by standard methods. Examples are the precipitation from aqueous solutions, followed by calcination (presence of air) or thermal treatment in the range of 100–600 °C. However, minor changes in preparation and storage of the powders can strongly influence the photocatalytic activity, so commercial and self-prepared samples may differ considerably in activity. Good reproducibility is reported for the *hydrothermal method* allowing controlled crystallisation.[27] Co-precipitation, doping and surface modification are also frequently employed. Most powders have specific surface areas between 40 and 300 m²/g and are often suspended in water or alcohols, rarely in aprotic solvents like ethers and acetonitrile. There they form micrometre aggregates of nano-sized crystallites. The latter are not small enough to exhibit a quantum size effect. Unless noted otherwise, all semiconductors mentioned in the book are of the *n*-type.

Titania is one of the most important large-scale industrial products. In 2020 about 4.6 million tonnes were produced worldwide [45]. The major users are the paint and coatings industry, but the white powder is also the most important semiconductor photo-

---

25 We recall that thermodynamic feasibility does not mean that the reaction rate is fast enough to observe appreciable product formation. In particular, multi-electron reactions have rather high activation energies.
26 The strongest elemental oxidising agent fluorine has a standard reduction potential of +2.8 V.
27 Hydrothermal reaction conditions imply heating an aqueous reaction system in an autoclave to temperatures between 100–1000 °C.

catalyst. This arises from its appropriate band edge positions, photostability, non-toxicity and easy availability. Absorption of UV light at the border to the visible region generates electrons and holes at about −0.5 V and +2.7 V. Accordingly, photoexcited titania is thermodynamically able to split water. The high oxidation potential of the hole enables the use in functional surface coatings for the photocatalytic removal of volatile organic compounds (VOC) and nitrogen oxides. Typical examples are water-based paints for interior and exterior house walls. However, pristine TiO_2 absorbs in the UV and modification is required if it is to be active with visible light as required for solar photocatalysis.

3.5.2 Phase change and crystal modifications

Phase changes (also described as crystal modifications) can have a major effect on electronic and photocatalytic properties. Vanadium dioxide (with a $3d^1$ configuration) undergoes a phase change from semiconductor to metallic conductivity on heating to 67 °C; below the transition there are alternating short and long V $\cdots$ V distances, while they are all equal in the metallic phase. Titanium dioxide ($3d^0$ configuration) exists in the three crystal modifications of anatase, rutile and brookite. Unless stated otherwise, the anatase modification is intended throughout this book. Commercially available titanium dioxide powder *P25* consists of about 80% anatase and 20% rutile and often is more active than the separate components (Figure 3.18). A likely explanation is based on EPR spectroscopy during irradiation. Since the signal of an electron depends on its surrounding solid structure, it is different for the two modifications. The results suggest that charge-recombination is partly inhibited by fast ICET from rutile to anatase (Figure 3.18) [46].

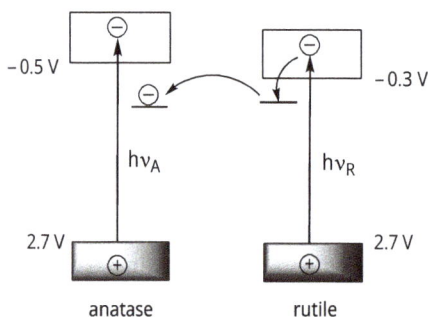

Figure 3.18: Model proposed for improved charge separation in titania P25 through ICET form rutile to anatase.

3.5.3 Solid solutions

When two components that only absorb UV light have the same crystal structure, homogeneous solutions may be formed which absorb visible light. A good example is the d^{10} system GaN–ZnO, easy obtainable by thermal treatment of a mixture of Ga_2O_3 and ZnO with NH_3 at 850 °C. The solid solutions exhibit bandgaps from 2.8 to 2.4 eV, corresponding to absorption onsets of 446–520 nm (Figure 3.19).

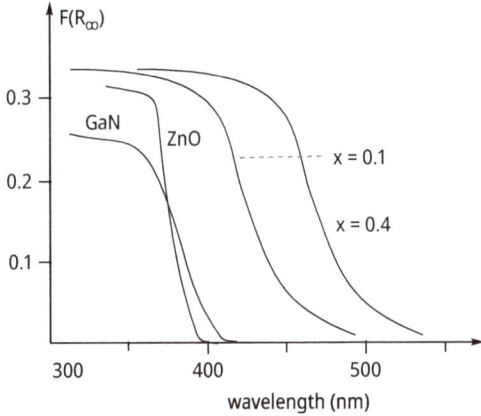

Figure 3.19: Schematically drawn diffuse reflectance spectra of GaN, ZnO and their solid solutions $(Ga_{1-x} Zn_x)(N_{1-x}O_x)$. Adapted from ref. [47].

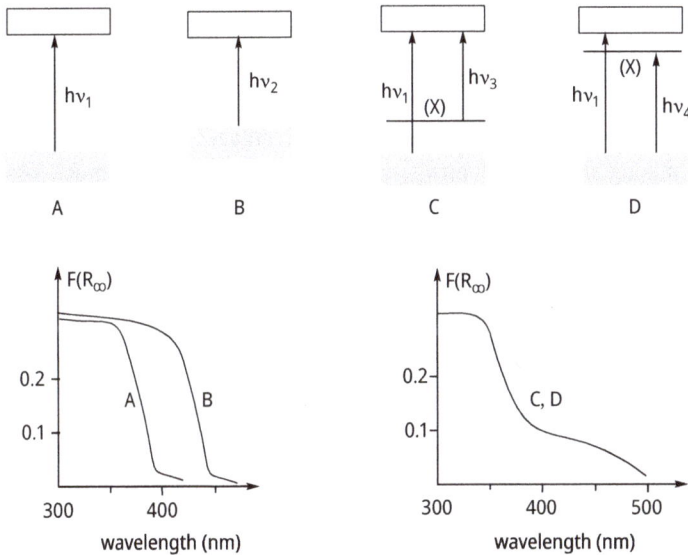

Figure 3.20: Influence of a strong (B) and weak (C) modifier or dopant (X) on the shape of the reflectance spectrum of the pure semiconductor A.

The bandgap narrowing may originate from an N(2p)-Zn(3d) repulsion resulting in a cathodic shift of the valence band [48]. A broadening of the latter is expected from strong electronic interaction between nitrogen and oxygen p-orbitals. Both effects rationalise (Figure 3.20B) the bandgap shift resulting in a steep absorption onset. The doping or modification of a semiconductor is different and in general leads to the formation of surface states within the bandgap. As a result, the onset displays a flat absorption shoulder resulting from new, localised electronic states (Figure 3.20C,D).

3.5.4 Surface modified semiconductor powders

As the surface of semiconductors used in photocatalysis contains under-coordinated metal ions and OH/SH/NH groups, chemical surface modification may affect the photocatalytic activity. One method is *grafting, i.e. forming a chemical bond between the two components.* It is recalled that the support is always present in a large excess over the grafting material. Positive effects were often found when the support is also a semiconductor, resembling a heterojunction.[28]

Grafting onto a semiconducting support
Grafting of 5% of CdS onto ZnS is easily performed by stirring an aqueous ZnS suspension containing various amounts of cadmium sulfate followed by addition of sodium sulfide as a reducing agent. Wavelength-selective irradiation of the CdS component of the resulting powder increases the rate of CO_2 reduction to formate about 40-fold [49]. Pristine CdS exhibits a negligible activity. Loading of higher percentages results in inactive powders. Mechanical mixing also leads to powders that are inactive or show much less photocatalytic activity.[29] This suggests that the two semiconductors are connected through Zn-S-Cd bonds enabling more efficient charge separation via an interparticle electron transfer.

Grafting of iron oxides onto titania exemplifies how experimental conditions influence electronic and photocatalytic properties. Impregnation of titania (0.05% Fe relative to rutile) with an aqueous iron(III) chloride solution of pH 2 at 90 °C and subsequent calcination at 110 °C generates amorphous FeO(OH) particles on titania [51]. The diffuse reflectance spectrum shows an absorption shoulder at 410–580 nm, assignable to a CT transition from the titania valence band to an FeO(OH) localised surface state located at about 0.5 eV below the conduction band edge. The bandgap of 3.00 eV is not changed (see Figure 3.20D). When using $Fe(acac)_3$ in an analogous procedure but calcining at 500 °C, an (FeO_x)-O-TiO_x surface complex seems to be formed.

---

28 In general, a heterojunction is the interface between two different semiconductors.
29 Similar "coupled" systems like CdS/TiO_2 were prepared just by grinding and are generally much less active [50].

Unlike the example above, this material does not exhibit an absorption shoulder but a shift of the bandgap by 0.40 eV. Probably, overlap of a small iron-localised energy band with the titania valence band leads to valence band broadening (Figure 3.20B) [52]. Besides iron, many other transition metals including cobalt and copper have also been employed in this type of surface modification. In most cases, better photocatalytic activity requires chemical bonding between the two components. Research on such types of heterojunctions greatly increased in the last two decades with the aim of improving visible-light photocatalytic activity [53].

A different type of surface modification is based on OH, NH and SH groups present on the surface of the corresponding metal oxides, nitrides and sulfide photocatalysts [54]. Complexation with appropriate organic or inorganic molecules leads to supramolecular hybrids that may improve the efficiency of formation of the reactive electron-hole pair and IFET. A good example is titania containing about 3–6 OH groups per nm^2 and under-coordinated titanium atoms. Therefore TiO_2 may act both as coordination centre (Figure 3.21A) or as a mono- and bidentate ligand (Figure 3.21B,C). Classic examples are 1,2-diols like catechol forming a red surface complex (Figure 3.21A) originating from a broad visible absorption band of ligand-to-metal character.

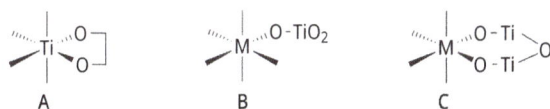

Figure 3.21: Titania surface complexes with Ti as a central metal (A) or a ligand (B, C) in a coordination complex (M = Ti, Rh, Pt).

Surface transition metal complexes

Surface transition metal complexes are also photo-active [44b,55]. Stirring a suspension of anatase hydrate in an aqueous solution of $H_2[PtCl_6]$ or rhodium(III) chloride and subsequent heating at 200 °C affords a yellowish and dark-red powder, respectively. There is good evidence for the presence of a surface metal complex containing a M-O-TiO_2 structural fragment (M = Pt(IV), Rh(III), Figure 3.21B). In the case of the rhodium complex, the red colour probably originates from a rhodium-to-titania charge transfer (MMCT) transition assigned to an intense absorption shoulder at 550 nm. In agreement with the presence of chemical bonding between the titania "ligand" and the platinum or rhodium complex, the electronic structure of titania is significantly changed. For both materials the quasi-Fermi potential, i.e. the Fermi potential under irradiation, is shifted from –0.54 V for anatase to about –0.24 V.[30] Both materials exhibit high photocatalytic activity in the field for complete aerial oxidation of pollutants in air and water. Thus, the ubiquitous pollutant 4-chlorophenol is oxidised to CO_2 and HCl by visible light. In the

---

30 Such a controlled shift of the quasi-Fermi level is very rarely possible.

proposed mechanism MMCT excitation results in charge separation to a Rh(IV)-O-Ti(III) intermediate, the molecular equivalent of a reactive electron-hole pair. The rather long distance between charges in the pair may slow down recombination and promote efficient IFET reactions to the primary redox products (Scheme 3.4).

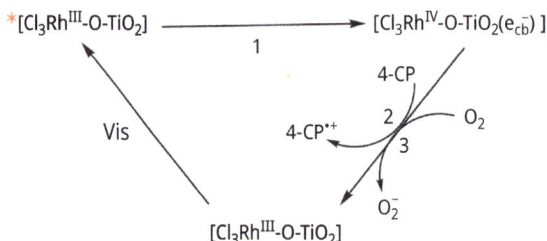

$*[Cl_3Rh^{III}-O-TiO_2] \longrightarrow [Cl_3Rh^{IV}-O-TiO_2(e_{cb}^-)]$

1

4-CP

Vis

4-CP$^{\bullet+}$

2

3

O_2

O_2^-

$[Cl_3Rh^{III}-O-TiO_2]$

Scheme 3.4: Primary reaction steps in the visible light oxidation of 4-chlorophenol photocatalysed by a two-centre surface complex with titania as a ligand.

Sensitisation by electron transfer

Solar photovoltaic cells containing Ru(bpy')$_2$(NCS)$_2$ and related complexes operate by electron injection into titania. The excited state potentials of the complexes with unsubstituted bipyridine ligands are sufficiently negative to inject an electron into the titania conduction band (see Chapter 2.2.6). However, only when the bpy ligand is modified in the 4,4′-positions by two carboxyl groups (bpy-CO$_2^-$) is the forward electron transfer fast enough to overcome the back-reaction. As often observed in surface modification, covalent attachment, here formation of a Ti(OCO$_2$–bpy)Ru fragment structure, is essential [56].

Metal-loaded semiconductors

In photoelectrochemistry the photogenerated charges are located at a light-sensitive semiconductor electrode and a metal counter electrode like platinum, connected by a conducting wire. The two IFET processes occur quite distantly from one another. It seems therefore likely that depositing platinum as "counter electrode" onto the semiconductor surface may accelerate the reductive IFET.[31] Loading a few weight percent of metallic nanoparticles onto the semiconductor surface may increase the reaction rate. A classic example is the platinisation of titania through irradiating a titania suspension in the presence of 1–2 weight % of hexachloroplatinic acid and a reducing agent [57]. Applying the method to CdS, the activation energy of water reduction by sodium sulfite is decreased from 54 to 8 kJ mol^{-1} (Equations 3.19–3.21). Furthermore,

---

31 The resulting metal-loaded or metallised semiconductor particle may be viewed as a "short-circuited" photoelectrochemical cell.

the quasi-Fermi level is shifted from −0.30 V to −0.60 V leading to an increase of the driving force of the reductive IFET (Equation 3.20). A maximal reaction rate is observed with 1–2% platinum [58].

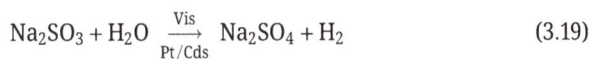

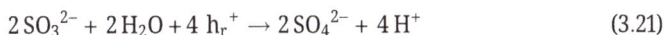

$$Na_2SO_3 + H_2O \xrightarrow[Pt/Cds]{Vis} Na_2SO_4 + H_2 \tag{3.19}$$

$$2 H_2O + 2 e_r^- \rightarrow H_2 + 2 OH^- \tag{3.20}$$

$$2 SO_3^{2-} + 2 H_2O + 4 h_r^+ \rightarrow 2 SO_4^{2-} + 4 H^+ \tag{3.21}$$

In the case that the metal particle size is much smaller than the wavelength of exciting light, the metal electrons may generate a local surface plasmon resonance (LSPR). Its resonant frequency strongly depends on the detailed particle shape (see Chapter 5.3.1) [59].

Carbon and nitrogen modified titania

In addition to metals and metal oxides, non-metals like carbon, nitrogen and sulfur have also been introduced into titania to achieve visible light activity. In most cases it is unknown if the latter originates from true doping, i.e. volume modification, or surface modification. A classic procedure consists of calcining titania in the presence of organic carbon or nitrogen compounds at temperatures of 200–600 °C.

Nitrogen-modified titania

TiO_2-N is prepared by three major methods: sputtering and implantation techniques, calcination of TiO_2 under N-containing atmospheres generated by nitrogen compounds like ammonia and urea, and by sol-gel methods employing N-containing reagents [60]. In most cases the nature of the surface species responsible for visible light activity was unknown, although species like NO_x [61], nitridic, amidic (NH_x) groups [62], oxygen vacancies and colour centres[32] were proposed [63]. In the case of urea or thiourea it was shown that titania initially catalyses formation of melamine, followed by condensation reactions with Ti(OH) surface groups to form a polytriazine-titania surface complex (Scheme 3.5). When high urea concentrations are employed, X-ray analysis reveals the presence of the stacked aromatic system of carbon nitride polymer (C_3N_4), known to be an organic semiconductor (vide infra). This surface complex is a very early example of a covalently bound heterojunction between an organic and inorganic semiconductor [64].

---

[32] Colour centres are crystal defects (mostly anion vacancies) occupied by one or more electrons, which are responsible for visible light absorption.

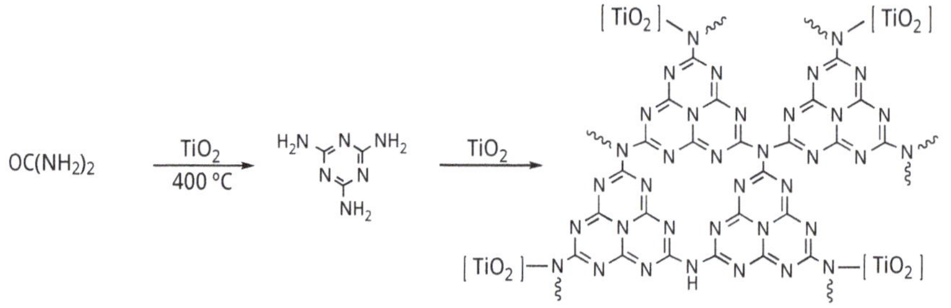

Scheme 3.5: Postulated structure of the titania-carbon nitride surface layer.

Carbon-modified titania

"Carbon"-modified titania (TiO_2-C) is easily accessible by calcining titania at 250–500 °C in the presence of an organic compound as carbon source [65]. Its diffuse reflectance spectrum exhibits a low-energy shoulder at 400–800 nm responsible for visible light activity for complete oxidation of air pollutants, even in diffuse daylight. It is therefore the basic component of commercially available wall paints. The chemical structure of the surface carbon molecule is unknown at present [66]. It may be a graphene derivative since heterojunctions of graphene with oxidic semiconductors can exhibit good visible light activity in photodegradation reactions [67].

(Poly)urethane-modified titania

The titania surface hydroxyl groups react with diisocyanate monomers with the formation of yellow titania-urethanes that are active in visible light. These hybrid materials photocatalyse aerial oxidation reactions (Scheme 3.6) [68].

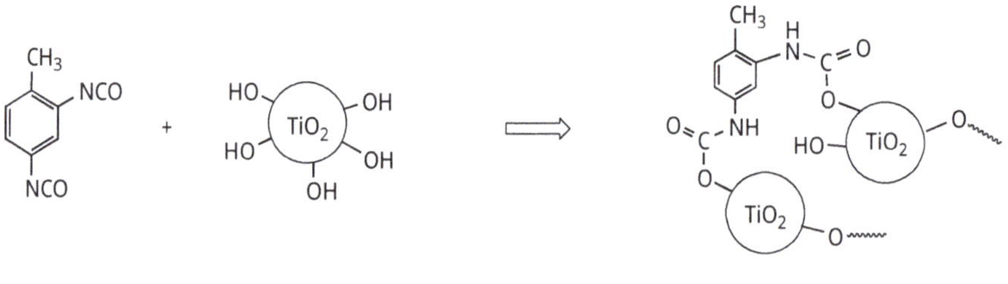

Scheme 3.6: Synthesis of titania modified polyurethanes.

Graphitic carbon nitride

Polymeric carbon nitrides were first obtained in the nineteenth century by Berzelius and Liebig as extremely thermally stable but insoluble polymers through thermal treatment of ammonium thiocyanate or urea. X-ray crystallography by Linus Pauling in 1937 proved the presence of heptazine cores (Scheme 3.5). In recent preparation methods heating of melamine or cyanamide leads to a graphitic phase (g-C_3N_4) able to act by itself as a semiconductor photocatalyst or as a sensitiser for wide-gap materials like titania (see above) [69]. Due to its bandgap of 2.7 eV the absorption onset is located at 463 nm, the conduction band edge at –1.2 V. Surface amino groups allow chemical bonding to various organic and inorganic semiconductors resulting in hybrid photocatalysts of increased activity in water splitting and small molecule activation [70].

Questions

1. How do the electronic absorption spectra of direct and indirect semiconductors differ?
2. Mark the position of the Fermi level in n- and p-semiconductors.
3. What is a diffuse reflectance spectrum?
4. What properties determine if a semiconductor is thermodynamically able to photocatalyse water splitting?
5. Explain the difference between photocatalytic and photoelectrochemical water splitting.
6. Explain the justification for the award of the 2023 Nobel Prize for the discovery and synthesis of quantum dots. What benefits do they bring to photophysics and photochemistry?
7. In 1857 Michael Faraday reported in a scientific paper how he had generated gold as small particles in suspension, finding that it was ruby-red in colour. Explain why the suspension is ruby-red and not the usually golden colour.
8. Cadmium sulfide is an n-type semiconductor with a bandgap of 2.4 eV. Loading with 1 wt% of colloidal platinum shifts the Fermi level from −0.9 to −1.2 V. Where is the valence band edge located, assuming that the Fermi level and the conduction band edge have the same energy?
9. AgBr crystallites are used in photographic films (see Chapter 4). What is the origin of the energy levels between the valence and conduction bands, and how does this relate to the photographic process?
10. At pH 7 the conduction band edge of TiO_2 is located at −0.5 V. The bandgap is 3.2 eV. (a) At what wavelength does light absorption of TiO_2 start? (b) What is the maximum oxidation potential of the light-generated hole assuming E_{CB} = -0.5 V?
11. In more than 90% of semiconductor-catalysed photoreactions two substrates are converted into an oxidised and a reduced product, in complete analogy with electrochemistry. Only in a very few cases is one unique addition product formed. Discuss the corresponding mechanism.
12. Formulate the effect of *photoredox amplification* by ethanol.
13. Discuss the principles of a silicon solar cell, of an LED and a diode laser.
14. What is electroluminescence and how does it relate to photoluminescence? It is applied in quantum dot and OLED displays. How do they work?

References

[1] a) A. L. Rogach, D. V. Talapin, H. Weller, in *Colloids and Colloid Assemblies* Wiley-VCH, Weinheim, Ed. F. Caruso, **2003**, p. 52–95; b) A. J. Nozik, *Next Generation of Photovoltaics* **2012**, *165,* 191–207; c) D. J. Norris, M. G. Bawendi, L. E. Brus, *Molecular Electronics*, International Union of Pure and Applied Chemistry, Ed. J. Jortner and M. Ratner, **1997**, p. 281–323.

[2] a) A. P. Alivisatos, *Journal of Physical Chemistry* **1996**, *100*, 13226–13239; b) F. P. Garcia de Arquer, D. V. Talapin, V. I. Klimov, Y. Arakawa, M. Bayer, E. H. Sargent, *Science* **2021**, *373*, 640.

[3] a) G. Kortuem, W. Braun, G. Herzog, *Angewandte Chemie* **1963**, *75*, 653–661; b) R. M. Edreva-Kardjieva, *Bulgarian Chemical Communications* **1992**, *25*, 166–192; c) B. M. Weckhuysen, R. A. Schoonheydt, *Catalysis Today* **1999**, *49*, 441–451.

[4] a) R. Beranek, *Advances in Physical Chemistry* **2011**, 1–20; b) P. Makuła, M. Pacia, W. Macyk, *Journal of Physical Chemistry Letters*, **2018**, *9*, 6814–6817.

[5] a) J. I. Pankove, *Optical Processes in Semiconductors*, Prentice-Hall Inc., New Jersey, **1971**; b) J. Tauc, R. Grigorovici, A. Vanuc, *Physica Status Solidi* **1966**, *15*, 627.

[6] S. M. Sze, *Physics of Semiconductor Devices*, John Wiley & Sons, New York, **1969**.

[7] a) B. Ohtani, O. O. P. Mahaney, F. Amano, N. Murakami, R. Abe, *Journal of Advanced Oxidation Technologies* **2010**, *13*, 247–261; b) E. S. Welter, S. Garg, R. Gläser, M. Goepel, *ChemPhotoChem* **2023**, *7*, e202300001.

[8] R. Kuenneth, G. Twardzik, G. Emig, H. Kisch, *Journal of Photochemistry and Photobiology A: Chemistry* **1993**, *76*, 209–215.

[9] a) K. Sooklal, B. S. Cullum, S. M. Angel, C. J. Murphy, *Journal of Physical Chemistry* **1996**, *100*, 4551–4555; b) L. Spanhel, H. Weller, A. Henglein, *Journal of the American Chemical Society* **1987**, *109*, 6632.

[10] J. Zhou, Q. Liu, W. Feng, Y. Sun, F. Li, *Chemical Reviews* **2015**, *115*, 395–465.

[11] D. J. Norris, Y. A. Vlasov, *Advances in Materials* **2001**, *13*, 371–376.

[12] J. E. G. J. Wijnhoven, L. Bechger, W. L. Vos, *Chemistry of Materials* **2001**, *13*, 4486–4499.

[13] J. I. Chen, E. Loso, N. Ebrahim, G. A. Ozin, *Journal of the American Chemical Society* **2008**, *130*, 5420–5421.

[14] a) W. Schmidt, *Optische Spektroskopie: Eine Einführung*, Wiley-VCH, Weinheim, **2014**; b) D. C. Dumitras, M. Petrus, A.-M. Bratu, C. Popa, *Molecules* **2020**, *25*, 1728.

[15] S. Manohar, D. Razansky, *Advances in Optics and Photonics* **2016**, *8*, 586–617.

[16] D. Brassington, *Journal of Physics D: Applied Physics* **1982**, *15*, 219.

[17] H. Kye, Y. Song, T. Ninjbadgar, C. Kim, J. Kim, *Sensors* **2022**, *22*, 5130.

[18] J. W. Nery, O. Pessoa, H. Vargas, F. de Am Reis, A. C. Gabrielli, L. C. Miranda, C. A. Vinha, *Analyst* **1987**, *112*, 1487–1490.

[19] K. Watanabe, J. Tokumine, A. K. Lefor, H. Nakazawa, K. Yamamoto, H. Karasawa, M. Nagase, T. Yorozu, *Scientific Reports* **2021**, *11*, 1–8.

[20] https://de.wikipedia.org/wiki/Photovoltaik. Accessed on April 7th 2024.

[21] https://en.wikipedia.org/wiki/Solar_power_by_country. Accessed on April 7th 2024.

[22] B. Wallikewitz, M. de la Rosa, D. Hertel, K. Meerholz, A. Falcou, H. Becker, *Frontiers in Optics*, **2005**. paper STuA2.

[23] a) H. Xu, R. Chen, Q. Sun, W. Lai, Q. Su, W. Huang, X. Liu, *Chemical Society Reviews* **2014**, *43*, 3259–3302; b) G. Hong, X. Gan, C. Leonhardt, Z. Zhang, J. Seibert, J. M. Busch, S. Bräse, *Advanced Materials* **2021**, *33*, 2005630.

[24] a) H. Shi, W. Yao, W. Ye, H. Ma, W. Huang, Z. An, *Accounts of Chemical Research* **2022**, *55*, 3445–3459; b) C. E. Housecroft, E. C. Constable, *Journal of Materials Chemistry C* **2022**, *10*, 4456–4482.

[25] H. Song, S. Luo, H. Huang, B. Deng, J. Ye, *ACS Energy Letters* **2022**, *7*, 1043–1065.

[26] a) J. J. Hopfield, *Physics and Chemistry of Solids* **1959**, *10*, 110–119; b) J. L. Birman, *Physical Review* **1959**, *115*, 1493–1505; c) M. Cardona, G. Harbeke, *Physical Review* **1965**, *137*, 1467–1476.

[27] H. Tang, K. Prasad, R. Sanilines, P. E. Schmid, F. Levy, *Journal of Applied Physics* **1994**, *75*, 2042–2047.

[28] a) H. Gerischer, *Surface Science* **1969**, *18*, 97–122; b) R. A. Marcus, *Angewandte Chemie International Edition* **1993**, *32*, 1111–1222.

[29] H. Kisch, D. Bahnemann, *Journal of Physical Chemistry Letters* **2015**, *6*, 1907–1910.

[30] R. Hayoun, K. M. Whitaker, D. R. Gamelin, J. M. Mayer, *Journal of the American Chemical Society* **2011**, *133*, 4228–4231.

[31] T. Berger, J. A. Anta, *Analytical Chemistry* **2012**, *84*, 3053–3057.

[32] N. Zeug, J. Buecheler, H. Kisch, *Journal of the American Chemical Society* **1985**, *107*, 1459–1465.

[33] S. Kozuch, J. M. Martin, *ACS Catalysis*, **2012**, *2*, 2787–2794.

[34] E. S. Welter, S. Kött, F. Brandenburg, J. Krömer, M. Goepel, A. Schmid, R. Gläser, *Catalysts* **2021**, *11*, 1415.

[35] a) A. V. Emeline, X. Zhang, M. Jin, T. Murakami, A. Fujishima, *Journal of Physical Chemistry B* **2006**, *110*, 7409–7413; b) L. Megatif, R. Dillert, D. W. Bahnemann, *Catalysis Today* **2019**, Ahead of Print.

[36] D. Bahnemann, P. Robertson, C. Wang, W. Choi, H. Daly, M. Danish, H. de Lasa, S. Escobedo, C. Hardacre, T. H. Jeon, *Journal of Physics: Energy* **2023**, *5*, 012004.

[37] H. Kisch, *Angewandte Chemie International Edition* **2010**, *49*, 9588–9589.

[38] H. Kisch, *Semiconductor Photocatalysis*, Wiley-VCH, Weinheim, Germany, **2015**.

[39] G. Haselmann, D. Eder, *Heterogeneous Photocatalysis: From Fundamentals to Applications in Energy Conversion and Depollution*, **2021**, Wiley-VCH, Weinheim, Ed. J. Strunk, pp. 187–219.

[40] a) H. H. Mohamed, D. W. Bahnemann, *Applied Catalysis B* **2012**, *128*, 91–104; b) P. V. Kamat, *Journal of Physical Chemistry Letters* **2012**, *3*, 663–672.

[41] P. C. Hiemenz, *Principles of Colloid and Surface Chemistry*, 2nd ed., Marcel Dekker, New York, **1986**.

[42] G. Horner, P. Johne, R. Kunneth, G. Twardzik, H. Roth, T. Clark, H. Kisch, *Chemistry: A European Journal* **1999**, *5*, 208–217.

[43] W.-J. Chun, A. Ishikawa, H. Fujisawa, T. Takata, J. N. Kondo, M. Hara, M. Kawai, Y. Matsumoto, K. Domen, *Journal of Physical Chemistry B* **2003**, *107*, 1798–1803.

[44] a) H. Ge, H. Tian, Y. Zhou, S. Wu, D. Liu, X. Fu, X.-M. Song, X. Shi, X. Wang, N. Li, *ACS Applied Materials & Interfaces* **2014**, *6*, 2401–2406; b) W. Macyk, G. Burgeth, H. Kisch, *Photochemical & Photobiological Sciences* **2003**, *2*, 322–328; c) S. Burnside, J.-E. Moser, K. Brooks, M. Grätzel, D. Cahen, *Journal of Physical Chemistry B* **1999**, *103*, 9328–9332; d) M. Nenadovic, T. Rajh, O. Micic, A. Nozik, *Journal of Physical Chemistry* **1984**, *88*, 5827–5830; e) Y. Cong, M. Chen, T. Xu, Y. Zhang, Q. Wang, *Applied Catalysis B: Environmental* **2014**, *147*, 733–740; f) E. Nurlaela, A. Ziani, K. Takanabe, *Materials for Renewable and Sustainable Energy* **2016**, *5*, 1–21; g) D. Meissner, R. Memming, B. Kastening, D. Bahnemann, *Chemical Physics Letters* **1986**, *127*, 419–423; h) J. F. Reber, K. Meier, *Journal of Physical Chemistry* **1984**, *88*, 5903–5913.

[45] https://www.grandviewresearch.com/industry-analysis/titanium-dioxide-industry. Accessed on April 7th 2024.

[46] D. C. Hurum, A. G. Agrios, S. E. Crist, K. A. Gray, T. Rajh, M. C. Thurnauer, *Journal of Electron Spectroscopy and Related Phenomena* **2006**, *150*, 155–163.

[47] K. Maeda, K. Domen, *Journal of Physical Chemistry C* **2007**, *111*, 7851–7861.

[48] S. H. Wei, A. Zunger, *Physical Review B: Condensed Matter* **1988**, *37*, 8958–8981.

[49] H. Kisch, P. Lutz, *Photochemical and Photobiological Sciences* **2002**, *1*, 240–245.

[50] K. R. Gopidas, M. Bohorquez, P. V. Kamat, *Journal of Physical Chemistry* **1990**, *94*, 6435–6440.

[51] H. Yu, H. Irie, Y. Shimodaira, Y. Hosogi, Y. Kuroda, M. Miyauchi, K. Hashimoto, *Journal of Physical Chemistry C* **2010**, *114*, 16481–16487.

[52] H. Tada, Q. Jin, H. Nishijima, H. Yamamoto, M. Fujishima, S.-i. Okuoka, T. Hattori, Y. Sumida, H. Kobayashi, *Angewandte Chemie International Edition* **2011**, *50*, 3501–3505. S3501/3501–S3501/3506.

[53] a) Q. Lin, S. Liang, J. Wang, R. Zhang, G. Liu, X. Wang, *ACS Sustainable Chemistry & Engineering* **2023**, *11*, 3093–3102; b) L. Zhang, J. Zhang, H. Yu, J. Yu, *Advanced Materials* **2022**, *34*, e2107668.

[54] G. E. Brown Jr., V. E. Henrich, W. H. Casey, D. L. Clark, C. Eggleston, A. Felmy, D. W. Goodman, M. Grätzel, G. Maciel, G. M. I. McCarthy, K. H. Nealson, D. A. Sverjensky, M. F. Toney, J. M. Zachara, *Chemical Reviews* **1999**, *99*, 77–174.

[55] a) W. Macyk, H. Kisch, *Chemistry: A European Journal* **2001**, *7*, 1862–1867; b) G. Burgeth, H. Kisch, *Coordination Chemistry Reviews* **2002**, *230*, 41–47.

[56] a) O. Kohle, M. Grätzel, A. F. Meyer, T. B. Meyer, *Advances in Materials* **1997**, *9*, 904–906; b) A. Hagfeldt, S.-E. Lindquist, M. Gratzel, *Solar Energy Materials and Solar Cells* **1994**, *32*, 245–257.

[57] J. R. White, A. J. Bard, *Journal of Physical Chemistry* **1985**, *89*, 1947–1954.

[58] M. F. Finlayson, B. L. Wheeler, N. Kakuta, K.-H. Park, A. J. Bard, A. Campion, M. A. Fox, S. E. Webber, J. M. White, *Journal of Physical Chemistry* **1985**, *89*, 5676–5681.

[59] S. Linic, U. Aslam, C. Boerigter, M. Morabito, *Nature Materials* **2015**, *14*, 567–576.

[60] a) X. Qiu, C. Burda, *Chemical Physics* **2007**, *339*, 1–10; b) C. Di Valentin, E. Finazzi, G. Pacchioni, A. Selloni, S. Livraghi, M. C. Paganini, E. Giamello, *Chemical Physics* **2007**, *339*, 44–56.

[61] a) S. Sato, *Chemical Physics Letters* **1986**, *123*, 126–128; b) S. Sakthivel, H. Kisch, *ChemPhysChem* **2003**, *4*, 487–490; c) Y. Cong, J. Zhang, F. Chen, M. Anpo, *Journal of Physical Chemistry C* **2007**, *111*, 6976–6982.

[62] H. Kisch, S. Sakthivel, M. Janczarek, D. Mitoraj, *Journal of Physical Chemistry C* **2007**, *111*, 11445–11449.

[63] a) N. Serpone, *Journal of Physical Chemistry B* **2006**, *110*, 24287–24293; b) V. N. Kuznetsov, N. Serpone, *Journal of Physical Chemistry C* **2007**, *111*, 15277–15288.

[64] D. Mitoraj, H. Kisch, *Chemistry: A European Journal* **2010**, *16*, 261–269.

[65] a) C. Lettmann, K. Hildenbrand, H. Kisch, W. Macyk, W. F. Maier, *Applied Catalysis B: Environmental* **2001**, *32*, 215–227; b) S. Sakthivel, H. Kisch, *Angewandte Chemie International Edition* **2003**, *42*, 4908–4911.

[66] P. Zabek, H. Kisch, *Journal of Coordination Chemistry* **2010**, *63*, 2715–2726.

[67] a) A. Mondal, A. Prabhakaran, S. Gupta, V. R. Subramanian, *ACS Omega* **2021**, *6*, 8734–8743; b) Y. Li, C. Gao, R. Long, Y. Xiong, *Materials Today Chemistry* **2019**, *11*, 197–216.

[68] W. Zou, J.-L. Zhang, F. Chen, M. Anpo, D.-N. He, *Research on Chemical Intermediates* **2009**, *35*, 717–726.

[69] a) M. J. Bojdys, J.-O. Mueller, M. Antonietti, A. Thomas, *Chemistry: A European Journal* **2008**, *14*, 8177–8182; b) Z. Ni, Q. Wang, Y. Guo, H. Liu, Q. Zhang, *Catalysts* **2023**, *13*, 579.

[70] a) X. Wang, K. Maeda, X. Chen, K. Takanabe, K. Domen, Y. Hou, X. Fu, M. Antonietti, *Journal of the American Chemical Society* **2009**, *131*, 1680–1681; b) C. M. You, C. Huang, S. Tang, P. Xiao, S. Wang, Z. Wei, A. Lei, H. Cai, *Organic Letters* **2023**, *25*, 1722–1726; c) L. Yang, Y. Peng, X. Luo, Y. Dan, J. Ye, Y. Zhou, Z. Zou, *Chemical Society Reviews* **2021**, *50*, 2147–2172.

4 Photochemistry

4.1 Light reactions and dark reactions

The difference between a dark reaction and a light reaction is shown schematically in Figure 4.1 through the energy changes along the reaction coordinate of an endothermic process.[1] By the end of the reaction the product **P** has a higher energy than the starting material **G** by an amount ΔE.

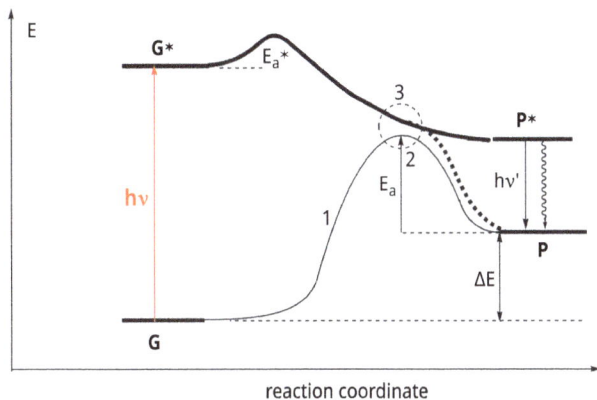

Figure 4.1: Energetic course of an endothermic reaction in the dark (—) and in the light (━, - - -).

As can be seen, the energy changes in a characteristic way. In the dark reaction, the product is not formed by the shortest route from **G** to **P**, but must climb over an activation hill (path 1–2 in Figure 4.1). The required activation energy E_a lies in the region 40–80 kJ/mol and will normally be delivered by heating. It is required to weaken the bond that is to be broken in a bond-breaking reaction or to bring a reaction partner to a favourable position in the case of a bond-making reaction. The smaller the activation energy, the faster the reaction. This relationship can be exploited by adding a substance that lowers the activation energy but is not consumed: a catalyst (cat). The catalyst forms weak bonds with the substrates and controls their spatial orientation such that the desired reaction is accelerated. The most important example is the Haber-Bosch process for the manufacture of ammonia from atmospheric nitrogen and hydrogen.[2] Without an iron catalyst the formation of product is much too slow

---

1 The diagram shows a two-dimensional cut through the multidimensional potential energy hypersurface along the reaction coordinate that best describes the progress of the reaction (e.g. change in bond length).
2 Ammonia is subsequently converted into fertilisers such as ammonium sulfate and ammonium nitrate.

https://doi.org/10.1515/9783111029375-004

even at 400 °C and high pressure (300 atm). In the presence of metallic iron, the formation of ammonia is so fast that the most important chemical process can proceed on a huge scale.[3]

In a photoreaction (or light reaction), in contrast to the dark reaction, an energy-rich, short-lived, electronically excited state G^* is formed instantaneously through absorption of a photon. The energy contributed by absorption of visible light lies in the range 170–300 kJ/mol. In this case, the activation energy E_a^* is so small that the required energy can be taken from the surroundings. In the subsequent course of the reaction to form product **P**, the potential energy curve of the light reaction and the dark reaction almost cross one another at point 3. The nearer that the two curves come to one another, the greater is the probability that the photochemical reaction crosses onto the thermal curve and forms product **P** directly (adiabatic reaction). Alternatively, G^* converts to P^* and then **P** through one or more additional potential energy surfaces (non-adiabatic reaction) with emission of light (hv′) or release of heat. As explained in Chapter 2.2.1, a photoreaction occurring on a single potential energy surface is described as adiabatic, whereas a non-adiabatic reaction requires a change of potential energy surface.

The crossing point in the reaction coordinate makes the well-known phenomenon of chemiluminescence understandable: many exothermic dark reactions proceed with emission of light. As can be seen from Figure 4.1, the reaction can proceed through the sequence $G \rightarrow 1 \rightarrow 2 \rightarrow 3$ to P^* which is converted to **P** by luminescence or by a radiationless process. In many cases, the initial product P^* is itself labile and is converted to a stable end-product by light emission. The very first example of that reaction type is the aerial oxidation of white phosphorus (see textbox) [1].

Cold fire – light instead of gold

The best known example of chemiluminescence is the reaction of white phosphorus (P_4) with oxygen in the air. This reaction generates such a bright light that the alchemist Hennig Brand used it as an attraction for the public. When searching for the "philosopher's stone", he discovered a white substance in 1696 that exhibited this fascinating light phenomenon when he heated urine with ash. He named this cold fire. Later the substance was named phosphorus (light carrier). This was the first chemical element to be discovered for which we know the name of the discoverer. In Brand's experiments, the carbon in the ash acts as the reducing agent for the phosphate (e.g. Na_2HPO_4) in the urine.

---

[3] Food shortages threatened at the end of the nineteenth century because of the rapidly increasing populations and the limited availability of natural fertilisers, such as guano. Increased production of cereal crops was not possible without fertiliser.

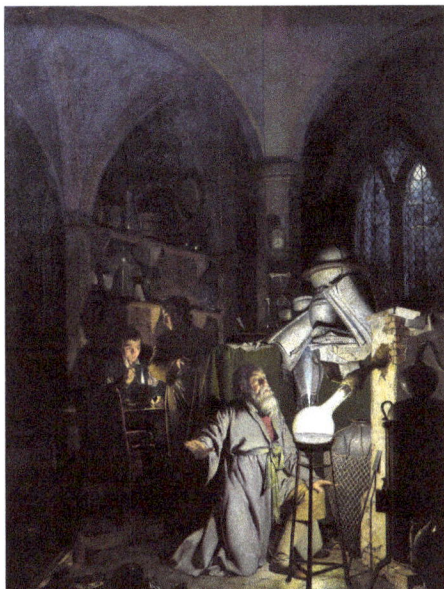

Painting of the discovery of phosphorus by Joseph Wright of Derby (1771), Derby Museum and Art Gallery.

A decisive difference between the thermal and photochemical reactions is that the heat energy in the thermal reaction is distributed evenly over the whole molecule, whereas the photon energy is localised in the chromophore groups. Consequently, photochemical reactions can lead to products that are different from those obtained by heating. Moreover, photochemistry can be performed at low temperature allowing the generation of thermally labile products (see Chapter 1.8). Individual chromophores can be excited selectively by tuning the wavelength of excitation. However, wavelength selective photochemistry is typically a characteristic of prompt photochemical reactions (see e.g. metal carbonyls). If an equilibrated excited state is formed by internal conversion and/or intersystem crossing, there is little dependence on the excitation wavelength.

Chemiluminescence is exploited in forensics as a very sensitive test for blood (glowing blood). The oxidation of the test substance luminol (A) by hydrogen peroxide occurs in the presence of haemoglobin (Hb) which is an extremely active catalyst. The product **P** is formed in its excited state which decays by blue luminescence to its ground state (Figure 4.2).

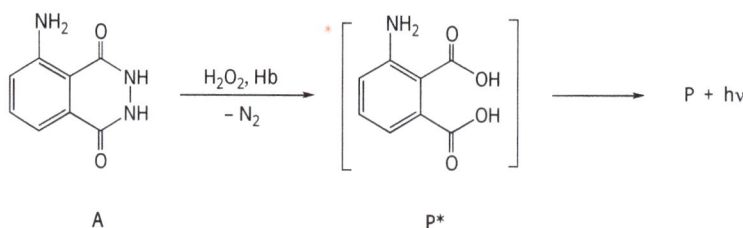

Figure 4.2: Reaction sequence of luminol chemiluminescence. A: 3-aminophthalic acid hydrazide, Hb: haemoglobin catalyst. Other iron compounds can also act as catalysts.

4.2 Non-catalytic photochemistry of organic molecules

The reactivity of an excited state can be roughly estimated from its electronic properties. If bonding MOs are depopulated or antibonding MOs populated, the bond is weakened and is likely to be cleaved [2]. This is the case for singlet excited states formed by π,π^*-transitions that undergo rearrangements, cycloadditions, cyclo-eliminations as well as proton- and electron-transfer reactions. The corresponding triplet states have diradical character and exhibit radical rearrangements and are especially prone to hydrogen abstraction reactions. $n\text{-}\pi^*$ states also have diradical characters and similar reactivity, but the difference between singlets and triplets is smaller.

4.2.1 Addition and substitution reactions

Addition of alcohols to olefins
The zwitterionic S_1 state of aryl-alkyl substituted olefins favours a nucleophilic addition of the alcohol across the double bond (Scheme 4.1).

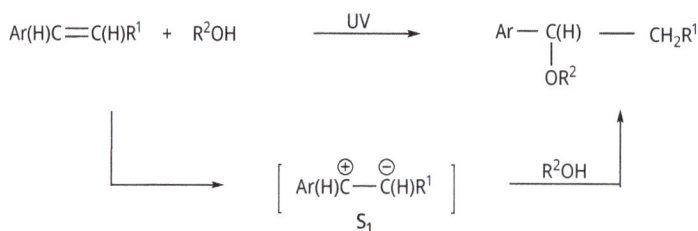

Scheme 4.1: Photoaddition of alcohols to olefins (R^1, R^2 = alkyl).

Photosubstitution in aromatics often exhibits different regiospecificity as compared to the thermal reaction. Thus, gentle heating of 4,4-dimethoxynitrobenzene in the presence of potassium hydroxide affords the para-substituted isomer as the major product, whereas only the meta-hydroxy product is formed upon UV-excitation (Scheme 4.2). MO-calculations reveal that the density of positive charge both in the singlet and triplet excited states is largest at the meta-position. Thus the nucleophilic attack of hydroxide occurs selectively at this carbon atom.

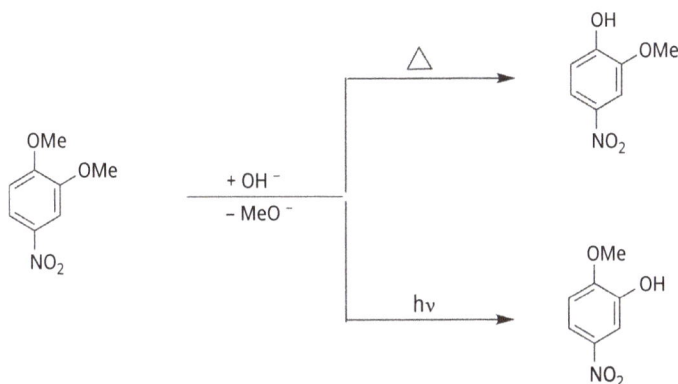

Scheme 4.2: Regiospecific substitution of methoxy groups in dimethoxynitrobenzene.

An example of a reaction that occurs from the $^3(n,\pi^*)$ excited state is the hydrodimerisation of benzophenone in propan-2-ol (Scheme 4.3). H-abstraction of the excited ketone gives the radical pair RP, initially within a common solvent cage. Surprisingly, the expected hetero-C-C coupling (Equation 4.1) occurs only to a minor extent, while benzpinacol as formed by homo-C-C coupling (Equation 4.3), is the main product. This suggests that radical diffusion out of the solvent cage and homo-coupling between fully solvated benzhydryl radicals (Equations 4.2 and 4.3) is faster than hetero-coupling within the cage. Accordingly, the latter is only preferred when a scavenger for solvated ketyl radicals such as α-camphor is present. Hydrogen transfer according to

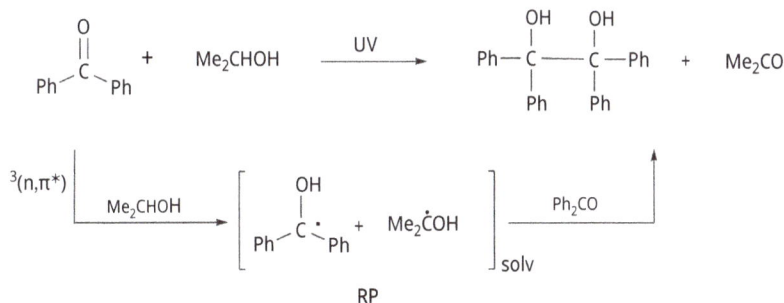

Scheme 4.3: Hydrodimerisation of benzophenone by isopropanol.

Equation 4.4 leads to acetone and a further benzhydryl radical. The intermediate radicals can be used as initiators for the polymerisation of alkenes.

$$[RP]_{solv} \rightarrow Ph_2C(OH) - C(OH)Me_2 \tag{4.1}$$

$$[RP]_{solv} \rightarrow Ph_2C^{\bullet}(OH) + C^{\bullet}(OH)Me_2 \tag{4.2}$$

$$2\,Ph_2C^{\bullet}(OH) \rightarrow Ph_2C(OH) - C(OH)Ph_2 \tag{4.3}$$

$$C^{\bullet}(OH)Me_2 + Ph_2CO \rightarrow Me_2CO + Ph_2C^{\bullet}(OH) \tag{4.4}$$

4.2.2 Cycloaddition reactions

The intermediate radicals postulated in Scheme 4.3 can be evidenced when the hydrodimerisation is conducted in a non-alcoholic solvent in the presence of an olefin (Paterno-Büchi reaction, Scheme 4.4).

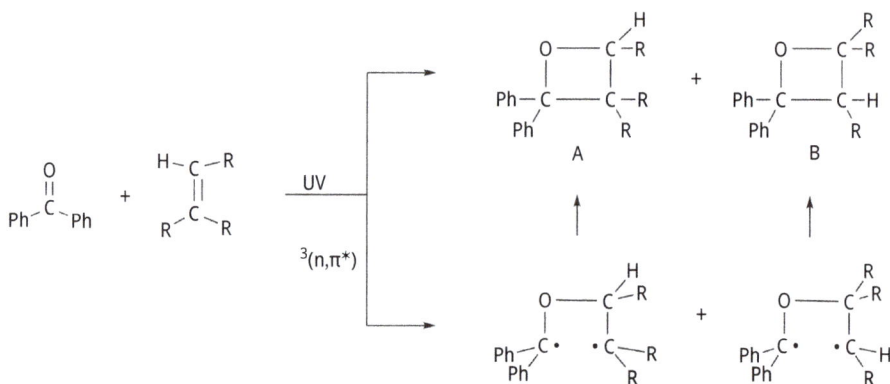

Scheme 4.4: [2 + 2] cycloaddition between benzophenone and an electron-rich alkene.[4]

If a phenyl group in benzophenone is replaced by p-H_2N-C_6H_4, the lowest excited state has CT character, and the photoreaction no longer occurs – a good example of the tuning of reactivity through a change in the nature of the excited state.

The development of the Woodward-Hoffmann [3] rules for pericyclic reactions was one of the biggest breakthroughs in understanding how reactivity of excited states differs

---

4 The numbers in brackets express how many atoms are involved in bond making.

from that of ground states.[5] In a pericyclic reaction making and breaking two or more bonds occurs at the same time. For such concerted reactions Hoffmann developed rules by considering how the occupied frontier MOs of the reactants map onto those of the product via MO correlation diagrams.[6] When the occupied orbitals of the ground states correlate with one another, the reaction is thermally allowed, but when the correlation requires occupation of the excited state orbitals, the reactions are photochemically allowed but thermally forbidden (Figure 4.3). From such frontier orbital correlation analyses, the rule was formulated as summarised in Figure 4.3 where n corresponds to the number of π-electrons involved in the bond breaking or bond forming step (n = 1, 2, 3, etc.). Accordingly, the photochemical ring opening of cyclobutene, n = 2, should occur through a disrotatory movement, but through a conrotatory way in the thermal reaction. Thus, the Woodward-Hoffmann rules allow the prediction of the stereochemical consequences of pericyclic reactions. In these reactions, the transition state consists of a cyclic array of all atomic centres involved in bond breaking and bond making. It is important that the rules apply only for concerted reactions, excluding formation of intermediates such as e.g. radicals.

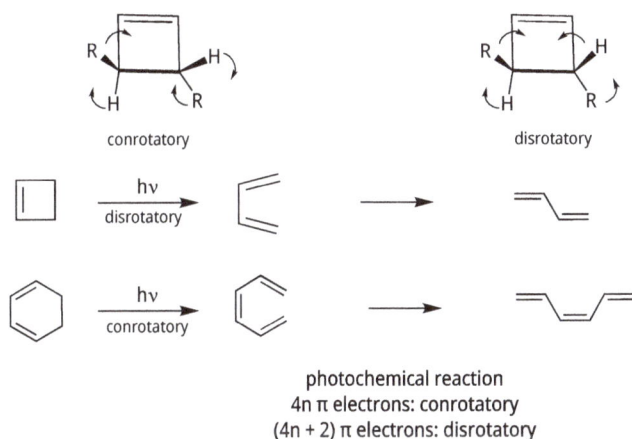

photochemical reaction
4n π electrons: conrotatory
(4n + 2) π electrons: disrotatory

Figure 4.3: Electrocyclic ring opening reactions.

To illustrate the role of frontier MOs in the application of the rules, Figure 4.4 describes in a simplified way the stereochemical consequences of orbital interactions. In the

---

5 Roald Hoffmann won the Nobel Prize in 1981 together with Kenichi Fukui "for their theories, developed independently, concerning the course of chemical reactions". Robert Burns Woodward had already won the Nobel Prize in 1965.
6 Molecular orbitals are said to correlate if they have common symmetry characteristics between reactant and product, hence the name of the original book by Woodward and Hoffmann: *The Conservation of Orbital Symmetry.*

photoreaction the LUMO must be considered since it contains one electron. As symbolised by the arrows, only a disrotatory movement leads to a positive orbital overlap for the ring closing of butadiene. In the case of a thermal reaction the HOMO is relevant and bond making should occur through a conrotatory movement.

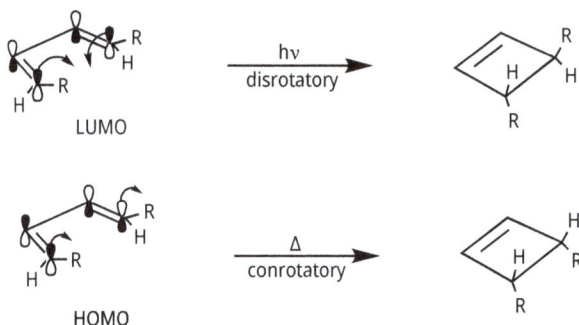

Figure 4.4: Stereochemical consequences of the electrocyclic ring closure of 1,3-butadiene in a photochemical and thermal reaction. The MOs are symbolised by the contributing atomic orbitals.

According to the rules, a [2 + 2]-cycloaddition is thermally forbidden, but photochemically allowed. These rules apply to ring-opening, the reverse of cycloaddition. DNA damage by UV radiation in the 280–300 nm-region forms an example of [2 + 2]-cycloaddition. If two thymine bases are close enough in the DNA-chain, a cyclobutane ring can form leading to distortion of the tertiary structure. This reaction occurs predominantly from the singlet excited state and is complete within picoseconds (Figure 4.5) [4]. The MO diagram shows the correlation between the MOs of the precursor and the product demonstrating that the reaction requires photochemical initiation. The orbitals of the reactant and the product that correlate with one another are shown with a box of the same colour. The correlation addresses whether the orbitals are symmetric or antisymmetric with respect to the two mutually perpendicular planes shown in the figure.

Intramolecular cycloaddition reactions can lead to very strained molecules that are not accessible by thermal reactions. For example, UV-irradiation of norbornadiene generates quadricyclane with an energy content increased by 92 kJ mol^{-1} (Figure 4.6). This energy can be readily released by regeneration of norbornadiene in the presence of a catalyst. Since there are almost no side-reactions, this system would be ideal for thermal storage of solar energy. To achieve such storage, the absorption spectrum of norbornadiene must be shifted into the visible region. In part, this can be achieved by variation of the substituents R^1 and R^2 but the amount of stored energy is greatly reduced [5].

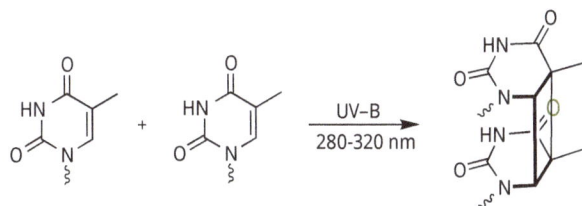

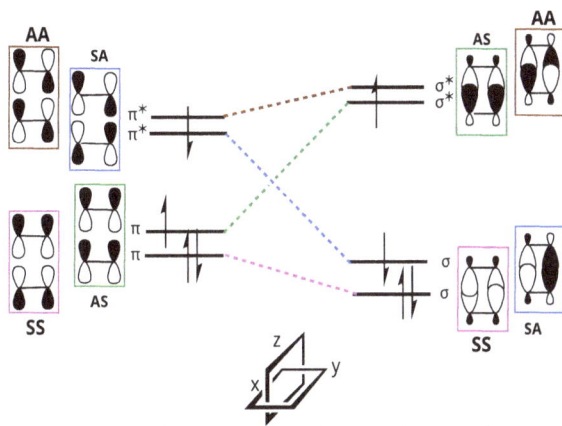

symmetry planes: S = symmetric, A = antisymmetric

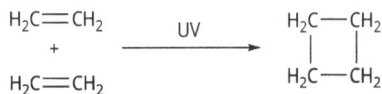

Figure 4.5: Above: cyclodimerisation of two neighbouring thymine bases in a DNA chain. Bonds without labels symbolise methyl groups. Below: MO diagram showing the correlation of the MOs and the symmetry planes conserved in the correlation. The diagram displays the singlet excited state of the reactant and product.

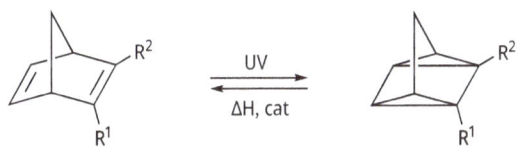

Figure 4.6: Valence isomerisation of norbornadiene ($R^1 = R^2 = H$) and its derivative ($R^1 = CO_2Et$, $R^2 = Ph$).

The conversion of bisnorbornadiene (A) into the highly strained molecular cage (B) is an even more spectacular example of a [2 + 2]-cycloaddition that proceeds in good yield (Figure 4.7).

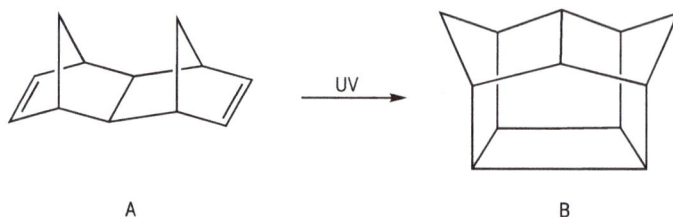

A   B

Figure 4.7: Intramolecular cycloaddition leading to a cage-compound.

4.2.3 Isomerisation and rearrangement reactions

Cis-trans isomerisations

The simplest intramolecular isomerisation reactions are the cis-trans (Z-E) rearrangements of alkenes and acyclic azo compounds (Figure 4.8). Such reversible processes lead to a significant change in the spatial structure together with other chemical and physical properties. Since the absorption spectrum changes, one of the most important effects is photochromism [6].

The hindrance of rotation around a C=C and N=N double bond forms the basis for the existence of geometric isomers at room temperature. Geometric isomerisation is observable therefore only upon absorption of thermal or photochemical energy. In the case of the colourless and orange solids 1,2-diphenylethene (stilbene) and 1,2-diphenyldiazene (azobenzene), respectively, UV and visible light can deliver this energy. In the solid state at room temperature the cis-isomers return slowly to the more stable trans-compounds. Introduction of appropriate substituents into the phenyl groups can change the stability considerably. Obviously, sterically demanding groups in the ortho-positions favour fast cis-trans isomerisation.

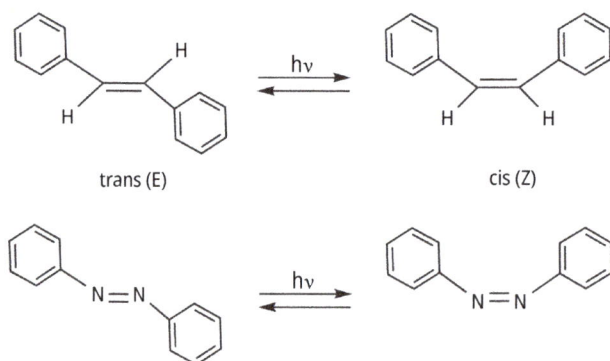

trans (E)   cis (Z)

Figure 4.8: Photochemical (E) to (Z) isomerisations of stilbene and azobenzene.

Such isomerisation processes play a huge role in biology, most notably in vision (see Chapter 6.3). One of the first abiotic examples was the tuning of selectivity of the complexation of group 1 metal ions by crown ethers. Thus, the smaller ring of the E-isomer prefers lithium and sodium ions, while the bigger ring of the Z-isomer selects potassium and rubidium (Figure 4.9) [7]. Since the effect is thermally reversible, the more stable E-isomer is regenerated after a time in the dark.

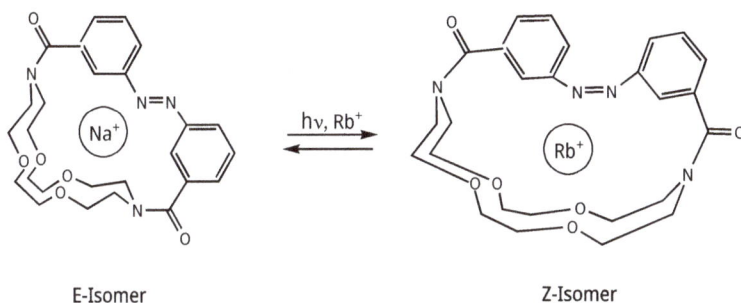

E-Isomer   Z-Isomer

Figure 4.9: Tuning of selectivity of metal-ion coordination on *E-Z* photoisomerisation.

When photosensitive functional groups are included into much bigger structures, the term supramolecular photochemistry is used to describe the general phenomenon; interconversion of isomers as in this example is called a photoswitch [2a]. Such reactions can be used to switch biological reactions on or off if the light is restricted to local regions of an organism by careful focussing. This new subject of photopharmacology is developing rapidly now.

Similar effects are used in construction of molecular machines. Stoddart pioneered the idea of a molecular shuttle based on a rotaxane in which a molecular ring is held on an axle by mechanical bonds and cannot slip off the axle because of the steric bulk of the end-groups.[7] In a shuttle, the ring can occupy two positions and can be induced to move from one to the other by a chemical signal, by electron transfer or by a light signal. An example of a photo-induced shuttle is shown schematically in Scheme 4.5, where *E-Z* interconversion of an azobenzene acts as the switch and the direction of shuttling can be changed by altering the irradiation wavelength [8].

A rotary motor consists of a rotor and a stator linked by an axle such that rotation occurs in one direction only. Feringa and colleagues achieved the feat of synthesising a motor on a molecular scale by using the power of photo-induced isomerisation of an overcrowded alkene that includes a stereogenic centre as illustrated in Scheme 4.5. The light reaction induces rotation of the rotor to a sterically unstable form that con-

---

7 Ben Feringa and Fraser Stoddart won the Nobel Prize in 2016 together with Jean-Pierre Sauvage for the design and synthesis of molecular machines.

B: ⌇⌇N⌇⌇ ... ⌇⌇NH⌇⌇

R = iPr

stable

less stable

less stable

stable

hν

hν

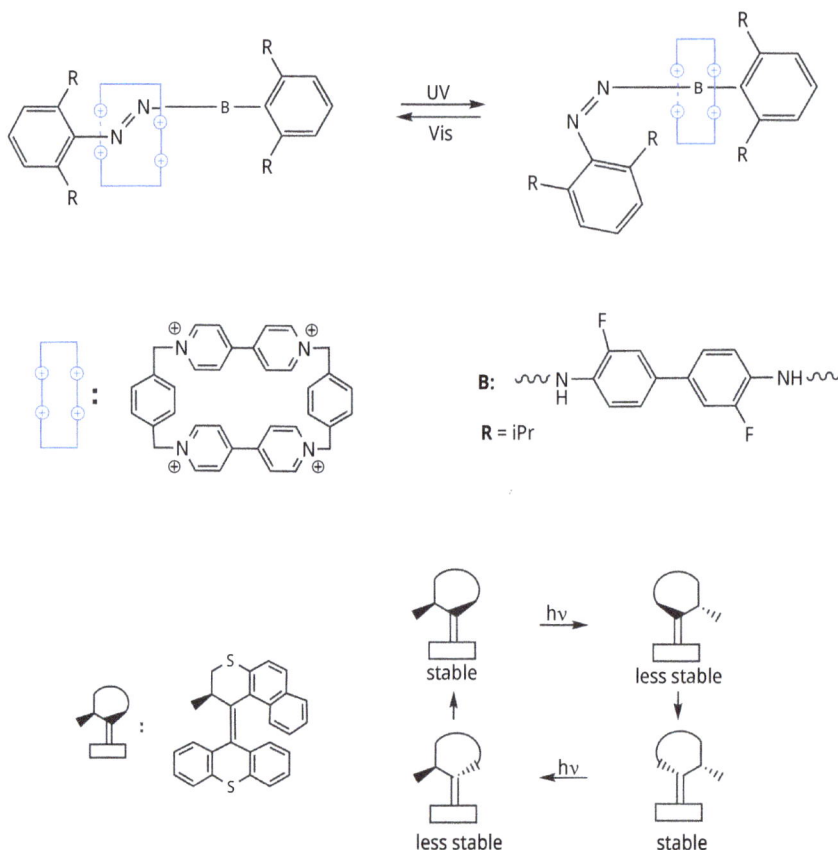

Scheme 4.5: Above: principle of light-driven molecular shuttle based on E-Z isomerisation of an azobenzene unit [9]. Below: Feringa's second generation photochemically driven molecular motor based on a very crowded and chiral alkene [10].

verts to the mirror image of the original molecule thermally. In a second photochemical step, this product undergoes a photo-induced rotation to the mirror image of the initial unstable product. In turn, this species isomerises thermally back to the original molecule (Scheme 4.5). The steric requirements of the molecule force these isomerisations to proceed in one direction only [11].

Sigmatropic reactions

In a sigmatropic reaction a σ-bond of a π-system migrates within the molecule to a new position, usually along a conjugated chain of atoms. To concisely describe such processes, a sigmatropic rearrangement of order [i,j] is a reaction in which a σ-bond that is flanked by one or more π-bonds migrates to a new position whose termini are i-1 and j-1 atoms removed from the original bonded position (Figure 4.10).

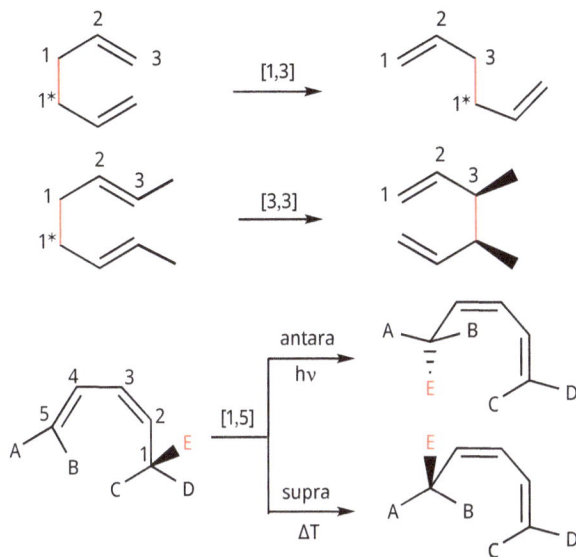

Figure 4.10: Nomenclature for sigmatropic C,C- and C,H- shift reactions.

According to the Woodward-Hoffmann rules [3], the stereochemical consequences can be predicted, as exemplified for a [1,5]-hydrogen shift. Recalling that in the cyclic transition state all carbon atoms are in a plane, the hydrogen atom may migrate within one side of the system (suprafacial path) or pass from the top to the bottom face (antarafacial). According to the rules the presence of 4n π–electrons induces a photoreaction in an antarafacial course while for (4n + 2) π–electrons a suprafacial movement is predicted. The opposite is predicted for the corresponding thermal reaction versions. It is recalled that these predictions are valid only for *concerted reactions*, but not for stepwise reactions via intermediates such as radicals.

In the industrial synthesis of vitamin D, photo-induced electrocyclic ring opening, followed by a [1,7]-sigmatropic hydrogen shift are key mechanistic steps (Scheme 4.6).

Di-π-methane rearrangement

A 1,4-diene containing one cis-olefin group may undergo a [1,2]-shift to an intermediate diradical or zwitterion, depending on the nature of the substituents. Subsequent ring closure affords the cyclopropane product (Figure 4.11), retaining the cis-position of R^1 during the rearrangement. The reaction is a very useful preparation of vinyl cyclopropanes. It also works when the double bond in the 1,2-position is part of an aromatic ring like benzene or when it is replaced by a keto group (oxa-di-π-methane rearrangement), but it is essential that R^2 and R^3 are alkyl groups.

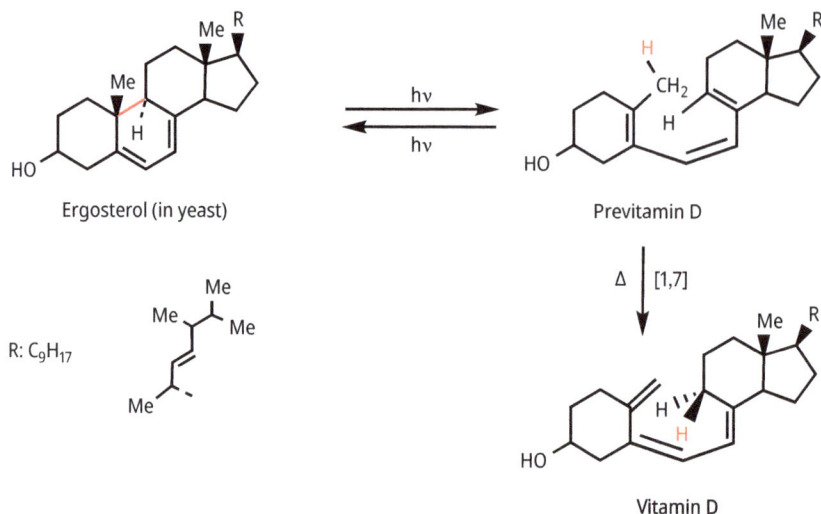

Scheme 4.6: Electrocyclic ring opening and [1,7]-H shift during synthesis of Vitamin D_2.

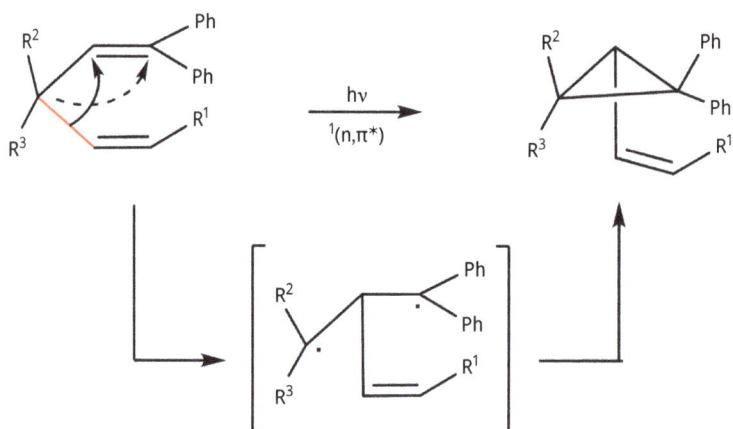

Figure 4.11: Synthesis of small ring compounds through the di-π-methane rearrangement.

Isomerisation of nitrite esters to δ-nitroso alcohols (Barton reaction)

UV-irradiation of a nitrous acid ester (alternatively named as an alkyl nitrite) with a hydrogen atom in the δ-position, generates nitric oxide and an alkoxy radical. Successive hydrogen abstraction leads to a hydroxyalkyl radical which undergoes a radical C-N coupling with NO to the final δ-nitroso alcohol (Scheme 4.7). The regioselectivity of abstraction is a consequence of formation of an energetically favourable six-membered transition state.[8]

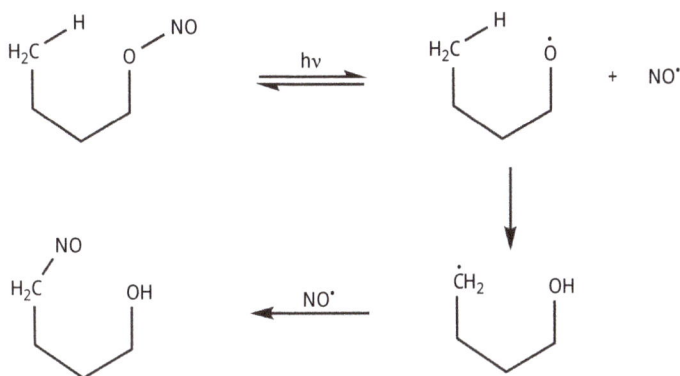

Scheme 4.7: Rearrangement of an alkyl nitrite into a nitroso alcohol (Barton Reaction).

The Barton reaction is a key step in the *in vitro* synthesis of aldosterone, a steroid hormone responsible for the regulation of blood pressure and of the sodium and potassium concentration in the plasma.

4.2.4 Fragmentation reactions

Elimination of dinitrogen

The photochemical reactions of diazo-compounds and diazirines lead to prompt loss of N_2, generating carbenes as short-lived intermediates. It is through these reactions that we know that the ground states of CH_2 and dialkylcarbenes are triplets whereas diphenylcarbene has a singlet ground state (Chapter 1.7). Similarly, the photochemical reactions of azido-compounds lead to nitrenes (Scheme 4.8).

The behaviour of phenylazide and its substituted analogues is particularly intriguing. Irradiation generates the singlet nitrene on the picosecond timescale. Two isomerisation reactions take place in competition with ISC to its ground triplet state, generating intermediates with either a 3-membered ring (benzazirine) or a 7-membered ring (cycloketenimine) that can be trapped to yield different products (Scheme 4.9). The formation of the 7-membered ring was established through the IR signature of its cumulated double bond in matrix isolation and TRIR experiments. In a remarkable demonstration of

---

8 Derek Barton was awarded the Nobel Prize in Chemistry in 1969 for his work on understanding conformations of organic molecules, a key for the finding of regiospecific radical reactions.

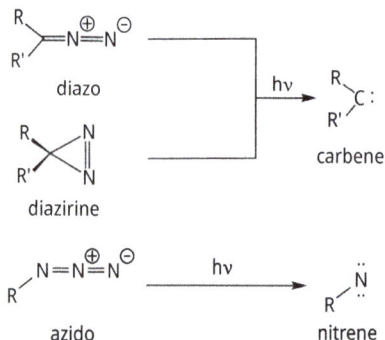

Scheme 4.8: Photochemical loss of N_2 from diazo-compounds, diazirines and azido-compounds (R = alkyl, aryl, etc.).

the power of wavelength-selective photochemistry, the three isomers have all been interconverted in an argon matrix at 10 K [12].

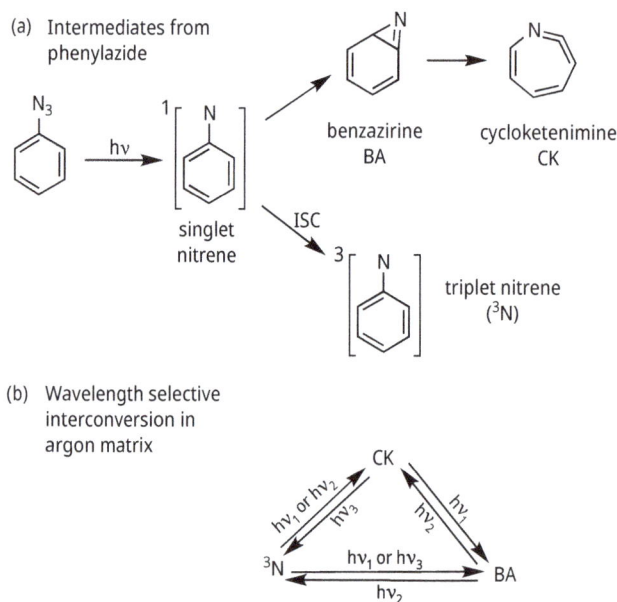

(a) Intermediates from phenylazide

benzazirine
BA

cycloketenimine
CK

singlet nitrene

ISC

triplet nitrene
(^{3}N)

(b) Wavelength selective interconversion in argon matrix

Scheme 4.9: (a) Photochemical products of phenyl azide observed by matrix isolation and time-resolved spectroscopy, (b) Interconversion of isomers of phenyl nitrene by wavelength-selective photochemistry.

C-C α-cleavage and intramolecular hydrogen abstraction of ketones (Norrish Types I and II)

The photochemical properties of ketones have been intensively investigated, both with respect to synthesis and mechanism. In most cases the reactive excited state has

$^3(n,\pi^*)$ character giving rise to the primary processes of bond cleavage, hydrogen abstraction, and elimination of carbon monoxide. Depending on the nature of the substituents attached to the keto group, all examples may be observable. In class I cleavage (Scheme 4.10) α-cleavage generates a radical pair (named also a diradical DR) within the solvent cage. Hydrogen atom transfer affords an aldehyde and olefin as final products. When the t-butyl group is exchanged by a propyl substituent, hydrogen atom transfer (HAT) affords a biradical (BR).[9] Subsequent C-C bond formation leads to a cyclobutyl alcohol.

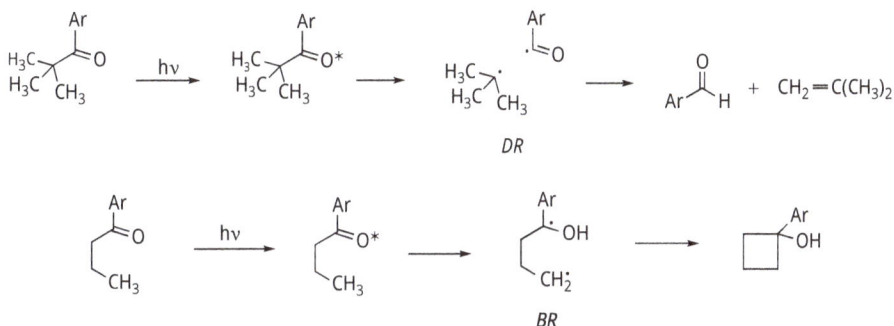

Scheme 4.10: Norrish Type I α-cleavage and Norrish Type II hydrogen abstraction form triplet excited states of ketones.

In the case of dibenzyl ketone, the primary diradical generates a benzyl radical through elimination of carbon monoxide. Subsequent radical C-C coupling generates $PhCH_2CH_2Ph$ as final product.

4.2.5 Reactions with O_2

Type I photooxidation
The oxygen molecule, which starts absorbing light at 240 nm, is one of the most important reactants for chemical conversions in natural and artificial systems. In air-saturated water its concentration is in the range of 0.2×10^{-3} M, in organic solvents about ten times higher. Unlike almost all other mutual reaction partners, which are usually present in a singlet ground state, oxygen has a triplet ground state ($^3O_2$) containing two unpaired electrons in its π^*-HOMO. One of the valence bond formulations, ·O-O·, signals the biradical property responsible for efficient coupling to other radicals.

According to the Wigner spin conservation rule, the reaction of a triplet with a singlet cannot generate two singlet products (see chapter 2.2.3). Consequently, oxygen can-

---

9 Thus, unlike a diradical, the two electrons are localised on *one* molecule (BR).

not react directly in its ground state to oxidise most organic molecules. These reactions are so slow that the biosphere of earth can survive in its present form. The presence of an initiator can accelerate aerial oxidations. The initiator may generate an organic radical (Type I oxidation) or sensitise singlet oxygen formation (Type II oxidations).

In Type I oxidations light absorption by the sensitiser S generates its triplet state $^3S_1$ via $^1S_1$ (Equation 4.5). Subsequently, a HAT reaction according to Equation 4.6 affords two radicals. One of them couples to O_2 forming a peroxyl radical, which in turn abstracts hydrogen from the sensitiser radical S-H$^\cdot$ (Equations 4.7–4.8). Since the peroxyl radical may abstract hydrogen from RH (Equation 4.9), a new starter radical R$^\cdot$ can induce a radical chain oxidation of RH without further involvement of the photosensitiser. Assuming S is benzophenone (Ph$_2$CO) and RH is isopropanol, the hydroxoperoxide Me$_2$C(OH)(O$_2$H) is produced according to Equation 4.8. This labile hydroperoxide transforms to acetone and hydrogen peroxide as final products. Termination of the photoinitiated radical chain reaction can happen according to Equation 4.10. Carbon-centred radicals are good reductants and may also generate the superoxide radical anion in a proton-coupled electron transfer according to Equation 4.11. Figure 4.12 shows the connections between the oxygen species.

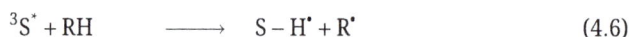

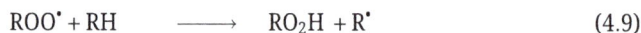

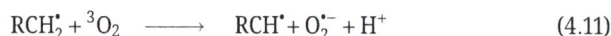

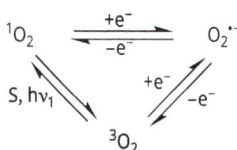

$$S \xrightarrow{\ h\nu\ } {}^3S^* \tag{4.5}$$

$$^3S^* + RH \longrightarrow S-H^\cdot + R^\cdot \tag{4.6}$$

$$R^\cdot + {}^3O_2 \longrightarrow ROO^\cdot \tag{4.7}$$

$$S-H^\cdot + ROO^\cdot \longrightarrow S + RO_2H \tag{4.8}$$

$$ROO^\cdot + RH \longrightarrow RO_2H + R^\cdot \tag{4.9}$$

$$2\,RCH_2OO^\cdot \longrightarrow RCH_2OH + RCHO + O_2 \tag{4.10}$$

$$RCH_2^\cdot + {}^3O_2 \longrightarrow RCH^\cdot + O_2^{\cdot-} + H^+ \tag{4.11}$$

Figure 4.12: Relationship between most relevant molecular oxygen species.

Type II photooxidation

While in Type I oxidation the inertness of $^3O_2$ is overcome by attack of a radical (Equation 4.7), in Type II triplet energy transfer generates singlet oxygen (Equations 4.12, 4.13). This lowest excited state of oxygen is located just 94 kJ/mol above the

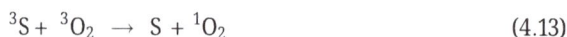

$$S \rightarrow {}^3S \tag{4.12}$$

$$^3S + {}^3O_2 \rightarrow S + {}^1O_2 \tag{4.13}$$

ground state and therefore much lower than the excited states of common organic compounds (see Chapter 2.2.4). Together with its rather long life-time of a few microseconds these properties make $^3O_2$ an efficient quencher for almost all other excited states. For that reason, photochemical experiments are generally conducted under dinitrogen or argon. Quenching may occur by energy and electron transfer generating $^1O_2$ and superoxide as summarised in Equations 4.7 and 4.11. Typical sensitisers are organic dyes and porphyrins. When these are excited at about 600 nm[10] and are able to selectively adsorb to a human cancer cell, they produce singlet oxygen *in vivo*, a cytotoxic agent. This is a commonly used method in cancer treatment named *photodynamic therapy* (PDT, Chapter 6.2) [13].

Organic syntheses with singlet oxygen

The microsecond lifetime of singlet oxygen is long enough to enable useful bimolecular reactions [14]. In the presence of a sensitiser the first step generates $^1O_2$ which is capable of adding to various unsaturated compounds. With electron-rich alkenes like tetramethylethylene 1,2-endoperoxides are formed (Figure 4.13, upper). These are thermally unstable and decompose under slight warming to ketones with efficient emission of light (chemiluminescence). Therefore 1,2-endoperoxides are used as a *light source* for conducting photoreactions "in the dark". With less substituted alkenes containing one allylic hydrogen atom, the synthetically useful *ene-reaction* is another typical reaction mode (Figure 4.12, middle).[11] It probably proceeds via a six-membered transition state (TS) favouring HAT to O_2 and a shift of the double bond.

---

10 At 600–800 nm light can penetrate 3–10 nm into human tissue.
11 Alkylhydroperoxides are in general highly explosive liquids. An exception is *t*-butylhydroperoxide, which is the reagent for the Sharpless-epoxidation of alkenes.

Through sun and spinach to a pharmaceutical

Professor G. O. Schenck in his private open-air laboratory in Heidelberg (1949).

1 kg of nettle or spinach leaves are suspended in 10 L of EtOH and 5 L of a 1% ethanol solution of α-terpinene. After leaving for 2–3 days in sunlight, 10 g of ascaridole could be isolated.

With 1,3-dienes like α-terpinene a [4 + 2] cycloaddition to the bicyclic peroxide ascaridole is observed [15]. The reaction was applied practically in 1943, using spinach leaves as sensitiser (chlorophyll) and sunshine as light source (see textbox) [16]. The

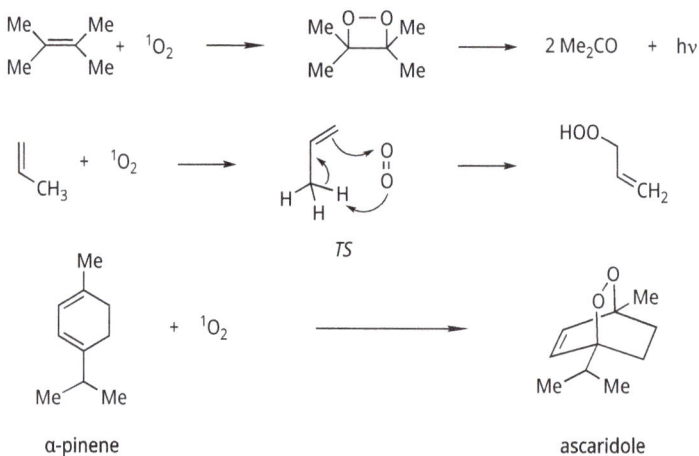

Figure 4.13: Some classical singlet oxygen reactions. S = triplet sensitiser.

product is an antihelminitic (antiparasitic) drug (Figure 4.13). In a much more recent industrial-scale application, singlet oxygen is used in the Sanofi process for the production of the antimalarial drug artemisinin [17] and in other routes to the drug [18].

4.2.6 Commercially relevant photoreactions

At a first glance it is disappointing that photochemical reactions are only used in industrial chemical processes in a few cases. For the most part large-scale reactions are conducted thermally in the presence of a catalyst. This is primarily because light is an expensive energy source, especially for high-intensity irradiation, necessary for obtaining economically relevant space-time yields. This is also the problem of the practical application of solar photochemistry [19]. Some of these issues can be overcome by using flow methods. Large-scale photochemical processes are therefore in industrial use only when the desired product is not accessible thermally or when light can act as an initiator for thermal chain reactions like polymerisation and aerial oxidation. On the other hand, the use of photochemistry in analysis for biochemistry and medicine is a rapidly growing field. Table 4.1 summarises the most important large-scale reactions including approximate product amounts.

Table 4.1: Some industrial photochemical processes [20].

Reaction / Product	Company	Tons per year
Photonitrosylation / Caprolactam	Tora	$> 10^4$
Photonitrosylation / Lauryllactam	Atochem	$> 10^4$
Photochlorination / Paraffins	Philips	$> 10^4$
Photochlorination / Chloromethane	Atochem	$> 10^5$
Photochlorination / Benzyl chloride	Atochem, Bayer, Monsanto	$> 10^5$
Sulfochlorination / CH_3SO_2Cl	Atochem	$> 10^3$
Sulfoxidation / Alkyl sulfonic Acid	Hoechst	$> 10^4$
Photooxidation / Rose Oxide	Dragoco	> 10
Electrocyclic Reaction / Vitamins D, A	Hoffmann-La Roche, Duphar, BASF	?

Photonitrosylation

Irradiation (600–200 nm) of nitrosyl chloride in liquid cycloalkanes generates NO and Cl radicals in the first reaction steps. The latter abstracts a hydrogen atom from the alkane generating HCl and a cycloalkyl radical, which adds to NO forming the oxime hydrochloride as intermediate (Scheme 4.11). Acid-catalysed Beckmann rearrangement produces caprolactam, which is hydrolysed to aminocaproic acid. Subsequent base-catalysed polycondensation affords nylon-6. Analogously, cyclododecane produces nylon-12.

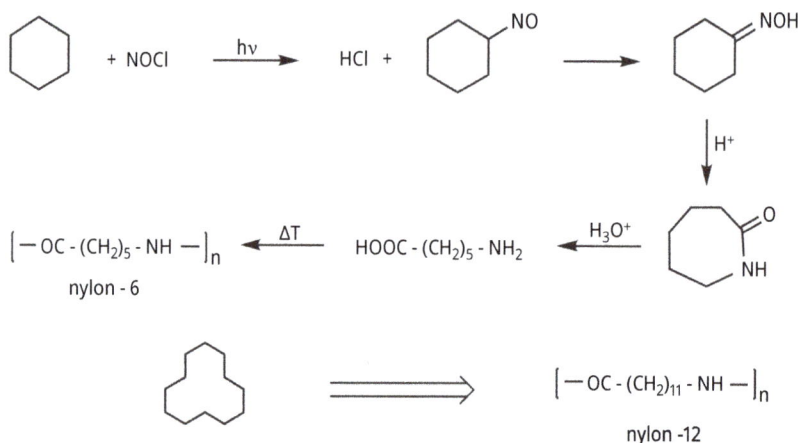

Scheme 4.11: Visible light photo-oximination of cycloalkanes followed by thermal rearrangement and polycondensation.

Photochlorination, photosulfochlorination, photosulfoxidation

Unlike nitrosylation, all the transformations in this section are photo-induced chain reactions. In photochlorination, visible light homolysis of chlorine generates chlorine atoms (Equation 4.14) which abstract hydrogen from the alkane producing an alkyl radical (Equation 4.15). That starts the chain reaction through the generation of a new chlorine atom (Equations 4.15, 4.16). The reaction is stopped by radical side reactions like recombination (Equation 4.17).

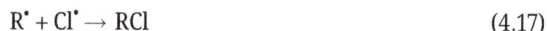

$$Cl_2 + Vis \rightarrow 2\,Cl^\bullet \tag{4.14}$$

$$Cl^\bullet + RH \rightarrow R^\bullet + HCl \tag{4.15}$$

$$R^\bullet + Cl_2 \rightarrow RCl + Cl^\bullet \tag{4.16}$$

$$R^\bullet + Cl^\bullet \rightarrow RCl \tag{4.17}$$

When sulfur dioxide is present in the above reaction, it is attacked by the intermediate alkyl radical generating an alkylsulfonyl radical which abstracts a chlorine atom producing chlorosulfonic acid ($R = C_{10-15}$, Equation 4.18). After alkaline hydrolysis the resulting sodium sulfonates are used in detergents. When converted to alkyl sulfonic acid esters, they become plasticisers for synthetic materials.

$$RSO_2^\bullet + Cl_2 \rightarrow RSO_2Cl + Cl^\bullet \tag{4.18}$$

Sulfonic acids may be produced more directly by running the above process with liquid alkanes ($C_{16} - C_{20}$) in the presence of air but omitting chlorine (Equation 4.19).

$$R - H + SO_2 + 1/2O_2 + UV \rightarrow RSO_3H \tag{4.19}$$

In the first reaction step, $^3SO_2$ undergoes a HAT with RH generating the starting alkyl radical (Scheme 4.12). The latter adds to sulfur dioxide and oxygen to form a peroxyl radical, which decomposes through further intermediates with production of further alkyl radicals.

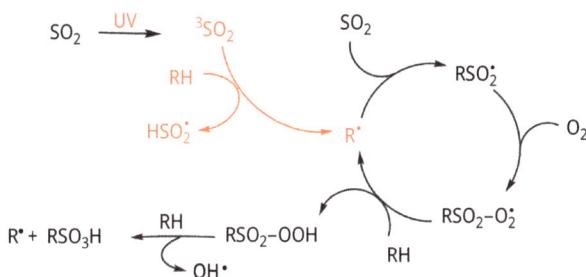

Scheme 4.12: UV induced sulfoxidation of alkanes.

Photochemical synthesis of perfumes

In the case of some expensive chemicals, photochemical methods may provide the only route to the desired product. This is true for the monoterpene *rose oxide*, which may be formed as cis- and trans-isomers, each existing as (+)- and (–)- stereoisomers, but only the (–)-cis isomer exhibits the distinctive olfaction of roses. The key reaction step is the dye-sensitised formation of aerial singlet oxygen (Scheme 4.13, first step). A subsequent *ene*-reaction generates an allyl hydroperoxide, followed by reduction to the diol and cyclisation to a mixture of cis- and trans-rose oxide. Its characteristic fragrance is also generated by grapes and lime-tree blossoms.

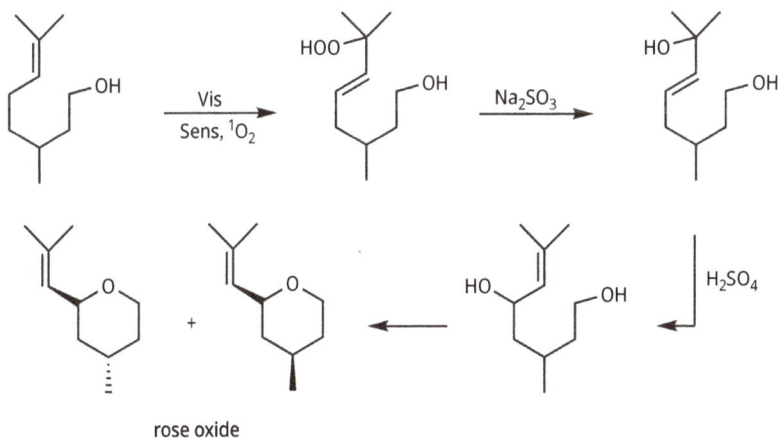

rose oxide

Scheme 4.13: Synthesis of rose oxide involving an *ene*-reaction in the first reaction step.

Photopolymerisation

In general, photopolymerisation is initiated by highly reactive radicals that are generated by photocleavage of O–O–, C–N– and C–C bonds (Equations 4.20–4.22). They start the radical chain by

$$R-O-O-R + UV \rightarrow 2\,RO^{\bullet} \tag{4.20}$$

$$R-N=N-R + UV \rightarrow 2\ R^{\bullet} + N_2 \tag{4.21}$$

$$Ar-CH(OH)-C(O)-Ar \rightarrow ArCH(OH)^{\bullet} + ArCO^{\bullet} \tag{4.22}$$

adding to the monomer (Equation 4.23, E = COOEt). After complete monomer consumption (Equations 4.24, 4.25), the chain is finished through radical recombination or disproportionation (Equation 4.26). In addition to organic compounds metal complexes and inorganic semiconductors may also function as radical initiators.

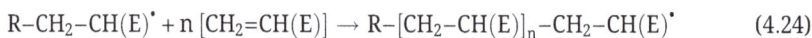

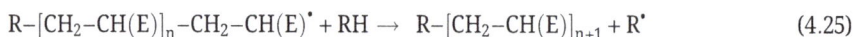

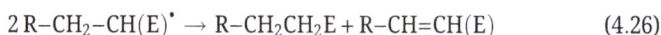

$$R^{\bullet} + CH_2=CH(E) \rightarrow R-CH_2-CH(E)^{\bullet} \tag{4.23}$$

$$R-CH_2-CH(E)^{\bullet} + n\,[CH_2=CH(E)] \rightarrow R-[CH_2-CH(E)]_n-CH_2-CH(E)^{\bullet} \tag{4.24}$$

$$R-[CH_2-CH(E)]_n-CH_2-CH(E)^{\bullet} + RH \rightarrow R-[CH_2-CH(E)]_{n+1} + R^{\bullet} \tag{4.25}$$

$$2\,R-CH_2-CH(E)^{\bullet} \rightarrow R-CH_2CH_2E + R-CH=CH(E) \tag{4.26}$$

In many applications, 2-methylacrylate derivatives are used as monomers, for example in dental filling treatments. Usually, yellow camphorquinone (2,3-bornanedione) is used as the photoinitiator.

Photochemical curing

Photochemical curing refers to the hardening of a thin film or paste containing a photoinitiator and monomeric and/or oligomeric unsaturated compound capable of inducing photopolymerisation and/or cross-linking. In addition to dental fillings, curing by UV light is used for coating wood, paper, plastics and metals. Unlike thermal curing, photochemical curing also enables imaging since it is localised to irradiated parts of the material. An interesting example is the photochemical cross-linking of an Si–H terminated polymer with an alkene-terminated polymer called "hydrosilylation" (see below).

Photolithography

Photolithography is a process that generates patterns in the range of macro- to nanoscale sizes by photoirradiation. The working principle is summarised in Figure 4.14. The pattern is generated by a so-called photoresist that is a material which upon irradiation becomes soluble (positive mode) or more insoluble (negative mode). Typical resists are thin films of specially designed polymers allowing the nano-structuring of computer chips. The smaller the wavelength of exciting light, the higher the lateral resolution. In very recent developments a laser-generated tin plasma radiates ex-

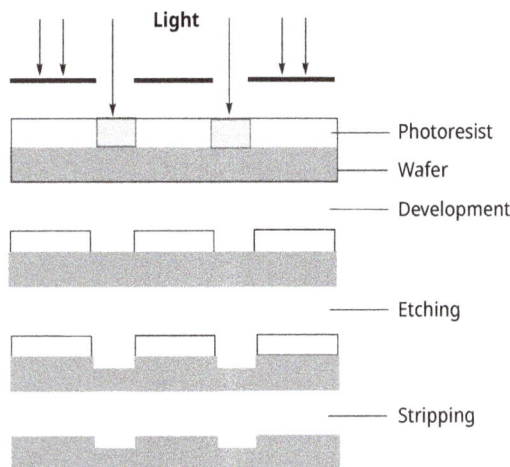

Figure 4.14: Structuring of a wafer surface by photolithography with a photoresist of positive mode.

treme UV light of $\lambda = 13.5$ nm [21], that allows the "printing" of structures down to below 30 nanometres on the most advanced microchips.

In industry, phenol-formaldehyde polymers are typically employed as resists working in the positive mode. They contain diazonaphthoquinone as a dissolution inhibitor (Scheme 4.14). Upon irradiation it decomposes to nitrogen and a short-lived α-ketocarbene derivative, which undergoes a Wolff rearrangement to an intermediate ketene. The latter is finally hydrolysed to the corresponding carboxylic acid. Addition of a basic developer leads to efficient dissolution of the irradiated film sections.

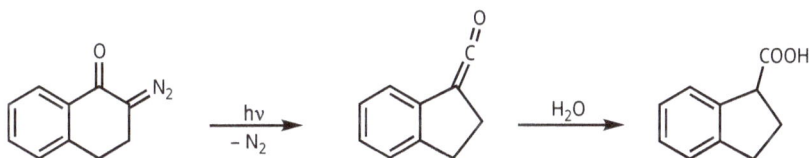

Scheme 4.14: A photogenerated acid accelerates dissolution of the irradiated part of the positive photoresist.

Optical brighteners and photostabilisers

Optical brighteners, also called *fluorescent brightening agents* (FBAs) absorb in the UV and violet region and emit in the visible. They are used as whitening agents in laundry detergents. In most cases they possess a highly substituted stilbene structure (Figure 4.15). Upon light absorption in the near UV (~360 nm), they emit an intense fluorescence in the blue region (~470 nm). This mixes with the often slightly yellow, complementary colour of the laundry, resulting in a bright white. Therefore, an extremely white-looking shirt does not prove that is very clean but only that it is freshly laundered.[12]

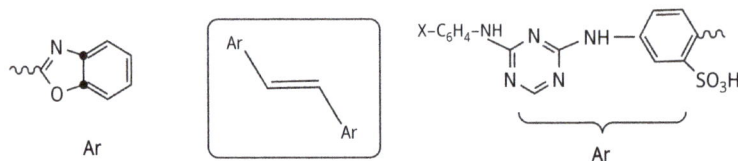

Figure 4.15: Structures of optical brighteners.

Photostabilisers

It is not surprising that almost all polymers suffer photo-induced damage. Although the UV part of solar light arriving at the earth's surface amounts to only about 3% of the total solar energy, the corresponding photon energies are high enough for bond break-ing (see Table 1.1). The resulting radicals act as initiators for oxidative radical chain deg-radation. Two methods of stabilisation are employed. In one, absorption by the polymer is prevented through addition of e.g. 2-hydroxybenzophenone or salicylic acid esters (Figure 4.16). At wavelengths below 370 nm these molecules have high molar absorption coefficients, and the singlet state undergoes very fast radiationless deactivation to the ground state. The presence of the hydroxy group in the 2-position enables efficient and reversible intramolecular hydrogen transfer on a picosecond time scale. In the second method the start of the oxidative degradation is inhibited by intercepting the generated radicals through addition of organic scavengers such as sterically hindered amines.

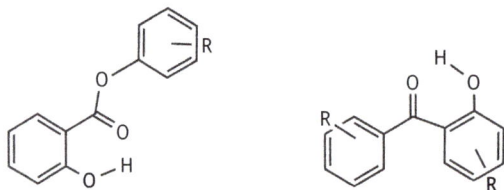

Figure 4.16: Photostabilisers undergoing efficient radiationless deactivation.

Photochromism

When a molecule can undergo a reversible reaction between two isomers having dif-ferent absorption spectra, and one or both steps occur only upon light absorption, the phenomenon is called photochromism (see Chapter 1.1). Scheme 4.15 depicts the basic reaction profile for the molecule A undergoing an endergonic photoreaction to isomer

---

12 The amount of FBA in a washed shirt is high enough to exhibit the strong blue/purple fluorescence when exposed to Disco light.

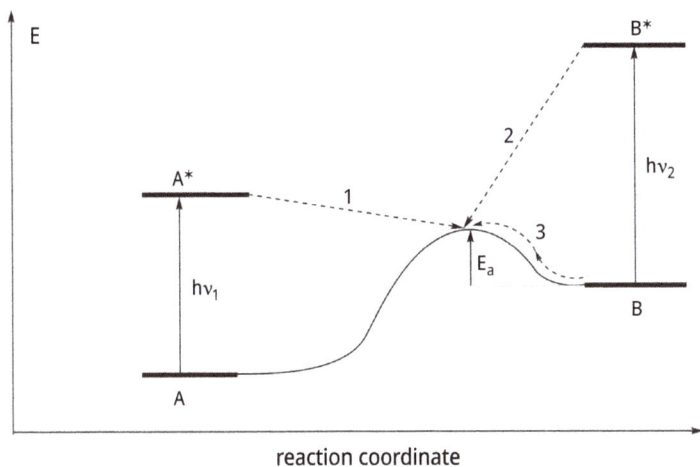

Scheme 4.15: Energy profile of a photochromic system consisting of A and its isomer B.

B (path 1). Molecule B may differ from molecule A also in its dipole moment or another physical property. Molecule B can reform molecule A either photochemically (path 2) or thermally (path 3). The latter process must require an activation energy E_a, big enough to ensure thermal stability.

Classical photochromic systems are based on cis-trans photoisomerisation and ring opening/closing reactions [22]. Figure 4.17 gives some characteristic examples. Naphtho-pyrans [23] have the advantage that, depending on the nature of substituents, they re-convert to the closed form either thermally (T-type) or photochemically (P-type) [24].

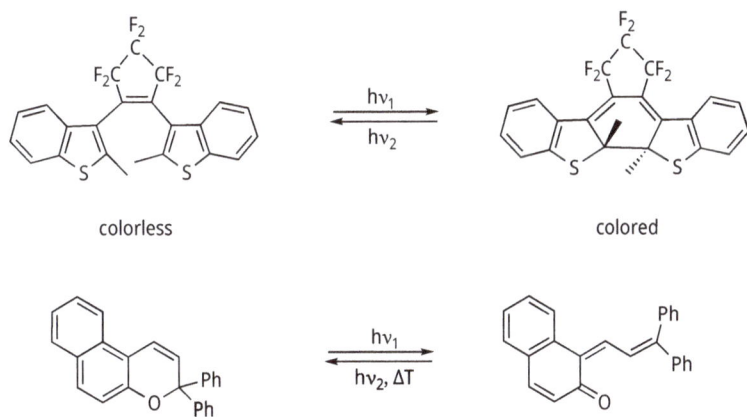

Figure 4.17: Diarylethylenes and naphthopyrans.

Applications of photochromism include photochromic sunglasses, optical filters, and smart windows. Photochromic inks may hide security markers in documents, banknotes, and transport packages. In nanotechnology they may significantly increase the information density of memory storage devices. Figure 4.18 exemplifies the principle. Due to the photochromic properties, reading and writing of information can be managed with one single compound (Figure 4.18a). As indicated, reading is coupled to erasing. This can be avoided by converting state B (locking) into a third state C (Figure 4.18b) [25].

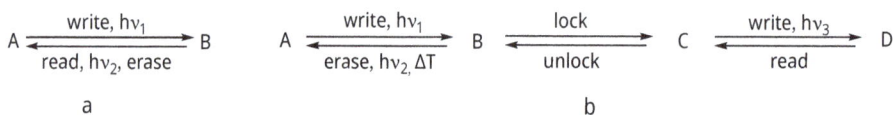

$$A \underset{\text{read, } h\nu_2, \text{ erase}}{\overset{\text{write, } h\nu_1}{\rightleftharpoons}} B \qquad\qquad A \underset{\text{erase, } h\nu_2, \Delta T}{\overset{\text{write, } h\nu_1}{\rightleftharpoons}} B \underset{\text{unlock}}{\overset{\text{lock}}{\rightleftharpoons}} C \underset{\text{read}}{\overset{\text{write, } h\nu_3}{\rightleftharpoons}} D$$

a   b

Figure 4.18: Simplified description of information storage through a photochromic system.

Luminescent sensors

Fluorescent chemosensors, more often called by the name *luminescence sensors,* have become important tools in analytical chemistry both in artificial and biological systems [2a]. They consist of a receptor for selective binding of an analyte, of a luminophore whose luminescing properties change upon analyte binding, and of a spacer linking the two former components together (Figure 4.19). Thus, the analyte generates a luminescence signal that can be easily measured with high accuracy. The most common analytes are protons, metal ions, and oxygen. Classical luminophores employ emissive aromatic molecules, but highly emissive metal complexes may also be used (see Chapter 2.3.3 for use of lanthanide complexes). When the luminophore is located on the tip of a fibre-optic, it is called an *optode, or optrode.* A commercial instrument, available as single-use cartridge, can measure concentrations of critical blood components (alkali and alkaline earth ions, oxygen, glucose, urea) within half a minute. This is deployed in ambulances and intensive-care units. Figure 4.20 displays the structure of a molecule with a 4-aminophthalimide as fluorophore. In the absence of an alkali metal ion the crown ether substituent quenches a green emission through electron transfer to the phthalimide group. Upon coordination of sodium ions to the ether part, the phthalimide emission is strongly amplified.

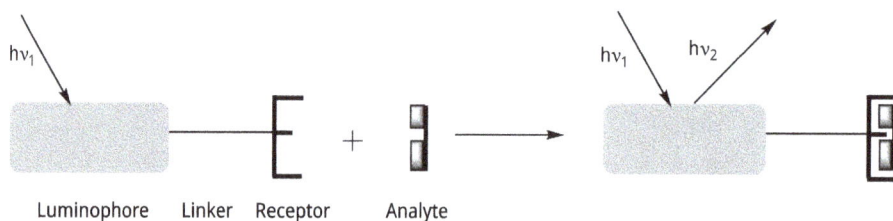

Luminophore   Linker   Receptor   Analyte

Figure 4.19: Principle of a luminescence sensor.

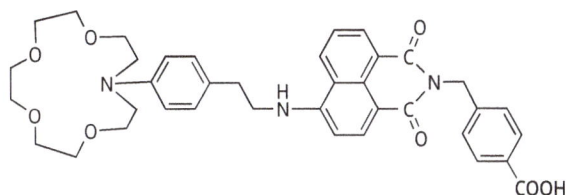

Figure 4.20: Fluorescence sensor for sodium ions.

Logic gates

The simplest molecular logic gate corresponds to a luminescent sensor. YES is generated when a chemical input activates a luminescence output. NOT is generated when a chemical input quenches the luminescence output [26]. The naphthalene-bridged crown ether depicted in Figure 4.21 satisfies this definition. It does not exhibit the well-known intense fluorescence of the naphthalene unit, but only a weak emission band at 438 nm, assigned to a crown ether-to-naphthalene CT interaction. Upon complete protonation of the amino groups, the weak band completely disappears and the intense naphthalene fluorescence appears at 342 nm. This behaviour can be transformed to the "truth table" of YES and NOT operations (Table 4.2). Since in general luminescence is easily quenched by a chemical input, NOT behaviour is quite common.

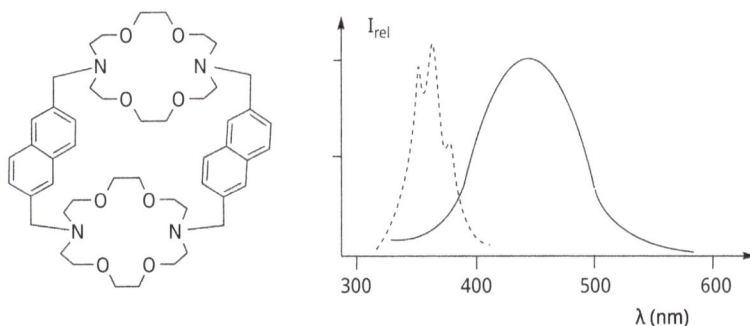

Figure 4.21: A simple molecular logic gate and its luminescence spectra in the neutral (—) and protonated (---) form. Redrawn from reference [26].

Table 4.2: Logic truth table for the gate function of the photoswitch shown in Figure 4.21.

	YES (H$^+$)	NOT (H$^+$)
Input (H$^+$)	Output (342 nm)	Output (438 nm)
0	0	1
1	1	0

Depending on the nature of the chemical input transition metal complexes may also act as the luminophore. In the last decade research in this field underwent rich development leading to many different types of logic gates (OR, NAND, NOR, XAND, etc.) [27].

Photochemical hydrosilylation

The addition of a molecule H–X across a C=C double bond is a commonplace reaction, but addition of a silane, H–SiR$_3$ (R=H, alkyl, aryl, alkoxy) requires catalysis, most commonly by platinum complexes. Its principal commercial application is in the cross-linking of silicones to form flexible, elastic polymers used in adhesives, insulation, resins and in medical applications (Scheme 4.16a, b). Photocatalytic methods have advantages when thin films are required or when the timing or spatial position of the reaction needs to be controlled precisely. The best established photosensitive pre-catalysts are (η^5-C$_5$H$_4$Me)PtMe$_3$) and Pt(acac)$_2$ (Me = CH$_3$, acac = acetylacetonate, Scheme 4.16c). A few seconds near UV irradiation is necessary to convert (η^5-C$_5$H$_4$Me)PtMe$_3$ to its active catalytic form, probably Pt nanoparticles. The reaction then proceeds without further irradiation [28].

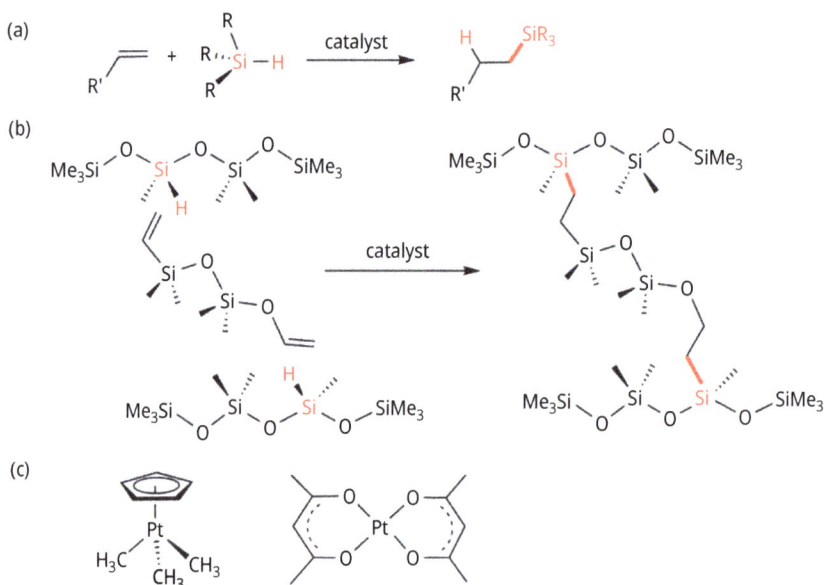

Scheme 4.16: (a) Hydrosilylation: addition of Si-H across C = C double bond; (b) cross-linking of silicones by hydrosilylation; (c) platinum photocatalysts that initiate hydrosilylation.

Silver halide photography

Silver halides formed the basis of most photographic processes from the 1840s to early 2000s and they still remain significant. Photographic film contains small grains of silver

(a)

(b)

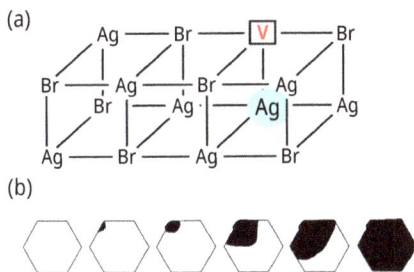

Figure 4.22: (a) Diagram of part of the unit cell of AgBr showing one silver ion that has moved to an interstitial position and the resultant vacancy **V** in the lattice position. (b) Growth of silver particle during development of latent image [29].

bromide, usually including a little silver iodide, embedded in gelatine. Silver bromide has the same cubic structure as sodium chloride with each silver ion surrounded by six bromide ions and vice versa. Another description of the structure is a face-centred cubic array of bromide ions with silver ions occupying the octahedral holes between the bromide ions. However, it contains "defects" that usually consist of silver ions that have moved from the octahedral holes to the interstitial tetrahedral holes, leaving vacancies at the octahedral sites. This type of defect is called a Frenkel defect (Figure 4.22a). Silver bromide contains far more "defects" than sodium chloride, but even then they are in range of parts per billion. The defect concentration may be 100 times greater, however, at the surface of the grain than in the interior.

Both these defects and the doping with small quantities of silver iodide affect the electronic and ionic properties of silver bromide. As a consequence, there are extra vacant energy levels for the silver 5s orbitals between the valence band and the conduction band. Excitation with visible light generates electrons that can migrate through the solid – photoconduction – and are liable to be trapped at the interstitial silver ions forming a silver atom. The source of the electron is a bromide ion – the resultant bromine atoms are eventually released from the surface as Br_2. Moreover, there is an equilibrium between the silver bromide and bromine vapour. A further consequence of the presence of defects is that silver ions can move through the solid as they hop from an octahedral site to an adjacent vacancy, giving it properties as an ionic conductor. This process assists the next stage of forming the photographic image. The newly formed silver atom is unstable with respect to loss of an electron. This helps it act as a nucleus for growth of particles of silver containing a few atoms. The particle grows by a process of ionisation of the Ag atom, migration of the silver ions and trapping of photoelectrons to form a bigger silver particle. Indeed, the negative charge associated with the silver cluster helps to attract more silver ions. These particles constitute the "latent image" in the exposed film.[13]

---

13 The theory of the photographic process was developed by R.W. Gurney and N. F. Mott in 1938. Nevill Mott won the Nobel Prize for Physics in 1977 for his work on the electronic structure of disordered systems.

Now we move from the camera to the dark room. Application of a reducing agent or "developer" reduces all the silver ions within the crystallite containing the silver particles to silver (Figure 4.22b). The development of the latent image is stopped by transfer of the film to an acidic solution, traditionally acetic acid. Those crystallites in the paper or film that do not contain silver particles are removed by a fixer consisting of a thiosulfate salt such as $Na_2(S_2O_3)$. Silver ions react with thiosulfate to form soluble $[Ag(S_2O_3)_2]^{3-}$ and $[Ag(S_2O_3)_3]^{5-}$ leaving the stable negative image on the film in black and white [30].

The subsequent processes of transfer of the image onto paper or reversal to form a positive image have essentially the same physical and chemical basis. Colour photography also uses the same fundamentals of silver chemistry, but in conjunction with dyes – see ref. [31].

4.3 Non-catalytic photochemistry of transition metal complexes

In previous sections on transition-metal complexes in Chapter 2, we saw examples of prompt photodissociation (e.g. $Cr(CO)_6$) and of formation of equilibrated charge-transfer excited states (e.g. $[Ru(bpy)_3]^{2+}$) that proceed to electron-transfer reactions. Although the course of photoreaction is very different in these two cases, the *ground* electronic states of these two molecules have some features in common: they are both d^6 low-spin complexes and consequently are both inert to substitution. Our ability to identify and study photochemical reactions of transition metal complexes requires that they are inert to the breaking of metal-ligand bonds under the conditions of reaction. The rate of substitution of transition metal complexes varies by about 16 orders of magnitude from extremely labile to extremely inert and is intimately linked to the electron configuration. High-spin configurations are usually very labile because of occupation of antibonding orbitals, while d^3, d^6 low-spin and d^8 square-planar complexes are remarkably inert. Thus a large proportion of the examples we will see, have these three electron configurations. The total valence electron count is also a good indicator of inert configurations for organometallic compounds. An 18-electron configuration around a particular metal corresponds to occupation of all bonding orbitals and no antibonding orbitals as in $(C_5H_5)Rh(C_2H_4)_2$ or $Mn_2(CO)_{10}$. Both of these molecules are substitution inert but photochemically active.

4.3.1 Photo-induced bond cleavage

Bond cleavage, one of the fundamental photoprocesses for metal complexes, can take the form of cleavage of M–L bonds, reductive elimination, or homolytic bond cleavage either of MM or MR bonds (Scheme 4.17, M = metal, L = 2-electron donor).

Examples of 2-electron ligands that can be photodissociated are CO, N_2, C_2H_4, NO, NH_3 and phosphine. Reductive elimination of H_2 from a cis-metal dihydride is another

Photodissociation

$$[M]-L \underset{+L}{\overset{h\nu,\,-L}{\rightleftharpoons}} [M]$$

with branching to:

R'H →	$[M]\overset{R'}{\underset{H}{\diagup}}$	oxidative addition
L' →	$[M]-L'$	ligand binding
[M] →	$[M]-[M]$	dimerisation
[M]L →	$[M]-[M]$ $\searrow\diagup$ L	reaction with precursor

Photo-induced reductive elimination

$$[M]\overset{H}{\underset{R}{\diagup}} \underset{+RH}{\overset{h\nu,\,-RH}{\rightleftharpoons}} [M]$$

with branching to:

L' →	$[M]-L'$	ligand binding
R'H →	$[M]\overset{R'}{\underset{H}{\diagdown}}$	oxidative addition

Photo-induced MM bond cleavage

$$[M]-[M] \overset{h\nu}{\rightleftharpoons} 2[M]^{\cdot} \xrightarrow{RY} 2[M]-Y + R^{\cdot}$$

radical abstraction, coupling, etc.

Homolytic M-R cleavage

$$[M]-X \underset{+X^{\cdot}}{\overset{h\nu,\,-X^{\cdot}}{\rightleftharpoons}} [M]^{\cdot} \xrightarrow{RY} [M]-Y + R^{\cdot}$$

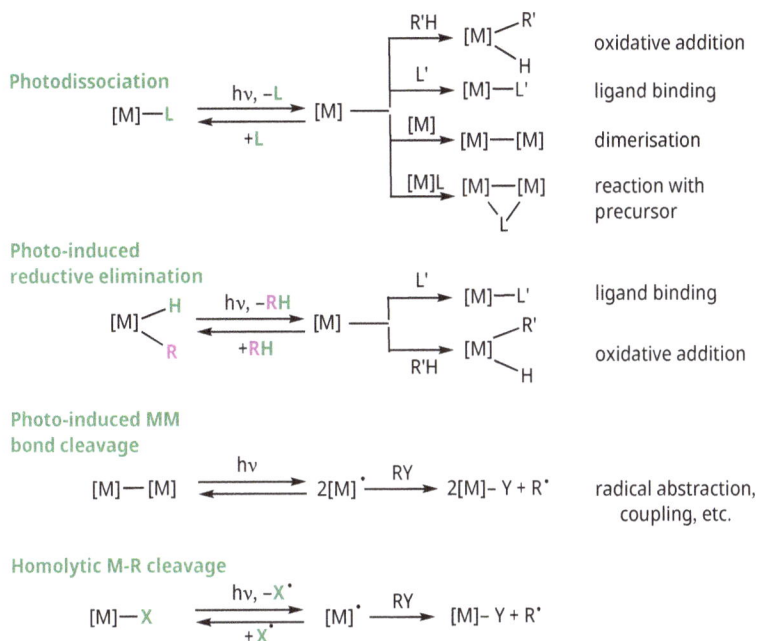

[M] = metal-ligand complex. L, L' = CO, N_2, Xe, NO, alkene, amine, phosphine, ether, etc
R, R' = H, alkyl, aryl. X = H, alkyl, OH, N_3

Scheme 4.17: Principal photochemical processes at transition metal centres involving initial bond cleavage.

common process in which two M–H bonds are broken with concomitant reduction in oxidation state of the metal by two units and formation of an H–H bond. Reductive elimination of alkane from an alkyl hydride complex is rarer. Compounds with M–M single bonds are usually photosensitive and undergo M–M bond cleavage. Homolytic cleavage of M–X (X = H, alkyl, OH, N_3) bonds can sometimes be observed. The occurrence of these photoprocesses depends on the nature of the absorption bands, the photolysis wavelength and the availability of deactivation pathways.

As has been described in Chapter 2, many of these reactions occur by prompt photodissociation and generate an extraordinarily reactive fragment. This fragment, shown as [M], may react with solvent within picoseconds, or may react with dissolved substrates at astonishing rates, even up to the diffusion-controlled limit. Scheme 4.17 illustrates the general pathways of these reactions. Since the initial steps of these reactions are typically thermally reversible, recombination with the expelled ligand(s) competes with reactions with added substrates.

In Chapters 1 and 2, we already encountered examples of metal carbonyl photochemistry, notably substitution at $Cr(CO)_6$ (Equations 2.36–2.38), formation of a methane complex from $[CpOs(CO)_3]^+$ (Scheme 1.5) and $Mn_2(CO)_{10}$ where CO dissociation and Mn–Mn bond cleavage are in competition (Chapter 2.3.2.9) [32]. Scheme 4.17 shows the various processes that can follow CO dissociation. An example of metal-metal bond formation is shown in Scheme 4.18. In addition to applications in synthesis, metal carbonyl photochemistry is used for photopolymerisation, photocatalysis and modification of surfaces. Since the discovery that CO at very low concentrations acts as a signaller in humans, chemists have searched for effective ways to introduce CO. One method is to photolyse metal carbonyls (photochemical CO-releasing molecules or Photo-CORMs) [33]. A current issue is the design of complexes that lose CO on irradiation with visible light rather than UV.

In solution, no more than one CO is dissociated from metal carbonyls per photon absorbed. However, several CO's may be ejected in the gas phase following the absorption of a single photon and the number ejected increases with the photon energy. The reason for this difference is that the photon may carry more energy than is needed for dissociation. In solution, the excess energy is removed by the solvent, but in the gas phase, the excess energy remains in the photofragment until it is dissipated by dissociation of further CO molecules.

Metal carbonyls have often been used to form complexes photochemically with two or more metals. The fragment formed by CO loss may react with the metal carbonyl precursor as in the conversion of $Fe(CO)_5$ to $Fe_2(CO)_9$ or may dimerise if it has a sufficiently long lifetime, as happens in the formation of $[CpRh(CO)]_2$ from $CpRh(CO)_2$ (Scheme 4.18). The latter subsequently reacts with ejected CO to yield $[CpRh(CO)]_2(\mu$-CO), but the C_5Me_5 analogue is stable.

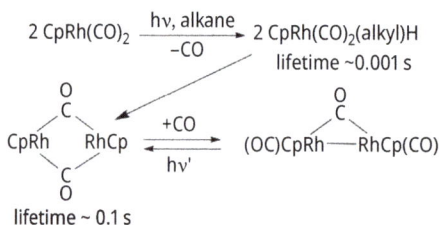

Scheme 4.18: Photochemical formation of dinuclear complexes of rhodium (Cp = η^5-C_5H_5).

Not all metal carbonyls eject CO on irradiation. Metal monocarbonyls such as $CpIr(CO)(H)_2$ often eject another ligand (see below). If the complex includes an electron-accepting ligand such as bipyridine, charge transfer to that ligand occurs leading to electron-transfer reactions as in $Re(CO)_3(bpy)Cl$.

Metal nitrosyl complexes also undergo photodissociation, though dissociation competes with linkage isomerisation – see below. Since NO is a well-known biological signaller, photodissociation of NO is also of biomedical interest. (In the biological literature, photodissociation is termed photo-uncaging). Examples of NO photodissociation from nitrosyls include $[Fe_2(\mu$-SR$)_2(NO)_4]^{2-}$ that releases all the NO groups on irradiation in aerated aqueous solution. The release can be made more efficient by using R groups that absorb strongly in the visible region such as porphyrins. Several simple ruthenium nitrosyls such as $[RuCl([15]aneN_4)(NO)]^{2+}$ ($[15]aneN_4$ is a macrocycle coordinating through four nitrogen atoms) undergo NO photodissociation; the latter has a quantum yield of 0.61 at 355 nm. Complexes with suitable absorbing ligands lose NO on visible irradiation [33, 34].

The 18-electron complex $CpRh(C_2H_4)_2$ forms an example of a thermally inert but photosensitive metal-ethene complex that undergoes a wide variety of substitution and oxidative addition reactions (Scheme 4.19a). The reaction with Et_3SiH forms a complex containing ethene, silyl and hydride ligands as proposed in the typical mechanisms for hydrosilylation (the process of addition of SiH bonds across an alkene, used in the manufacture of silicones, Chapter 4.2.6). The related complex $CpRh(PMe_3)(C_2H_4)$ has been used to study C-H bond activation. It reacts photochemically with benzene to form $CpRh(PMe_3)(Ph)H$ via a transient $Rh(\eta^2\text{-}C_6H_6)$ complex. With fluorinated benzenes, the product distribution can be controlled to give either an aryl hydride or an η^2-arene complex or an equilibrium between the two types (Scheme 4.19b). The aryl hydride products observed with fluorinated benzenes were the first to show the ortho-fluorine effect decisively, in which there is a thermodynamic drive to place the fluorine substituents ortho to the metal-carbon bond.

A variety of photosensitive cis-dihydride metal complexes are illustrated in Scheme 4.20 [36]. All of these molecules undergo photochemical loss of H_2 in a concerted process. The resulting fragments are often extraordinarily reactive toward intermolecular oxidative addition reactions with H_2 (i.e. recombination), C–H, Si–H and B–H bonds. Examples are found that react with the C–H bonds of alkanes, arenes and alkenes (Scheme 4.21). In contrast to selectivity of radicals for tertiary C–H bonds, the fragments formed by H_2 loss show selectivity for primary, over secondary, over tertiary C–H bonds. Even methane itself is reactive. The mechanism of these reactions has been studied in detail for rhodium complexes: the interconversion of the alkane complex and the isomeric alkyl hydride complex enables a mechanism that allows the metal to "walk" along the chain until it reaches the primary C–H bond. The reactivity of these species toward alkanes has excited hopes, so far unfulfilled, that a low energy catalytic route from methane to methanol would result.

In Chapter 1, we described the photochemical methods of time-resolved spectroscopy (both with UV/visible and IR detection) and matrix isolation that have revealed numerous highly reactive coordinatively unsaturated metal complexes, such as 16-electron $Fe(CO)_4$, $Cr(CO)_5$ and $CpRh(C_2H_4)$. Their structures have been determined by use of isotopic substitution. Their reactivity toward unusual substrates including alkanes and noble gases has been revealed by these methods and spectroscopy in liquefied noble gases. In this way, labile molecules such as $Cr(CO)_5Xe$ or $Cr(CO)_5(H_2)$ can be characterised and the kinetics of their reactions monitored. Of particular interest, have been the reactions of alkanes to form $M(\eta^2\text{-alkane})$ and M(alkyl)H complexes. Infrared methods revealed indirect evidence for the $M(\eta^2\text{-alkane})$ complexes and identified $CpRe(CO)_2(\eta^2\text{-alkane})$ as a target be-

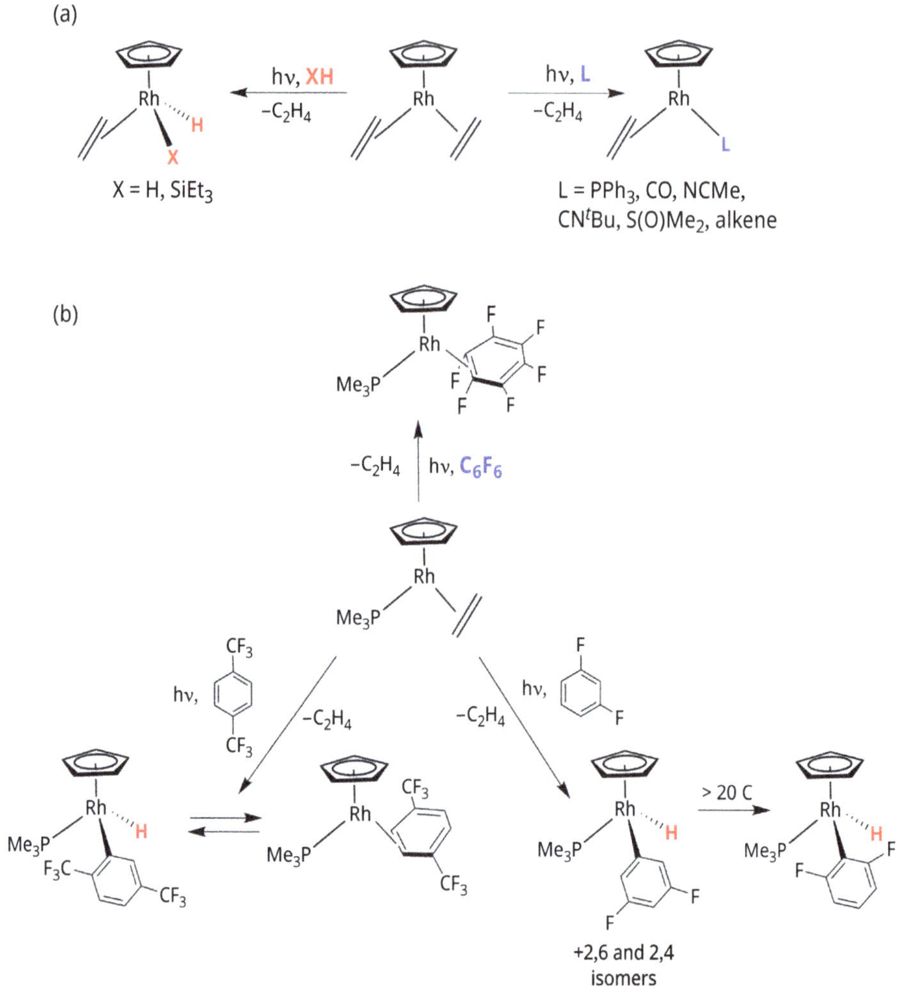
(a)
hν, XH
−C2H4
hν, L
−C2H4
Rh
H
X
X = H, SiEt3
L
L = PPh3, CO, NCMe,
CNtBu, S(O)Me2, alkene
(b)
Me3P
F
−C2H4
hν, C6F6
CF3
hν,
−C2H4
hν,
> 20 C
F3C
+2,6 and 2,4
isomers

Scheme 4.19: Photochemistry of ethene complexes: (a) $CpRh(C_2H_4)_2$ substitution and oxidative addition (b) control of products of $CpRh(PMe_3)(C_2H_4)$ with fluorinated benzenes [35].

cause of its long lifetime. As a result, it proved possible to irradiate samples of $CpM(CO)_3$ (M = Mn, Re) in alkane solvents within the probehead of an NMR spectrometer and obtain direct characterisation of $CpM(CO)_2(\eta^2\text{-alkane})$ by NMR spectroscopy. The first example with methane as the alkane was $[CpOs(CO)_2(\eta^2\text{-}CH_4)]^+$ (Scheme 1.5).

Homolytic cleavage of bonds to a metal is also possible. As we have seen M–M bond cleavage competes with M–CO cleavage for $Mn_2(CO)_{10}$ and many metal-metal bonded species. This reaction has been used for photopolymerisation with $Mn_2(CO)_{10}$ and other metal-metal bonded carbonyls such as $Cp_2Mo_2(CO)_6$. The metal carbonyl radical abstracts iodine from iodoalkanes and the resulting alkyl radical initiates polymeri-

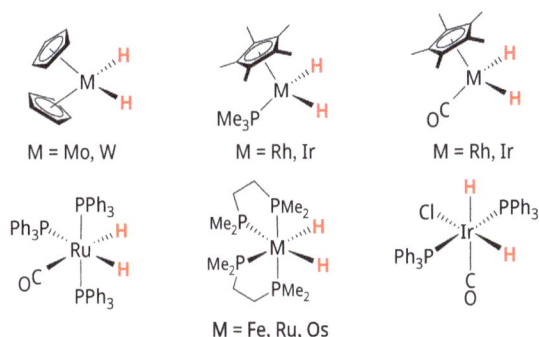

Scheme 4.20: Photosensitive metal dihydride complexes.

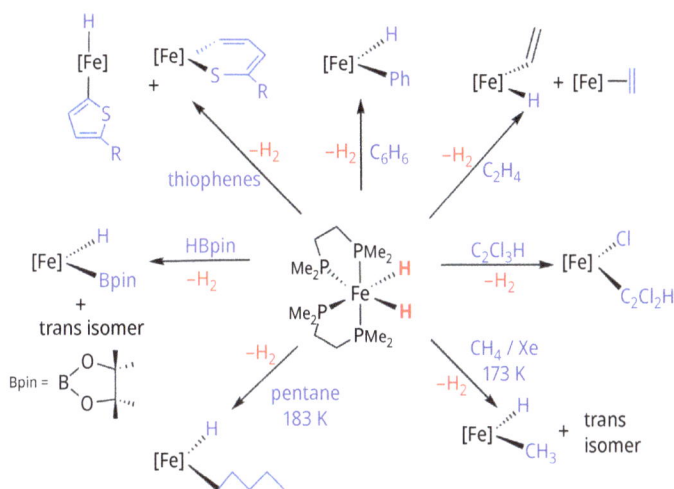

Scheme 4.21: Photochemistry of an iron diphosphine dihydride complex. [Fe] = $Fe(Me_2PCH_2CH_2PMe_2)$. (Adapted from *Chem. Rev.* **2016**, *116*, 8506, Copyright 2016 American Chemical Society, reference [36]).

sation of $CH_2{=}CF_2$ [37]. Photochemical homolytic cleavage of metal hydride and metal alkyl bonds does not occur systematically, but there are important examples nonetheless: for metal hydrides, $HMn(CO)_5$ and $Re(\eta^5\text{-}C_5Me_5)_2H$; for metal methyls, methylcobalamin (vitamin B_{12}) and its analogues, $(\eta^5\text{-}C_5H_4Me)Pt(Me)_3$ used for photo-induced hydrosilylation. We encountered the homolytic cleavage of the Fe-OH bond in Fenton's reagent (Chapter 2.3.2.3). Metal-azide bonds can also be broken homolytically on LMCT excitation. An example of interest as a pro-drug for cancer therapy is all trans-$[Pt(N_3)_2(OH)_2(py)_2]$ that loses azide radicals on excitation with visible light [38].

Homolysis of a bond within a ligand has also been reported: a simple chromium nitrito complex loses NO on irradiation with visible light generating a Cr=O moiety.

This reaction provides an alternative route to NO release instead of using a nitrosyl precursor [33].

4.3.2 Photo-induced isomerisation

In addition to the four classes of photo-induced bond cleavage, four types of photo-isomerisation are encountered: configurational isomerisation (cis to trans or facial to meridional), linkage isomerisation, oxidative cleavage and spin-state isomerisation (Scheme 4.22).

Configurational isomerisation from cis to trans is typical of square planar Pt^{II} complexes, for example Pt(PCy$_3$)$_2$(SiEt$_3$)H (Cy = cyclohexyl). Such reactions are thought to proceed via a pseudo-tetrahedral transition state. A particularly interesting example of facial to meridional isomerisation is fac-Re(bpy)(CO)$_3$Cl (bpy = 2,2'-bipyridine). This complex is well known for its charge transfer photochemistry induced by long-wavelength irradiation, but it isomerises on short-wavelength (313 nm) photolysis.

In the example of linkage isomerisation in Scheme 4.22, a nitro ligand is converted to a nitrito ligand, but other examples involve nitrosyls, N$_2$, sulfur dioxide, and Me$_2$SO (Scheme 4.23) [39]. Photoisomerisation is often studied in the crystalline state by single-crystal X-ray diffraction. This method, introduced by Coppens to the study of coordination complexes [40], allows variable temperature and time-resolved measurement. Often, the irradiation results in partial conversion of isomers, but there are instances where full conversion is observed while maintaining crystallinity, implying that the crystal environment is sufficiently flexible to allow these changes. Although the products may be unstable at ambient temperature, they may be metastable at low temperature.

Photoisomerisation

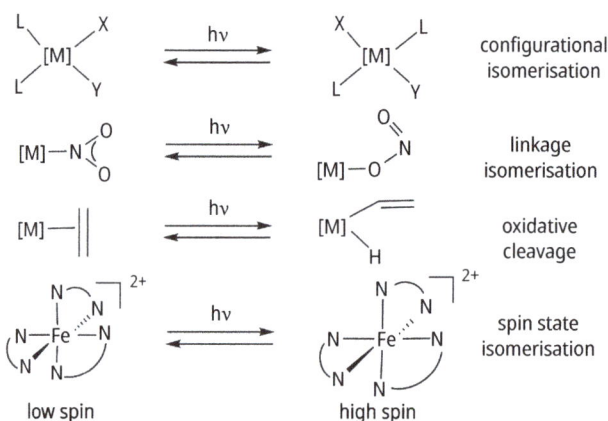

Scheme 4.22: Photoisomerisation processes at transition metal complexes.

Photocrystallography has revealed coordination modes that had not been documented previously such as side-bound NO and N_2 coordinated to a single metal. A classic example was the discovery by Coppens et al that the nitrosyl ligand of sodium nitroprusside $Na_2[Fe(CN)_5(NO)] \cdot 2H_2O$ could be converted from the familiar N-bound form to the O-bound isomer on photolysis at 488 nm at low temperature and in turn to the side-bound isomer with 1064 nm irradiation. Photocrystallography is often complemented by IR spectroscopy since the different coordination modes have characteristic vibrational signatures. Indeed, the conversion of the NO group of CpNi(NO) from linear Ni–N–O to sideways was observed by matrix isolation revealing a huge 447 cm^{-1} shift in the NO stretching frequency that could be reversed by a change in photolysis wavelength. It wasn't until 20 years later that the photoisomer was shown to be side-on bound by photocrystallography.

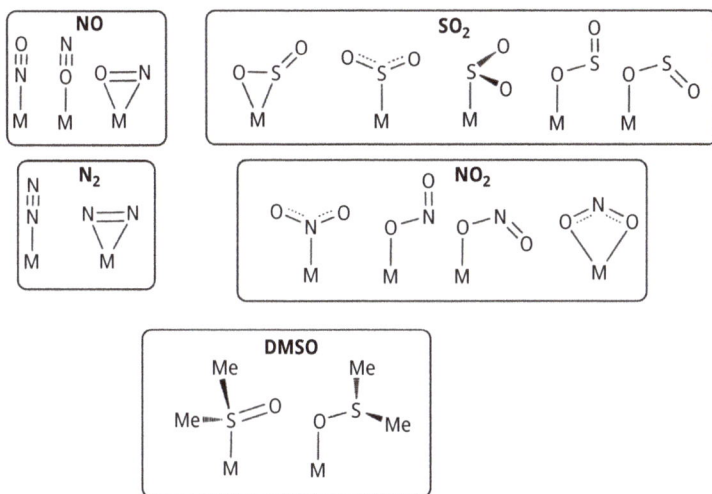

Scheme 4.23: Linkage isomers investigated photochemically.

Isomerisation by oxidative cleavage (Scheme 4.22) describes a process in which the metal is oxidised and a bond in the ligand cleaves. Thus a metal ethene complex may sometimes be isomerised to a metal vinyl hydride as is observed for $CpRh(PMe_3)(C_2H_4)$ at low temperature, but the precursor is regenerated already at 253 K (see also Scheme 4.19). However, the corresponding iridium vinyl hydride reverts to the $Ir(C_2H_4)$ isomer only at 393 K.

The fourth group of reactions involving photo-induced isomerisation concerns spin-crossover. For numerous Fe^{II} and Fe^{III} complexes, as well as some manganese and cobalt complexes, low spin and high spin forms lie very close in energy. Consequently, equilibria between the two forms are observed in solution while crystalline forms may switch spin states with temperature or pressure. Since high spin forms populate the higher-lying antibonding orbitals (d_{z^2} and $d_{x^2-y^2}$) in an octahedral geometry), the high-spin isomers have metal-ligand bonds as much as 0.2 Å longer than the low-spin forms. The magnetic properties also change drastically with spin state. Considering all that we said in Chapters 1 and 2 about transitions between spin states being forbidden, it might be thought that light absorption could not alter the spin state. On the contrary, it was discovered in 1984 that the spin states of a crystal at low temperature could be switched entirely through irradiation with visible light [41]. The phenomenon could also be observed in solution and, astonishingly, the spin state can switch within < 50 femtoseconds (5×10^{-14} s) following light absorption [42]. The changes are described as Light-Induced Excited Spin-State Trapping (LIESST) and the changes in magnetic properties as photomagnetism. The first example was found in [Fe(ptz)$_3$][BF$_4$]$_2$ (ptz = 1-propyl tetrazole) which is in the high-spin form (spin quintet) at room temperature and colourless. On cooling to 80 K or lower, the crystal switches to the low spin form (spin singlet) which is red, due to spin-allowed d-d absorption bands. On irradiation into these bands, it switches back to the high-spin form which is metastable at low temperature.

The dynamics of the light-induced spin crossover have been studied in the very simple complex, [Fe(bpy)$_3$]$^{2+}$ in aqueous solution by UV/Vis absorption and by X-ray absorption [42a]. In this case the absorption spectrum is dominated by MLCT bands. The mechanism of isomerisation has been established as intersystem crossing (ISC) from the singlet MLCT state to the triplet MLCT state followed by a second ISC to the ground quintet state of the spin-state isomer. The great advantage of the X-ray absorption method is that the elongation of the Fe–N bonds is observed directly; the time-resolution of this method led to a timescale for the change of ca. 150 femtoseconds, but more recent optical experiments have reduced the timescale yet further to ca. 50 femtoseconds. The Fe–N bond length is measured to be 0.2 Å longer in the high spin state than in the low spin state due to the occupation of the antibonding orbitals (Figure 4.23).

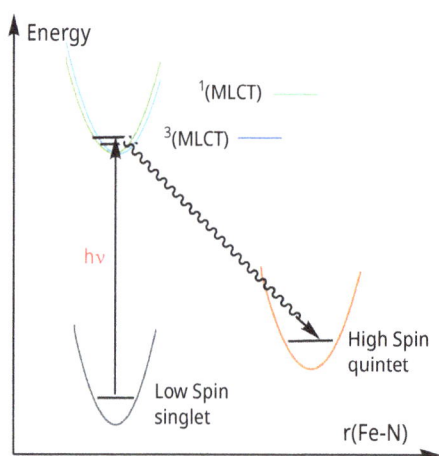

Figure 4.23: Schematic energy level diagram for spin-state isomerisation of [Fe(bpy)$_3$]$^{2+}$.

4.3.3 Photo-induced electron transfer (PET)

The ability of transition-metal complexes to shift charge through light absorption was demonstrated in Chapter 2.3.2 with as many as six different classes of charge-transfer: metal-to-ligand (MLCT), ligand-to-metal (LMCT), ligand-to-ligand (LLCT), charge-transfer to solvent (CTTS), metal-to-metal (MMCT) and ion-pair charge transfer (IPCT). Following excitation, the charge can shift back to where it started either radiatively or non-radiatively. This is back-electron-transfer and is unproductive as a chemical reaction. Alternatively, chemical reaction can ensue – the slower the back-electron-transfer, the greater is the time available for chemical reaction. We have already seen a few examples. For ferrioxalate, the LMCT excitation leads to decomposition with release of CO_2 (Figure 2.27) while LMCT irradiation of $[Fe(OH)(H_2O)_5]^{2+}$ generates $OH^\bullet$ radicals (Equation 2.39). We also saw how a CTTS transition can generate solvated electrons. Here we illustrate how other types of charge-transfer excitation can lead to redox reactions.

The paradigm for MLCT photochemistry is $[Ru(bpy)_3]^{2+}$ (Chapter 2.2.6). As we showed, its excited state is both a stronger oxidising agent and a stronger reducing agent than its ground state. An example of its action as a reducing agent is its photoreaction with peroxodisulfate that generates $[Ru(bpy)_3]^{3+}$ and sulfate (Equation 4.27). In the presence of reducing agents such as NEt_3, ascorbate, SO_3^{2-}, or Eu^{2+}, the excited state $[Ru(bpy)_3]^{2+*}$ is reduced to $[Ru(bpy)_3]^{+}$, i.e. $[Ru(bpy)_3]^{2+*}$ acts as an oxidising agent.

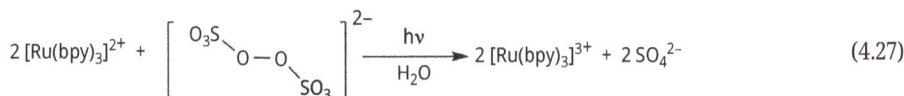

$$2\,[Ru(bpy)_3]^{2+} + \left[\begin{array}{c} O_3S \\[-2pt] \diagdown \\ O-O \\ \diagup \\ SO_3 \end{array} \right]^{2-} \xrightarrow[H_2O]{h\nu} 2\,[Ru(bpy)_3]^{3+} + 2\,SO_4^{2-} \tag{4.27}$$

Photochemistry may also be observed on irradiation into the MMCT band of mixed valence complexes. Scheme 4.24 shows an example where the electron transfer causes reduction of inert Co(III) to form labile Co(II), such that the ligands on cobalt are replaced by solvent water and the reaction is irreversible.

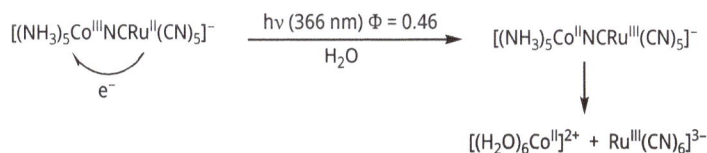

$$[(NH_3)_5Co^{III}NCRu^{II}(CN)_5]^{-} \xrightarrow[H_2O]{h\nu\ (366\ nm)\ \Phi = 0.46} [(NH_3)_5Co^{II}NCRu^{III}(CN)_5]^{-}$$

$$\underset{e^-}{\curvearrowright}$$

$$\downarrow$$

$$[(H_2O)_6Co^{II}]^{2+} + Ru^{III}(CN)_6]^{3-}$$

Scheme 4.24: MMCT photochemistry.

Salts of metal carbonyl anions with metallocene cations such as $[Cp_2Co][Co(CO)_4]$ exhibit visible absorption bands due to IPCT in regions where the parent salts such as $Na[Co(CO)_4]$ and $[Cp_2Co]Cl$ do not absorb. When provided with a substrate such as PPh_3, electron transfer takes place leading to reduced and oxidised products (Equation 4.28). In the absence of another substrate, it appears that excitation in this region produces no reaction, but this is an illusion. The illusion is dispelled by time-resolved spectroscopy that reveals the immediate products of electron transfer and establishes that back electron transfer occurs by second order kinetics in ca. 50 μs [43].

$$2\,[Cp_2Co]^+[Co(CO)_4]^- + 2\,PPh_3 \xrightarrow[CH_2Cl_2]{h\nu} 2\,Cp_2Co + Co_2(CO)_6(PPh_3)_2 + 2\,CO \qquad (4.28)$$

In Scheme 2.16, we saw how excitation of a sensitiser can be used to reduce an electron acceptor and oxidise an electron donor. That principle provides an effective mechanism of separating the charge and prolonging the lifetime of the reduced acceptor and oxidised donor. The sensitiser, donor and acceptor components can be free in solution or incorporated in supramolecules as a dyad (sensitiser-acceptor) molecule or triad (donor-sensitiser-acceptor, Figure 4.24). This type of arrangement is inspired by natural photosynthesis where the donor, sensitiser and acceptor are strategically positioned within the chloroplast. Supramolecules have the advantage that electron transfer can occur much faster than in separated molecules in solution because there is no need to wait for the molecules to diffuse close to one another. The design is intended to separate the charge but, nevertheless, back electron transfer may be faster than for separated molecules and alternative reactions often compete with the desired process (see Chapter 5.1).

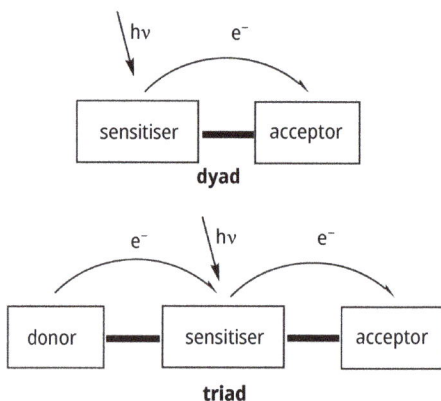

Figure 4.24: Supramolecules incorporating sensitiser, electron donor and electron acceptor (dyad above and triad below).

The sensitiser can be a derivative of $[Ru(bpy)_3]^{2+}$ with an MLCT excited state or can be, for instance, a zinc porphyrin derivative with an ILCT (π-π^*) excited state that acts as a powerful electron donor. This type of design is common in photocatalysis where the acceptor must be catalytically active.

4.4 Atmospheric photochemistry

4.4.1 The components of the atmosphere

Every schoolchild learns that Earth's atmosphere contains a mixture of gases, N_2, O_2, Ar, CO_2 and H_2O vapour, and that the CO_2 concentration has been rising, but that is only a small part of the story. It also contains much lower concentrations of other gases: Ne, Kr, Xe, O_3 (ozone), CH_4, other hydrocarbons, CO, SO_2, NO, NO_2, N_2O, chlorofluorocarbons and radical species such as $OH^{\bullet}$, $HO_2^{\bullet}$ and $ClO^{\bullet}$. They may be of biological, geological or anthropogenic origin. There are also a host of liquid or solid particles suspended in the atmosphere (aerosols) including water droplets, ice particles, dust from deserts containing SiO_2, Fe_2O_3 and other oxides as well as particulate hydrocarbons. Volcanic eruptions pour more gases and particles into the atmosphere. Trees emit high molecular mass hydrocarbons as gases or droplets. The water droplets may have been formed by spray off the oceans and therefore contain dissolved salts. Wildfires as well as human activities supply many more particles as smoke. These particles are transported round the globe by the winds. The temperature and pressure vary with latitude, season, weather and height above the surface. The atmosphere is subject to irradiation with 24-hour oscillation at wavelengths from the far UV to the IR. The wavelengths reaching the upper atmosphere are very different from those reaching Earth's surface. The different elements in the atmosphere, nitrogen, oxygen, carbon and sulfur are constantly being added to and removed from the atmosphere through cyclical processes (carbon cycle, etc.). While these cycles lead to a steady state in the fractions of N_2 and O_2, this is not necessarily the case for other species. An important feature of the atmosphere is that is not at chemical equilibrium – if it was at equilibrium, the nitrogen would be oxidised by the oxygen [44].

As we climb through the atmosphere, the pressure falls monotonically. The temperature falls to begin with but then starts to rise again. The boundary between these two regions is called the tropopause, the region below is the troposphere and the region above the stratosphere. The tropopause varies in height with latitude from about 6 to about 18 km. The troposphere contains most of the mass of the atmosphere. The stratosphere continues to a height of about 50 km when the temperature starts to fall again. These temperature boundaries are significant because the gases move within the troposphere rapidly but cross the tropopause very slowly.

Photochemistry is the principal driver of the chemical reactions of many of these atmospheric species and the source of all the oxygen (O_2) in the atmosphere is photochemical – it comes from photosynthesis. Of the principal components of the atmosphere only oxygen (O_2) absorbs parts of the solar spectrum between 200 and 300 nm and that absorption occurs only in the upper atmosphere. It is many of the minor species that absorb in the near UV and visible parts of the spectrum. Their photoreactions may take place in the gas phase, in solution in water droplets, in frozen ices, or in molecules adsorbed onto the surface of particles. They may be stoichiometric or catalytic.

Thus we encounter many of the principles of homo- and heterogeneous photochemistry that we described in previous chapters, but applied to small main-group molecules. As usual, if photochemistry is to be significant, the combination of absorption coefficient,[14] radiation intensity and quantum yield must be suitable. Additionally, the kinetics of the subsequent thermal reactions must be sufficiently fast, but we will not discuss kinetics here. Photochemistry in the stratosphere is substantially different from that in the troposphere because the wavelength distribution of solar energy differs enormously. The stratosphere is irradiated by wavelengths out to the far ultraviolet ~ 190 nm, that are capable of breaking bonds of at least 600 kJ/mol. In the troposphere, wavelengths less than 310 nm are cut off. Consequently, bonds stronger than ~ 380 kJ/mol cannot be broken. Although intensities are low in the region 310–380 nm, this region is significant because several atmospheric species absorb in this region and the photons are more energetic than at longer wavelengths.

4.4.2 Oxygen and ozone (O_2 and O_3) in the stratosphere

Oxygen in the stratosphere absorbs UV irradiation of wavelengths less than about 240 nm. Its absorption ensures that no radiation of wavelengths less than 200 nm reaches Earth's surface. In the region 200–240 nm, its absorption causes dissociation into oxygen atoms into their ground triplet states. They can recombine or react with O_2 in a spin-allowed reaction to form ozone in its ground singlet state. Ozone absorbs across the UV region from 200 nm to 350 nm and in the visible from 440–850 nm (at higher pressure it looks blue) with a dip at 400 nm. The absorption coefficient is so great in the region 250–310 nm that essentially all of this radiation is absorbed by the stratospheric ozone layer. However, the absorption coefficients at 400 and 600 nm are smaller than at 250 nm by a factor of a million and ten thousand, respectively, so visible radiation continues to reach Earth's surface. Ozone absorption in the stratosphere in the region 240–350 nm causes photodissociation to O and O_2, so shielding the Earth's surface from UV radiation in this region. Ozone can also be removed by reaction with oxygen atoms. The result is that we have two photochemical reactions and two thermal reactions (Scheme 4.25) that create a steady state concentration of ozone in the stratosphere. This is a cyclical process in which ozone is created and destroyed. Its maximum concentration in the stratosphere reaches about 10 parts per million (ppm). The energy released in these photodissociation and recombination reactions heats the stratosphere and gives rise to the temperature inversion between the troposphere and stratosphere.

---

14 Gas-phase chemists use the term absorption cross-section with units $cm^2\ mol^{-1}$ in place of absorption coefficient.

$$
\text{ozone formation} \begin{cases} O_2 \xrightarrow{\text{hv }(< 240 \text{ nm})} O + O \\ O + O_2 \longrightarrow O_3 \end{cases}
$$

$$
\text{ozone destruction} \begin{cases} O_3 \xrightarrow{\text{hv }(> 240 \text{ nm})} O + O_2 \\ O + O_3 \longrightarrow 2\,O_2 \end{cases}
$$

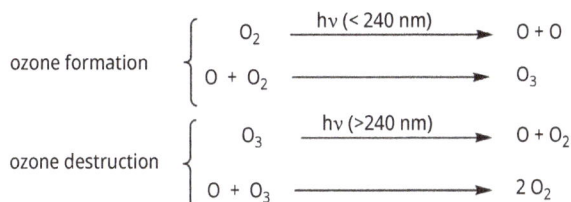

Scheme 4.25: Ozone formation and destruction in the stratosphere.

This steady state can be strongly perturbed by the action of species other than oxygen atoms that destroy ozone. Farman, Gardiner and Shanklin not only discovered the ozone hole over Antarctica in 1985 [45] but connected it with the earlier realisation by Molina and Rowland that chlorine atoms and chlorine oxides derived from chlorofluorocarbons (CFCs) could deplete ozone.[15] Chlorine oxides were detected over Antarctica shortly afterwards. UV photolysis of chlorofluorocarbons provides the source of anthropogenic chlorine atoms that can act as catalysts for the destruction of ozone. These molecules are too inert to react thermally and absorb at short wavelengths and are therefore liable to photodissociation only if they reach the stratosphere. Scheme 4.26 (CCl_2F_2 is one example of such a chlorofluorocarbon) shows the simplest catalytic cycle for ozone destruction by chlorine, but it is not just $ClO^{\bullet}$ but other chlorine oxides Cl_2O_2, $ClOO^{\bullet}$ that are involved. Chlorine monoxide reacts with NO_2 to from chlorine nitrate, $ClONO_2$, which acts as another source of $ClO^{\bullet}$ on photolysis. Species such as Cl_2O_2 and $ClONO_2$ act as reservoirs of chlorine atoms and $ClO^{\bullet}$ that are released with near UV or visible light, so broadening the range of wavelengths that induce ozone loss (see Scheme 1.6). Other "chain carriers" that promote related cycles photochemically include H_2O_2, $HOCl$, and $HONO_2$.

$$
\begin{array}{lrcl}
 & CCl_2F_2 & \xrightarrow{\text{hv (UV)}} & Cl^{\bullet} + CClF_2 \\
\text{reaction A} & Cl^{\bullet} + O_3 & \longrightarrow & ClO^{\bullet} + O_2 \\
\text{reaction B} & ClO^{\bullet} + O & \longrightarrow & Cl^{\bullet} + O_2 \\ \hline
\text{reactions A + B} & O + O_3 & \longrightarrow & 2\,O_2
\end{array}
$$

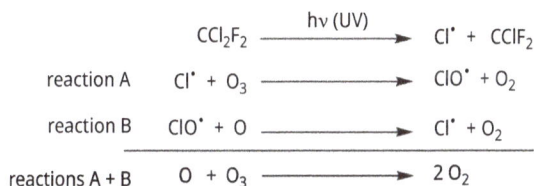

Scheme 4.26: Ozone destruction by chlorofluorocarbons exemplified by CCl_2F_2.

The reactions do not just occur homogeneously but on the surface of ice and dust particles. Polar stratospheric clouds can accumulate $ClONO_2$ and HCl on their surfaces (HCl is formed by reaction of methane with $Cl^{\bullet}$). These molecules can react together to

---

15 Mario Molina, Sherry Rowland and Paul Crutzen won the Nobel Prize in 1995 for their work on the formation and destruction of ozone, especially by CFCs and nitrogen oxides.

form chlorine gas and nitric acid. The chlorine is readily photodissociated to form chlorine atoms that can then react with ozone as above (Equations 4.29, 4.30).

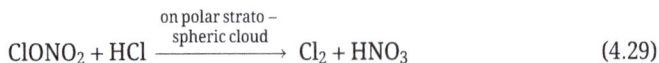

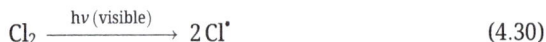

$$ClONO_2 + HCl \xrightarrow{\substack{\text{on polar strato-}\\\text{spheric cloud}}} Cl_2 + HNO_3 \tag{4.29}$$

$$Cl_2 \xrightarrow{h\nu\,(\text{visible})} 2\,Cl^{\bullet} \tag{4.30}$$

Proof of the role of ClO$^{\bullet}$ in ozone depletion came with the strong correlation between the growth in ClO$^{\bullet}$ to parts per billion concentrations and the loss of ozone in the Antarctic Spring as sunlight returned. These discoveries led to the Montreal Protocol, the most successful environmental treaty of all time, signed in 1987 and gradually strengthened since. As a result of the treaty, the use of ozone-depleting substances has declined drastically and the ozone levels in the stratosphere are gradually recovering [46].

Crutzen[15] had provided related evidence that nitrogen oxides could also destroy ozone. NO can react with ozone to form NO_2 that reacts in turn with oxygen atoms to form O_2 and more NO, perpetuating the cycle of destruction. This is one of several molecules that can be oxidised by ozone and create similar cycles.

4.4.3 Ozone in the troposphere

The role of ozone in the troposphere is very different from that in the stratosphere. Rather than being a protector of life, it is toxic to humans, damaging to plants, and an important greenhouse gas. In the troposphere, it cannot be formed by photolysis of O_2 because of the absence of far UV radiation. The principal source of ozone in the troposphere is photolysis of nitrogen dioxide, generating oxygen atoms that react with O_2 (Scheme 4.27). This process can become cyclical through the reaction of NO with the hydroperoxy radical $HO_2^{\bullet}$ generated in other processes. Ozone in the troposphere is subject to photolysis at all wavelengths since it absorbs across the near UV and visible spectrum. At any of these wavelengths, it undergoes photodissociation generating oxy-

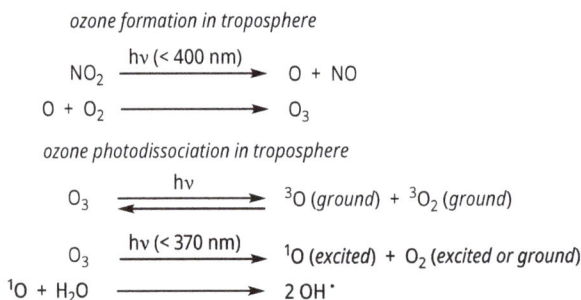

ozone formation in troposphere

$$NO_2 \xrightarrow{h\nu\,(<400\text{ nm})} O + NO$$

$$O + O_2 \longrightarrow O_3$$

ozone photodissociation in troposphere

$$O_3 \underset{}{\overset{h\nu}{\rightleftharpoons}} {}^3O\,(ground) + {}^3O_2\,(ground)$$

$$O_3 \xrightarrow{h\nu\,(<370\text{ nm})} {}^1O\,(excited) + O_2\,(excited\text{ or }ground)$$

$$^1O + H_2O \longrightarrow 2\,OH^{\bullet}$$

Scheme 4.27: Ozone formation and destruction in the troposphere. Note that only light > 310 nm is available in the troposphere.

gen atoms and oxygen molecules in their ground states which simply recombine. At shorter wavelengths than 370 nm, however, it can also produce oxygen atoms in their singlet excited state. The by-product can be O_2 in its singlet excited state (spin-allowed process) or in its triplet ground state (spin-forbidden). This reaction is of particular importance since singlet oxygen atoms react with water vapour to form $OH^•$ radicals. These $OH^•$ radicals are the principal oxidants in the lower atmosphere (Scheme 4.27).[16]

$OH^•$ radicals are capable of abstracting hydrogen from most hydrocarbons. The resulting hydrocarbyl radical reacts with O_2 to form alkylperoxy radicals that can proceed to more stable products such as CH_3CO_2H. If NO is present in addition to hydrocarbons, a cycle is formed in which further $OH^•$ is formed together with two molecules of NO_2 (Scheme 4.28). The latter can be photolysed leading to more ozone and $OH^•$ radicals. Since the $OH^•$ radicals are formed photochemically, their action takes place during daytime. The build-up of NO_2 via oxidation of hydrocarbons results, in turn, in the formation of more ozone (Scheme 4.27).

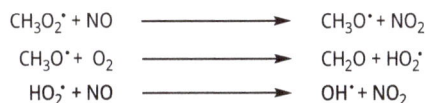

$$CH_3O_2^• + NO \longrightarrow CH_3O^• + NO_2$$
$$CH_3O^• + O_2 \longrightarrow CH_2O + HO_2^•$$
$$HO_2^• + NO \longrightarrow OH^• + NO_2$$

Scheme 4.28: Regeneration of $OH^•$ and NO_2 via reaction of methylperoxy radicals.

Photochemical smog is a major pollution problem in many cities, especially when the air doesn't mix well because of the geography or because of a temperature inversion. During the night-time and early morning, the concentration of NO and small particles from vehicle exhausts builds up. As the intensity of sunlight increases, concentrations of NO and $OH^•$ increase through the reactions in Scheme 4.27. The presence of hydrocarbons fuels the reactions in Scheme 4.28 leading to a build-up of peroxy radicals $RO_2^•$, NO_2, O_3 and peroxyacylnitrates (PAN). Reactions of ozone with hydrocarbons can also increase the concentration of particulates. The mix of NO_2 and particles results in the brown haze, while PAN causes eye irritation and ozone causes respiratory problems. Two of the photochemical reactions involved are shown in Scheme 4.29 together with the molecular structure of PAN. Formaldehyde may be formed by oxidation of methane by $OH^•$ and can undergo photodissociation to form more radical species at short wavelengths. Nitrous acid, HONO, is generated by reaction of NO with NO_2 and water. It is subject to photodissociation to $OH^• + NO$ at longer wavelengths. PAN is formed by reaction of acylperoxy radicals with NO_2.

Turning to sulfur compounds, $OH^•$ radicals are also able to oxidise H_2S to SO_2 (with the help of O_3 and O_2) and are critical in the oxidation of SO_2 to form SO_3 and sulfuric acid. The sulfuric acid may be deposited as acid rain.

---

16 Bimolecular reactions in the gas phase that yield a single molecule such as $O + O_2 \rightarrow O_3$ require a third body (usually designated M) to remove the excess energy. For simplicity, we omit M in this section.

CH$_2$O $\xrightarrow{\text{h}\nu\ (<340\ \text{nm})}$ HCO$^\bullet$ + H$^\bullet$

HONO $\xrightarrow{\text{h}\nu\ (<400\ \text{nm})}$ OH$^\bullet$ + NO

R = CH$_3$, C$_2$H$_5$, etc.

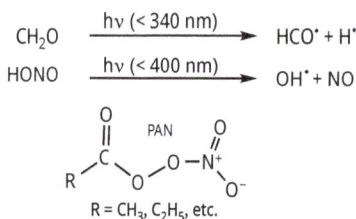

Scheme 4.29: Photochemical smog and PAN.

Photochemistry and ozone play a major role in the surface layer of the ocean, in the aerosols created above it and in the resultant gases [47]. The principal source of this reactivity is iodine, present as I$^-$ or IO$_3^-$ and dissolved organic matter. Iodide reacts with ozone in the surface layer to produce HOI and I$_2$ that are released to the atmosphere (Equations 4.31, 4.32).

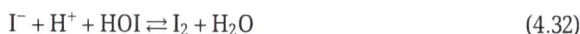

$$I^- + H^+ + O_3 \rightarrow HOI + O_2 \tag{4.31}$$

$$I^- + H^+ + HOI \rightleftharpoons I_2 + H_2O \tag{4.32}$$

Of lesser importance, organic iodine compounds such as CH$_3$I and CH$_2$ICl are also released. All of these gas-phase iodine compounds are photo-active generating iodine atoms. In a cyclical process iodine atoms react with ozone to form IO$^\bullet$ radicals and O$_2$, while the reverse process occurs by photodissociation of IO$^\bullet$ and subsequent reaction of O atoms with O$_2$ (Equations 4.33–4.35).

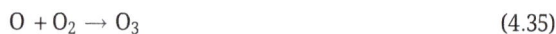

$$I^\bullet + O_3 \rightarrow IO^\bullet + O_2 \tag{4.33}$$

$$IO^\bullet \xrightarrow{\text{h}\nu} I^\bullet + O \tag{4.34}$$

$$O + O_2 \rightarrow O_3 \tag{4.35}$$

The IO$^\bullet$ radical also reacts with NO$_2$ to form IONO$_2$ and with HO$_2^\bullet$ to form HOI, both of which also generate iodine atoms photochemically. Dissolved organic matter can be oxidised via photosensitisation mechanisms as well as by OH$^\bullet$ radicals, yielding methyl radicals that react with iodine atoms to form CH$_3$I in the absence of O$_2$. One source of OH$^\bullet$ is the photolysis of nitrite in the presence of a proton source (see Equation 4.42) that may, in turn, be generated photochemically from nitrate. These sources of iodoalkanes are considered to be more important than biogenic formation in microorganisms. Although iodide is far less abundant than bromide or chloride, its reactions are of more importance than analogues for the other halides thanks to their much higher rates. The overall effect of these iodine compounds in the atmosphere is to reduce tropospheric ozone very substantially especially via equations 4.31 and 4.33. Scheme 4.30 summarises the compounds that can generate iodine atoms photochemically.

Photochemical sources of iodine atoms

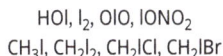

$HOI, I_2, OIO, IONO_2$

$CH_3I, CH_2I_2, CH_2ICl, CH_2IBr$

Scheme 4.30: Photoactive iodine compounds involved in ozone destruction.

4.5 Natural waters photochemistry

Solar photoreactions contribute substantially to the biological quality of natural waters. Recalling that about 70% of the earth's surface is covered by water, sun-driven processes should play an important role in global reactions such as the carbon cycle and the transformation of dissolved organic matter (DOM) [48].

As we have seen in Chapter 4.4, the photochemistry of the surface waters of the ocean has a major effect on the chemistry of the atmosphere, especially through its effects on the concentrations of ozone and $OH^{\bullet}$ radicals. The sources of the reactivity are iodine and DOM. In this section, we look at the effects of photochemistry on the natural waters themselves. Unless noted otherwise, the reactions discussed apply for the seawater surface layer.

Typical concentrations of the most relevant inorganic components of sea water are: chloride (0.5 M), bromide (0.8 mM), iodide (0.1 µM), sodium (0.4 M), magnesium (0.05 M), and calcium (0.01 M) [47b]. The iron concentration is much lower than that of the other metals because of the insolubility of Fe_2O_3. Among the iron halides, only iron bromide complexes can efficiently absorb sunlight due to their absorption onset at 415 nm and concentration. Nitrate and nitrite absorb at about 310 and 360 nm, respectively, just reaching the short wavelength part of sunlight. In contrast, iron(III) complexes and many chromogenic organic compounds (CDOM) exhibit medium to strong visible light absorption in the range of 400–700 nm. Both types of organic matter originate from aquagenic material such as carbohydrates and proteins, and from abiotic systems formed by soil erosion. To the CDOM group belong yellow to deep red coloured, nitrogen-containing carbohydrates and proteins, known by the umbrella names fulgic and humic acids. They make up the major part of CDOM and are able to photocatalyse solar reactions of DOM which itself cannot absorb sunlight. Figure 4.25 displays a comparison of the lake water absorption spectrum and spectral distribution of sunlight as a function of penetration depth. As a rule of the thumb, the solar flux above 350 nm can be memorised as 1 Einstein/m^2/h.

From the above it follows that the most relevant photoreactions in natural waters originate from light absorption by CDOM, Fe(III), nitrite and nitrate. The most reactive intermediates are ^{3}CDOM states. Strongly oxidising radicals such as $OH^{\bullet}$, $CO_2^{-\bullet}$, $^1O_2$. NO and $X_2^{-\bullet}$ (X = Cl, Br) radicals are often involved in the formation of persistent aromatic nitro- and bromo compounds [50]. The concentration of iodide is influenced by its biological equilibrium with iodate and by the photoreactions discussed in Chapter 4.4.3 [47b].

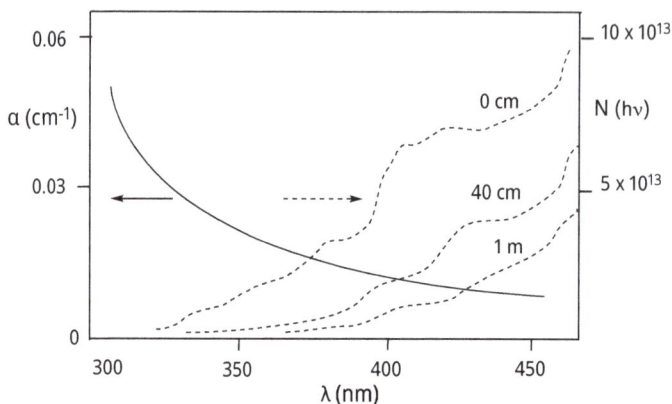

Figure 4.25: Solid line: Absorption spectrum of lake water (Lake Piccolo in Avigliana, Italy). Broken lines: depth dependence of absorbed sunlight. a corresponds to absorptivity which is equivalent to the absorption coefficient (see Equation 3.1). N corresponds to the number of photons absorbed per cm^2, nm and second. It is obtained from the product of spectral photon flux with the Avogadro constant. Redrawn from ref. [49].

4.5.1 Iron(III) complexes

Due to their ubiquitous presence in natural waters and their capability of visible light absorption, iron(III) species play a basic role in solar photochemistry. At neutral pH values $[Fe(OH)(H_2O)_5]^{2+}$ is the dominant species exhibiting a LMCT band maximum at 297 nm. Its low-energy tail overlaps with the UV part of sunlight inducing formation of Fe(II) and the strongly oxidising $OH^•$ radical,[17] which generates hydrogen peroxide and Fe(II). The latter is oxidised back to Fe(III) in the dark (Photo-Fenton reaction, see Equation 2.38 and Figure 2.26). In the presence of mono- or dicarboxylic acids, the corresponding iron(III) complexes undergo complete oxidation to carbon dioxide and inorganic salts such as sulfate and nitrate. This is true also for EDTA compounds, which are refractory to microbial decomposition.

Siderophores[18] form another class of organic ligands that bind very efficiently to Fe(III) and are generated by bacteria and fungi. Through hydroxycarboxylate and catecholate groups they form the corresponding Fe(III) complexes capable of generating Fe(II) via the Photo-Fenton reaction. It is significant that the bioavailability of iron for micro-organisms is high only for the oxidation state two.

Similar LMCT based photoreactions occur at the surface hydroxyl groups of iron(III) oxides. In the absence of iron compounds CDOM can also photocatalyse hydrogen peroxide generation via Equations 4.36–4.39, where D represents a weak re-

---

17 For the corresponding redox potential see Chapter 5.3.2.2.
18 Greek: "iron carrier".

ducing agent. The intermediate superoxide radical is converted to H_2O_2 via Equations 5.37–5.38.

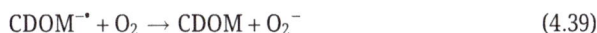

$$CDOM + h\nu \rightarrow {}^3CDOM \tag{4.36}$$

$$^3CDOM + D \rightarrow CDOM^{-\bullet} + D_{OX} \tag{4.37}$$

$$Fe(III) + CDOM^{-\bullet} \rightarrow Fe(II) + CDOM \tag{4.38}$$

$$CDOM^{-\bullet} + O_2 \rightarrow CDOM + O_2^{-} \tag{4.39}$$

As an example, the hydrogen peroxide concentration in freshwater versus sunshine time is given in Figure 4.26 for a New Zealand lake.

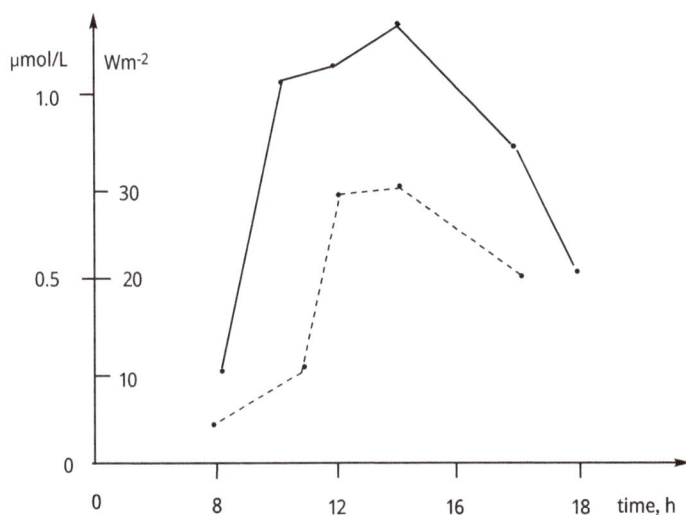

Figure 4.26: Variation of hydrogen peroxide concentration at 5 cm depth (–) and UV irradiance (---) with time for Lake Oxbow (within Waimakaririe River banks, New Zealand) [51].

Iron(III) complexes may also enable a link between global cycles of the elements of carbon and nitrogen (Equation 4.40) [52].

$$RCOOH + Fe(III) + 1/2O_2 + h\nu \rightarrow ROH + CO_2 + Fe(II) \tag{4.40}$$

4.5.2 $OH^{\bullet}$, $NO^{\bullet}$, NO_2, $CO_3^{-\bullet}$ radicals and singlet oxygen

In natural waters, the $OH^{\bullet}$ radical is the most active species. In addition to the Photo-Fenton reaction, it also generated from nitrite, nitrate, and CDOM (Equations 4.41–4.44).

$$NO_3^- + h\nu + H^+ \rightarrow OH^\bullet + NO_2^\bullet \qquad (4.41)$$

$$NO_2^- + h\nu + H^+ \rightarrow OH^\bullet + NO^\bullet \qquad (4.42)$$

$$CDOM + h\nu \rightarrow \; \rightarrow \; {}^3CDOM \qquad (4.43)$$

$${}^3CDOM + H_2O \rightarrow [CDOM - H]^\bullet + OH^\bullet \qquad (4.44)$$

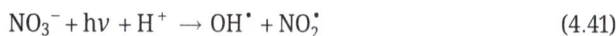

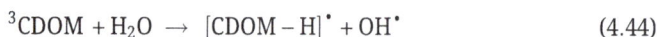

The generated nitrogen oxide radicals may be involved in environmental photonitrosylation reactions of chlorophenolic herbicides in the Rhone river delta [53].

Singlet oxygen is probably produced through energy transfer from a triplet excited CDOM state (Equations 4.43, 4.45). It may reach concentrations up to 10^{-13} M although radiationless conversion to the triplet ground state is very fast.

$${}^3CDOM + {}^3O_2 \rightarrow CDOM + {}^1O_2 \qquad (4.45)$$

The carbonate radical anion can be generated according to Equations 4.46–4.47 and has a reduction potential in the range of 1.60 V, about one Volt below that of the $OH^\bullet$ radical.

$$OH^\bullet + CO_3^{2-} \rightarrow OH^- + CO_3^{-\bullet} \qquad (4.46)$$

$$OH^\bullet + HCO_3^- \rightarrow H_2O + CO_3^{-\bullet} \qquad (4.47)$$

4.5.3 Dihalogen radicals $Cl_2^{-\bullet}$ and $Br_2^{-\bullet}$

The radical anion $Br_2^{-\bullet}$ is generally formed through oxidation of the halide by a hydroxyl radical to a bromine atom followed by addition to bromide (Equations 4.48–4.49). but Fenton-like oxidation processes are also likely (Equation 4.50). In addition to sea water, bromide occurs in considerable concentration in inland ephemeral rivers and lakes and is a very efficient scavenger of $OH^\bullet$ radicals.

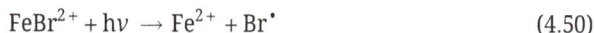

$$Br^- + OH^\bullet \rightarrow Br^\bullet + OH^- \qquad (4.48)$$

$$Br^\bullet + Br^- \rightleftharpoons Br_2^{-\bullet} \qquad (4.49)$$

$$FeBr^{2+} + h\nu \rightarrow Fe^{2+} + Br^\bullet \qquad (4.50)$$

In the case of the chloride reaction, the direct oxidation by an $OH^\bullet$ radical is a reversible process with an equilibrium constant which lies far to the left side at neutral pH. Furthermore, the reaction is endothermic, the opposite of bromide oxidation. Therefore reactions analogous to Equation 4.50 are more likely [54]. Scheme 4.31 summarises the most important halide-derived intermediate radical species. Ozone originates from the atmosphere through photocleavage into NO and oxygen atoms. The latter forms O_3 by reaction with O_2 (Scheme 4.27).

The dihalide radical anions absorb at longer wavelengths than the simple halides. As an example, chloride does not significantly absorb at wavelengths greater than 220 nm while $Cl_2^{-\bullet}$ has a broad band with $\lambda_{max} = 340$ nm and $\varepsilon(340$ nm$) = 9600$ M^{-1} cm^{-1} [55]. There is also a major role for release of iodine in seawater to the air (see Equations 4.31–4.32).

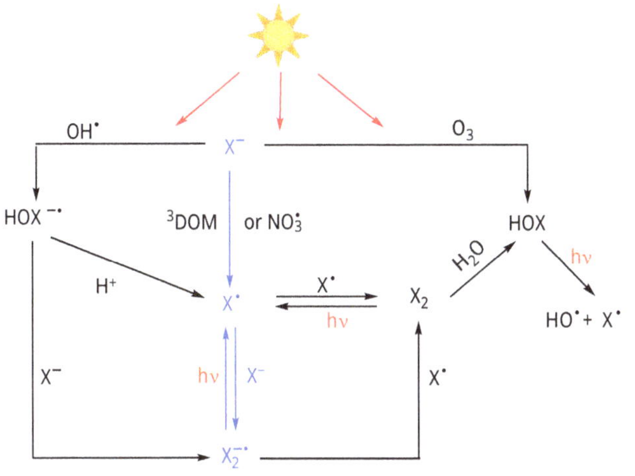

Scheme 4.31: Sunlight-induced formation of reactive halogen compounds [54].

4.5.4 Environmental aspects

As summarised above, a manifold of strong inorganic and organic oxidants is generated when solar light impinges on natural waters. In the presence of organic pollutants, omnipresent hydroxyl radicals may induce exhaustive oxidation to carbon dioxide and simple inorganic salts (see Equation 2.39 and Scheme 5.24). The small, but bactericidal, UV part of solar light enables disinfection even of untreated water to produce drinking water [56]. Mercury and simple mercury salts do not absorb solar radiation, but in the presence of ubiquitous DOM they form CT-complexes that are photoactive with visible light. The resulting sunlight-driven mercury cycle is summarised in Scheme 4.32.

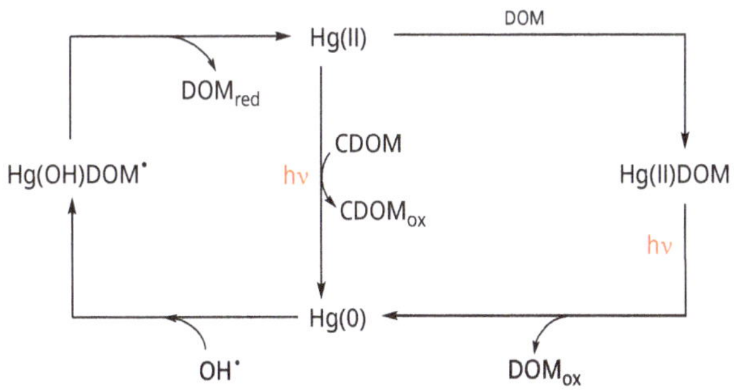

Scheme 4.32: Mercury photoredox cycle in presence dissolved organic matter (DOM) [57].

The toxic action of metallic mercury is boosted when converted to methylmercury chloride, CH_3HgCl, usually named methylmercury. It is formed from inorganic mercury by the action of microbes in aquatic systems and from anthropogenic sources and is soluble in water as the aqua complex $[CH_3Hg(H_2O)]^+$. Thus, very unlike most other metal-carbon bonds, it does not suffer hydrolytic cleavage. It has strong lipophilicity and high neurotoxicity. Exposure to humans generally occurs through eating fish from contaminated water as first observed in 1956 at Minamata Bay in Japan.[19] Methylmercury degradation occurs both biotically, and abiotically [57, 58]. The latter is much more efficient and very likely proceeds through solar CT excitation of $CH_3Hg \cdots DOM$ charge transfer complexes generating reduced mercury and oxidised DOM products [59].

Certain cyanobacteria produce cyanotoxins harmful to human and animal life. In the photic zone of natural waters solar degradation contributes to their overall degradation. Hydroxyl radicals, triplet DOM excited states and photosynthetic pigments induce their oxidative decomposition [60]. Analogous photooxidations contribute to the aging and removal of microplastics (defined as particles smaller than 5 mm) [61].

Questions

1. What is the difference between an adiabatic and non-adiabatic reaction?
2. What is the meaning of chemiluminescence? Chemiluminescence is used to determine concentrations of NO in the gas phase and in solution. Explain how this works.
3. What is the Paterno-Büchi reaction, and why is it named a [2 + 2] cycloaddition?
4. (a) What is the valence isomerisation of norbornadiene to quadricyclane, and why is it a model system for the thermal storage of solar energy. (b) What is its basic disadvantage?
5. Draw the Di-π-methane rearrangement of a 1,4-diene recalling that the first step is a [1,2] sigmatropic shift.
6. In the Barton reaction an alkylnitrite with δ–C-H bonds rearranges to a δ–NO alcohol. Formulate the mechanism.
7. What is a concerted reaction? How do the Woodward-Hoffmann rules demonstrate the differences between thermally and photochemically allowed electrocyclic reactions?
8. The key step in the activation of aliphatic C-H bonds is a hydrogen atom abstraction (HAT). Formulate this step for the primary photoreaction of alkane sulfoxidation.

---

[19] Minamata disease is a neurological disease. In extreme cases, paralysis and death can follow within a few weeks. It appeared first in 1956 in people eating fish from the Minamata Bay in Kumamoto, Japan. Fish from the bay were contaminated by methylmercury originating from a chemical plant producing acetaldehyde in the presence of a mercuric sulfate catalyst.

9. What is the meaning of photochemical curing? This method is used in dentistry. Explain the chemical basis of this application.

10. Explain the principle of photolithography and its application in the semiconductor industry.

11. What is the principle of an optical brightener?

12. Explain the principle and application of a fluorescence sensor. Fluorescent sensors have been devised for anions such as fluoride (see ref. [62]). How do they work, and what is the relation to logic gates?

13. Give some examples for the primary photoreactions of transition metal complexes having M-L, M(H)(R), M(H)(H) and M-M structural motifs; L = PPh_3, CO.

14. What is the difference between prompt photoreaction and formation of an equilibrated excited state? Give an example of each both in organic chemistry and transition-metal chemistry.

15. In the ground state $[Ru(bpy)_3]^{2+}$ is not oxidised by peroxodisulfate. Why does it occur in the excited state?

16. What properties of ruthenium and iridium complexes make them suitable for excited state redox reactions?

17. How can photochemical methods be employed to establish the existence of extremely labile metal complexes such as $M(CO)_n(alkane)$ or $M(CO)_nXe$?

18. Why are chlorofluorocarbons damaging to the ozone layer? How is ozone depleting potential defined?

19. How is ozone formed and removed photochemically in the lower atmosphere?

References

[1] https://upload.wikimedia.org/wikipedia/commons/thumb/9/9c/Joseph_Wright_of_Derby_The_Alche mist.jpg/462px-Joseph_Wright_of_Derby_The_Alchemist.jpg. Accessed on June 15 2024.

[2] a) V. Balzani, P. Ceroni, A. Juris, *Photochemistry and Photophysics*, Wiley-VCH, Weinheim, **2014**; b) N. J. Turro, V. Ramamurthy, J. C. Scaiano, *Modern Molecular Photochemistry of Organic Molecules*, University Science Books, Sausalito, CA, **2010**.

[3] R. Hoffmann, R. B. Woodward, *Accounts of Chemical Research* **1968**, *1*, 17–21.

[4] C. Rauer, J. J. Nogueira, P. Marquetand, L. González, *Journal of the American Chemical Society* **2016**, *138*, 15911–15916.

[5] a) H. Ikezawa, C. Kutal, *Journal of Organic Chemistry* **1987**, *52*, 3299–3303; b) T. Luchs, P. Lorenz, A. Hirsch, *ChemPhotoChem* **2020**, *4*, 52–58.

[6] H. D. Bandara, S. C. Burdette, *Chemical Society Reviews* **2012**, *41*, 1809–1825.

[7] S. Shinkai, M. Ishihara, K. Ueda, O. Manabe, *Journal of the Chemical Society, Perkin Transactions* **1985**, *2*, 511–518.

[8] B. Yao, H. Sun, L. Yang, S. Wang, X. Liu, *Frontiers in Chemistry* **2022**, *9*, 832735.

[9] T.-G. Zhan, M.-Y. Yun, J.-L. Lin, X.-Y. Yu, K.-D. Zhang, *Chemical Communications* **2016**, *52*, 14085–14088.

[10] N. Ruangsupapichat, M. M. Pollard, S. R. Harutyunyan, B. L. Feringa, *Nature Chemistry* **2011**, *3*, 53–60.

[11] B. L. Feringa, *Angewandte Chemie-International Edition* **2017**, *56*, 11059–11078.

[12] a) E. Leyva, M. S. Platz, E. Moctezuma, *Journal of Photochemistry and Photobiology* **2022**, *11*, 100126; b) H. Inui, K. Sawada, S. Oishi, K. Ushida, R. J. McMahon, *Journal of the American Chemical Society* **2013**, *135*, 10246–10249.

[13] a) D. Phillips, *Pure and Applied Chemistry* **1995**, *67*, 117–126; b) T. Maisch, R.-M. Szeimies, G. Jori, C. Abels, *Photochemical & Photobiological Sciences* **2004**, *3*, 907–917.

[14] D. R. Kearns, *Chemical Reviews* **1971**, *71*, 395–427.

[15] G. O. Schenck, K. Ziegler, *Naturwissenschaften* **1944**, *32*, 157.

[16] http://www.naomischenck.com/wp-content/uploads/2016/03/GartenLaborHeidelberg.jpeg. Accessed on April 22nd.

[17] J. Turconi, F. Griolet, R. Guevel, G. Oddon, R. Villa, A. Geatti, M. Hvala, K. Rossen, R. Göller, A. Burgard, *Organic Process Research & Development* **2014**, *18*, 417–422.

[18] Z. Amara, J. F. Bellamy, R. Horvath, S. J. Miller, A. Beeby, A. Burgard, K. Rossen, M. Poliakoff, M. W. George, *Nature Chemistry* **2015**, *7*, 489–495.

[19] J. S. Wau, M. J. Robertson, M. Oelgemoller, *Molecules* **2021**, *26*, 1685.

[20] D. Wöhrle, M. W. Tausch, W.-D. Stohrer, *Photochemie: Konzepte, Methoden, Experimente*, John Wiley & Sons, **2012**.

[21] R. Schupp, F. Torretti, R. Meijer, M. Bayraktar, J. Scheers, D. Kurilovich, A. Bayerle, K. Eikema, S. Witte, W. Ubachs, *Physical Review Applied* **2019**, *12*, 014010.

[22] S.-Z. Pu, Q. Sun, C.-B. Fan, R.-J. Wang, G. Liu, *Journal of Materials Chemistry C* **2016**, *4*, 3075–3093.

[23] B. Gierczyk, M. F. Rode, G. Burdzinski, *Scientific Reports* **2022**, *12*, 10781.

[24] L. Kortekaas, W. R. Browne, *Chemical Society Reviews* **2019**, *48*, 3406–3424.

[25] P. Ceroni, A. Credi, M. Venturi, *Chemical Society Reviews* **2014**, *43*, 4068–4083.

[26] R. Ballardini, V. Balzani, A. Credi, M. T. Gandolfi, F. Kotzyba-Hibert, J.-M. Lehn, L. Prodi, *Journal of the American Chemical Society* **1994**, *116*, 5741–5746.

[27] a) B. Daly, J. Ling, A. P. De Silva, *Chemical Society Reviews* **2015**, *44*, 4203–4211; b) T. Fukaminato, T. Doi, N. Tamaoki, K. Okuno, Y. Ishibashi, H. Miyasaka, M. Irie, *Journal of the American Chemical Society* **2011**, *133*, 4984–4990.

[28] a) L. D. Boardman, *Organometallics* **1992**, *11*, 4194–4201; b) R. J. Hofmann, M. Vlatković, F. Wiesbrock, *Polymers* **2017**, *9*, 534.

[29] N. F. Mott, R. W. Gurney, *Electronic Processes in Ionic Crystals*, 2nd ed., Oxford University Press, **1948**.

[30] J. Hamilton, *Advances in Physics* **1988**, *37*, 359–441.

[31] D. N. Rogers, *The Chemistry of Photography: From Classical to Digital Technologies*, Royal Society of Chemistry, **2006**.

[32] J. J. Turner, M. W. George, M. Poliakoff, R. N. Perutz, *Chemical Society Reviews* **2022**, *51*, 5300–5329.

[33] P. C. Ford, *Coordination Chemistry Reviews* **2018**, *376*, 548–564.

[34] T. R. deBoer, P. K. Mascharak, *Advances in Inorganic Chemistry* **2015**, *67*, 145–170.

[35] a) S. T. Belt, M. Helliwell, W. D. Jones, M. G. Partridge, R. N. Perutz, *Journal of the American Chemical Society* **1993**, *115*, 1429–1440; b) A. D. Selmeczy, W. D. Jones, M. G. Partridge, R. N. Perutz, *Organometallics* **1994**, *13*, 522–532.

[36] R. N. Perutz, B. Procacci, *Chemical Reviews* **2016**, *116*, 8506–8544.

[37] C. P. Simpson, O. I. Adebolu, J.-S. Kim, V. Vasu, A. D. Asandei, *Macromolecules* **2015**, *48*, 6404–6420.

[38] H. Shi, C. Imberti, P. J. Sadler, *Inorganic Chemistry Frontiers* **2019**, *6*, 1623–1638.

[39] L. E. Hatcher, J. M. Skelton, M. R. Warren, P. R. Raithby, *Accounts of Chemical Research* **2019**, *52*, 1079–1088.

[40] P. Coppens, I. Novozhilova, A. Kovalevsky, *Chemical Reviews* **2002**, *102*, 861–884.

[41] P. Gütlich, A. B. Gaspar, Y. Garcia, *Beilstein Journal of Organic Chemistry* **2013**, *9*, 342–391.

[42] a) C. Bressler, C. Milne, V.-T. Pham, A. ElNahhas, et al. *Science* **2009**, *323*, 489–492; b) G. Auböck, M. Chergui, *Nature Chemistry* **2015**, *7*, 629–633.

[43] T. Bockman, J. K. Kochi, *Journal of the American Chemical Society* **1989**, *111*, 4669–4683.

[44] A. M. Holloway, R. P. Wayne, *Atmospheric Chemistry*, Royal Society of Chemistry, **2015**.

[45] J. C. Farman, B. G. Gardiner, J. D. Shanklin, *Nature* **1985**, *315*, 207–210.

[46] F. S. Rowland, *Philosophical Transactions of the Royal Society B: Biological Sciences* **2006**, *361*, 769–790.

[47] a) L. J. Carpenter, P. D. Nightingale, *Chemical Reviews* **2015**, *115*, 4015–4034; b) L. J. Carpenter, R. J. Chance, T. Sherwen, T. J. Adams, S. M. Ball, M. J. Evans, H. Hepach, L. D. J. Hollis, C. Hughes, T. D. Jickells, A. Mahajan, D. P. Stevens, L. Tinel, M. R. Wadley, *Proceedings of the Royal Society A: Mathematical, Physical and Engineering Sciences* **2021**, *477*, 20200824.

[48] G. R. Helz, *Aquatic and Surface Photochemistry*, CRC Press, **2018**.

[49] D. Vione, M. Minella, V. Maurino, C. Minero, *Chemistry* **2014**, *20*, 10590–10606.

[50] D. Vione, *Applied Photochemistry: When Light Meets Molecules*, vol. 92, Springer, **2016**.

[51] R. Herrmann, *Environmental Toxicology and Chemistry: An International Journal* **1996**, *15*, 652–662.

[52] a) R. G. Zepp, B. C. Faust, J. Hoigne, *Environmental Science & Technology* **1992**, *26*, 313–319; b) B. C. Faust, *Aquatic and surface photochemistry* **2018**, 3–38.

[53] P. Reddy Maddigapu, D. Vione, B. Ravizzoli, C. Minero, V. Maurino, L. Comoretto, S. Chiron, *Environmental Science and Pollution Research* **2010**, *17*, 1063–1069.

[54] Z. Yang, Z. Mao, Z. Xie, Y. Zhang, S. Liu, J. Zhao, J. Xu, Z. Chi, M. P. Aldred, *Chemical Society Reviews* **2017**, *46*, 915–1016.

[55] P. Caregnato, J. A. Rosso, J. M. Soler, A. Arques, D. O. Mártire, M. C. Gonzalez, *Water Research* **2013**, *47*, 351–362.

[56] C. Chu, E. C. Ryberg, S. K. Loeb, M.-J. Suh, J.-H. Kim, *Accounts of Chemical Research* **2019**, *52*, 1187–1195.

[57] L. Si, P. A. Ariya, *Atmosphere* **2018**, *9*, 76.

[58] H. Du, M. Ma, Y. Igarashi, D. Wang, *Bulletin of Environmental Contamination and Toxicology* **2019**, *102*, 605–611.

[59] a) H. Luo, Q. Cheng, X. Pan, *Science of the Total Environment* **2020**, *720*, 137540; b) X. Shi, Z. Chen, X. Liu, W. Wei, B.-J. Ni, *Science of The Total Environment* **2022**, *846*, 157498.

[60] T. Kurtz, T. Zeng, F. L. Rosario-Ortiz, *Water Research* **2021**, *192*, 116804.

[61] P. Liu, H. Li, J. Wu, X. Wu, Y. Shi, Z. Yang, K. Huang, X. Guo, S. Gao, *Water Research* **2022**, *214*, 118209.

[62] S. Xiong, M. N. Kishore, W. Zhou, Q. He, *Coordination Chemistry Reviews* **2022**, *461*, 214480.

5 Photocatalysis

In the last two decades photocatalysis has raised tremendous interest among synthetic chemists because visible light excitation of catalysts can generate highly reactive intermediates such as radicals. In most cases the latter are produced through a reductive or oxidative single electron transfer (SET) followed by proton exchange. In a few cases an energy transfer may be the primary reaction of the excited photocatalyst. One of the biggest challenges of photocatalysis is the exploitation of solar energy, mimicking photosynthesis to produce fuels or chemicals or to recycle waste – variously described as artificial photosynthesis, solar fuels or solar chemicals. This problem is unsolved, although huge progress has been made. In the following we shall summarise some typical examples for homogeneous und heterogeneous systems.

5.1 Homogeneous photocatalysis

There are various ways of achieving homogeneous photocatalysis. The approaches discussed in this chapter are summarised in the textbox below.

Methods of homogeneous photocatalysis
1. Dissociation of a metal-to-ligand bond
2. Single electron transfer without photosensitiser
3. Single electron transfer induced by photosensitiser
   (a) Resulting in atom transfer radical addition
   (b) Combined with thermal organometallic catalysis
   (c) Combined with organocatalysis to achieve asymmetric synthesis
4. Hydrogen atom transfer catalysis (concerted proton-coupled electron transfer)
5. Energy transfer by photosensitiser

In Chapter 2.2.9 we defined photosensitisation as a process by which a molecule, the photosensitiser (S) absorbs the light, but another molecule, the substrate, undergoes a change. Sensitisation may take place by energy or electron transfer (EnT, ET). While in an EnT process the sensitiser returns to its ground state without suffering any chemical change, the ET case produces the oxidised or reduced sensitiser as depicted in Scheme 5.1 for the conversion of A to a reduced or oxidised product P. Accordingly, a reductant or oxidant is required to reconvert the sensitiser back to its original state. We note that the overall reaction is a sensitised reduction or oxidation of the substrate A according to Equations 5.1 and 5.2. Therefore, the sensitiser plays the role of a photocatalyst. When the same products are formed even in the absence of *Red* or *Ox*, the sensitisers become stoichiometrically consumed reaction partners of a standard photoreaction in which one reaction partner is the light absorbing species (Equations 5.3 and 5.4).

https://doi.org/10.1515/9783111029375-005

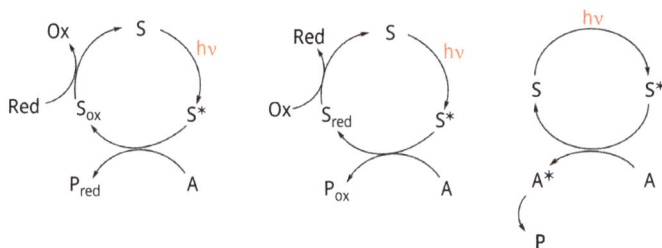

Scheme 5.1: Sensitisation of the reaction A to P. Left: by reductive electron transfer. Centre: by oxidative electron transfer. Right: by energy transfer.

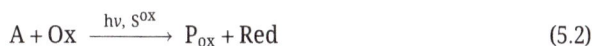

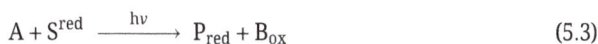

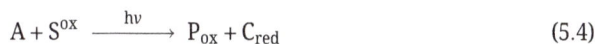

$$A + Red \xrightarrow{h\nu, \, S^{red}} P_{red} + Ox \tag{5.1}$$

$$A + Ox \xrightarrow{h\nu, \, S^{ox}} P_{ox} + Red \tag{5.2}$$

$$A + S^{red} \xrightarrow{h\nu} P_{red} + B_{ox} \tag{5.3}$$

$$A + S^{ox} \xrightarrow{h\nu} P_{ox} + C_{red} \tag{5.4}$$

Thus, the terms *sensitiser* and *photocatalyst* describe the same phenomenon. Usually, the former expression is used in the case of an energy transfer, the latter for an electron transfer mechanism. In this book we follow the same practice although there is no general agreement on this topic [1]. Irrespective of such definition problems, the fact that visible light activation of organic or inorganic molecules generates highly reactive compounds at ambient temperatures has led to tremendous interest in basic and industrial research.[1] In many cases the primary products act as photoredox catalysts opening convenient synthetic access to valuable compounds that are not obtainable or only very difficult to obtain by thermal reactions. Together with the development of novel photoreactors, photoredox catalysis has become a major strategy in natural product synthesis, relevant to medicine and biology [3].

5.1.1 Catalysis initiated by photodissociation of a metal-to-ligand bond

5.1.1.1 Regioselective 1,4-hydrogenation of 1,3-dienes

Transition metal catalysis is an important part of research and industrial applications. In addition to the one-electron processes of Scheme 5.1, coordinatively unsaturated metal centres can activate a substrate by the classical two-electron steps of oxidative addition of the substrate and reductive elimination of the product generated in the metal coordination sphere [4]. These unsaturated centres can also coordinate new two-electron donor

---

1 For a recent summary see a collection of articles in Chemical Reviews [2].

ligands such as alkenes, alkynes and H_2. The regioselective 1,4-hydrogenation of 1,3-dienes nicely illustrates the principles.

UV irradiation of catalytic amounts of chromium hexacarbonyl in *n*-hexane containing equal amounts of *cis*- and *trans*-1,3-pentadiene results in regioselective 1,4-hydrogenation of the *cis*-isomer (Scheme 5.2).

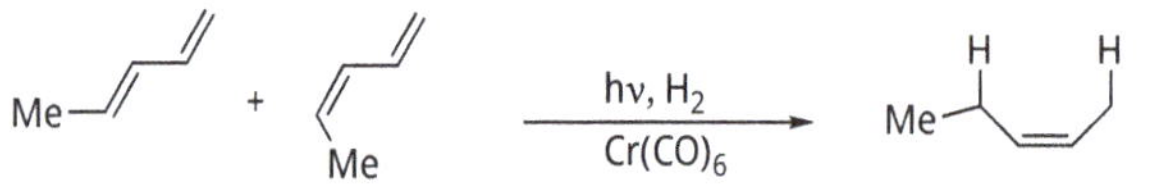

Scheme 5.2: Regioselective addition of hydrogen to a cis-1,3 diene.

In the first reaction step (Scheme 5.3) the excited carbonyl selectively reacts with *E*-**1** producing the true photocatalyst **3** [5]. Light absorption by the latter de-coordinates carbon monoxide to form a 16 valence electron (VE) intermediate **3a** and generates the non-classical dihydrogen complex[2] **a**, from which a stepwise 1,4-hydrogen transfer may occur via the allyl hydride Cr(II) intermediate **b** and the alkene Cr(0) intermediate **c**. Decomplexation of the product *Z*-**2** and complexation of *E*-**1** regenerates the intermediate **a** directly or via intermediate **3**. The latter possibility is more likely since continuous irradiation is necessary for substantial product formation. All the intermediates shown in Scheme 5.3 have 16- or 18-valence electron (VE) configurations.

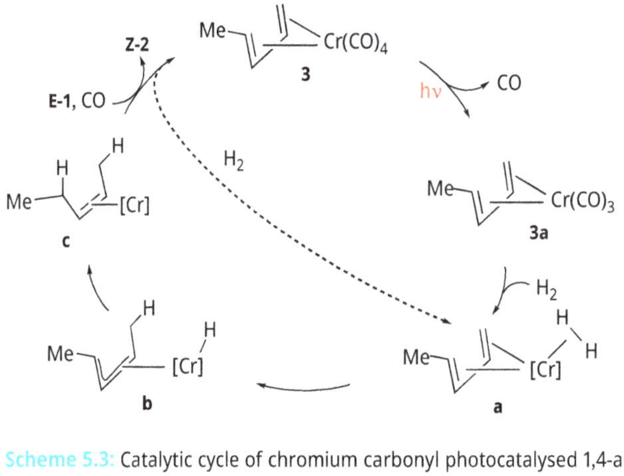

Scheme 5.3: Catalytic cycle of chromium carbonyl photocatalysed 1,4-addition of H_2 to 1,3-pentadiene. Bold lower case letters in this scheme and in the following scheme refer to postulated intermediates and [Cr] stands for Cr(CO)$_3$ in complexes **a–c**.

---

2 Classical dihydride complexes do not contain a hydrogen ligand with an H-H bond (M-H_2) but instead two M-H bonds (H-M-H).

5.1.1.2 Co-cyclisation of alkynes with nitriles

Similar photogeneration of a catalyst is the first step of the synthetically very useful pyridine preparation through a cyclisation of alkynes with nitriles (Scheme 5.4). Visible light irradiation of catalytic amounts of CpCo(CO)$_2$, Cp = η^5-cyclopentadienyl[3] in the presence of an alkyne and excess nitrile affords pyridine derivatives **4** [6]. The reaction also occurs upon heating [7]. Similar to the photocatalytic hydrogenation, loss of carbon monoxide to form 16 VE CpCo(CO) is the primary photoreaction and is followed by coordination of alkyne generating the alkyne complex **a** (Scheme 5.5). The latter is photoconverted to the bis-alkyne intermediate **b**. Subsequent steps are nitrile induced C–C coupling to a cobaltacyclopentadiene followed by insertion of the nitrile into a Co–C bond. A final alkyne/CO induced reductive elimination affords pyridine **4** and catalyst **a**.

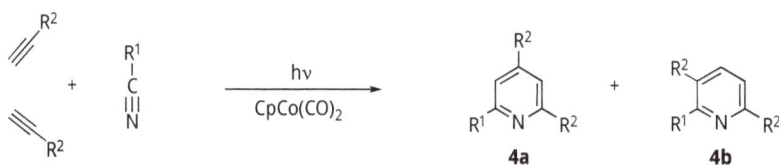

Scheme 5.4: Cocyclisation of alkynes with nitriles.

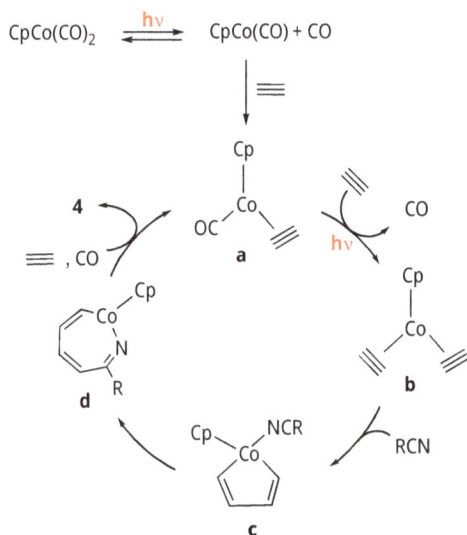

Scheme 5.5: Catalytic cycle of the CpCo(CO)$_2$ photocatalysed pyridine synthesis.

---

3 The affix η gives a topological indication of the bonding between a π-electron ligand and the central atom in a coordination entity. The superscript indicates the number of ligand atoms, which bind to the central metal.

5.1.2 Ligand substitution by photo-induced chain reaction

In the examples presented above continuous irradiation is required for product formation. Next we discuss substitution reactions, in which a short initial irradiation leads to a reaction that continues in the dark. The simple looking photosubstitution of the acetonitrile ligand in the d^6-rhenium complex 5^+ depicted in Equation 5.5 exhibits three unusual features [8].

- Firstly, no substitution is observable when PPh$_3$ is replaced by pyridine.
- Secondly, a value of $\Phi(6^+) = 20$ points to a photo-induced chain reaction.
- And thirdly, triphenylphosphine quenches the emission of 5^+, whereas pyridine has no effect.

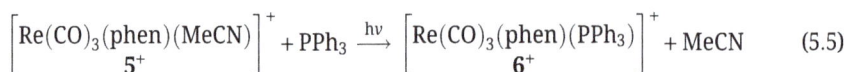

$$\left[\text{Re(CO)}_3(\text{phen})(\text{MeCN})\right]^+ + \text{PPh}_3 \xrightarrow{h\nu} \left[\text{Re(CO)}_3(\text{phen})(\text{PPh}_3)\right]^+ + \text{MeCN} \quad (5.5)$$
$$\mathbf{5^+} \qquad\qquad\qquad\qquad\qquad\qquad \mathbf{6^+}$$

According to the energies of the excited states involved, quenching by energy transfer is not thermodynamically feasible. However, according to the corresponding redox potentials, quenching by electron transfer is possible (Scheme 5.6). Thus, after excitation of 5^+, the primary step is a photo-induced electron transfer (PET). The resulting 19 VE complex **5** contains the accepted electron in an antibonding MO leading to a fast substitution reaction of the labile 19 VE complex **6**. The latter undergoes a fast single electron transfer to 5^+ regenerating **5** and producing the final product 6^+. This step is the growth reaction of an *electron transfer chain catalysis* process. Termination is due to side reactions such as the back-electron transfer from **5** to the triphenylphosphine radical cation. Rapid substitution is a characteristic of 17 VE and 19 VE species, in contrast to slow substitution of 18 VE species.[4] Thus this example illustrates single electron transfer without an added photosensitiser leading to a chain reaction.

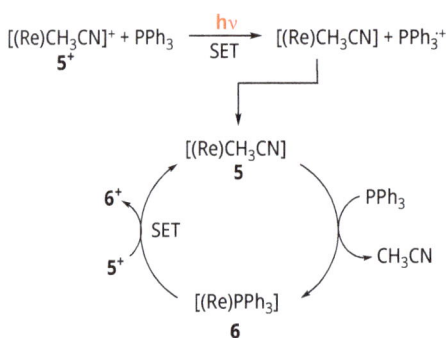

Scheme 5.6: Catalytic cycle of a radical chain substitution reaction initiated by photo-induced electron transfer. (Re): Re(CO)$_3$(phenanthroline).

---

[4] Reactions like this where light absorption occurs outside the catalytic cycle are sometimes referred to as photo-induced catalysis, as opposed to reactions where light absorption occurs within the cycle that are denoted as photoassisted catalysis.

Our next example illustrates the lability of a 17 VE intermediate. The tungsten hydride $(\eta^5\text{-}C_5H_5)W(CO)_3H$ undergoes thermal substitution of CO by PBu_3 in the dark very slowly (<1% conversion after 17 h at room temperature with 10 fold excess of PBu_3). However, irradiation into its absorption band at 311 nm for 87 min led to much higher conversion with a quantum yield exceeding 30 as a result of cleavage of the W–H forming the 17 VE radical $(\eta^5\text{-}C_5H_5)W(CO)_3^{\cdot}$ (Equation 5.6) [9]. The metal-metal bonded dimer $[(\eta^5\text{-}C_5H_5)W(CO)_3]_2$ was not formed in this reaction but has a dramatic effect if added deliberately because it splits into the same radical on irradiation into its visible absorption band at 500 nm (Equation 5.7). Addition of 9 mol% of the dimer to the solution of $(\eta^5\text{-}C_5H_5)W(CO)_3H$ and PBu_3 and irradiation at 500 nm resulted in rapid photosubstitution and a quantum yield of at least 1000. This reaction therefore represents an electron-transfer chain initiated either by homolytic W–H cleavage of the hydride or W–W cleavage of the dimer (Equations 5.8, 5.9).

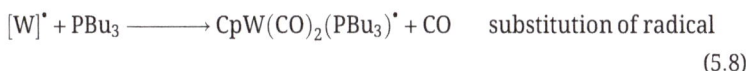

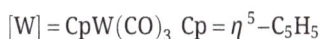

$$[W]H \xrightarrow{h\nu(311\,nm)} H^{\cdot} + [W]^{\cdot} \qquad\qquad \text{radical formation} \quad (5.6)$$

$$[W]-[W] \xrightarrow{h\nu(500\,nm)} 2\,[W]^{\cdot} \qquad\qquad \text{radical formation} \quad (5.7)$$

$$[W]^{\cdot} + PBu_3 \longrightarrow CpW(CO)_2(PBu_3)^{\cdot} + CO \quad \text{substitution of radical}$$
$$(5.8)$$

$$CpW(CO)_2(PBu_3)^{\cdot} + [W]H \longrightarrow CpW(CO)_2(PBu_3)H + [W]^{\cdot} \,\text{propagation} \qquad (5.9)$$

$$[W] = CpW(CO)_3 \quad Cp = \eta^5\text{-}C_5H_5$$

5.1.3 Single electron transfer (SET) photocatalysis

Due to their high reactivity, radicals play an important role in synthetic organic chemistry since they can often start the desired reaction, but their generation usually requires hazardous materials or UV irradiation. In view of these disadvantages, the role of radical intermediates was largely neglected for a long time. This changed significantly when it was found that visible light excitation of organic or inorganic compounds generated redox-active excited states capable of producing radicals through one-electron transfer (SET). In contrast, a two-electron transfer generates two substrate radicals at the heterogeneous semiconductor surface (see Chapter 5.3).

Unlike most organic photocatalysts, inorganic/metal organic compounds generally possess longer excited state lifetimes (microseconds) and have higher reducing and oxidising potentials in the excited state. Values of the latter can be estimated from the ground state potentials and absorption/emission spectra (see Chapter 2.2.6). The excited state potentials allow an estimate of the thermodynamic feasibility of the SET reaction step. Figure 5.1 summarises some classical examples, all of which require absorption of visible light. The ruthenium and iridium complexes (and their analogues) have found especially widespread application since they have both good reductive and oxidative properties enabling SET with donor and acceptor substrates (see Chapters 2.2.6 and 2.2.9). Anionic organic photocatalysts can have very negative excited state reduction potentials allowing efficient primary SET steps [10]. Together with the progress in photoreactor design (see Chapter 1.8), preparative photochemistry based on SET and to a minor part also on energy-transfer processes has grown immensely within the last two decades [1, 11]. Through combination with thermal catalysis and asymmetric syntheses the field has become a new technology for the preparation of highly functionalised molecules relevant to industry and biology [3a, 3c]. In the following we illustrate the basic mechanistic principles with a few characteristic examples.

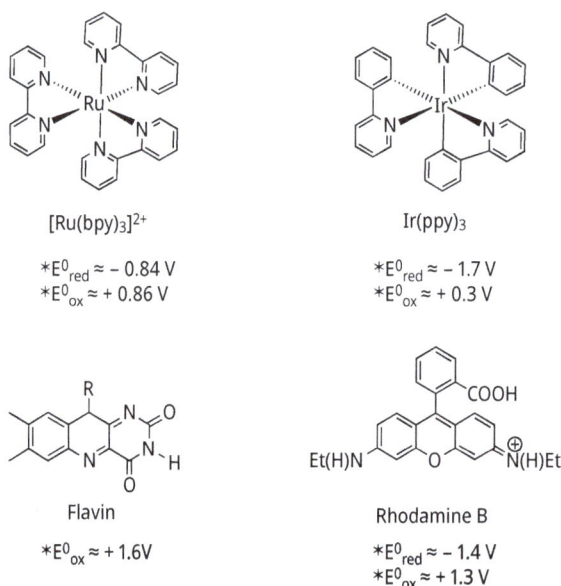

$[Ru(bpy)_3]^{2+}$

$*E^0_{red} \approx -0.84$ V
$*E^0_{ox} \approx +0.86$ V

$Ir(ppy)_3$

$*E^0_{red} \approx -1.7$ V
$*E^0_{ox} \approx +0.3$ V

Flavin

$*E^0_{ox} \approx +1.6$V

Rhodamine B

$*E^0_{red} \approx -1.4$ V
$*E^0_{ox} \approx +1.3$ V

Figure 5.1: Excited state redox potentials of some organometallic and organic photocatalysts. For nomenclature of excited state potentials, see Chapter 2.2.6.

5.1.3.1 Atom transfer radical addition (ATRA)

In medicinal, agricultural and industrial chemistry the selective introduction of fluorinated groups is a basic part of the field. As an example we mention the visible-light photocatalysed addition of fluoroalkyliodides to alkenes enabled by ruthenium(II) complexes or rhodamine B as summarised in Equation 5.10.

$$R{-}I \;+\; {=}\!\!\diagup^{R^1} \quad \xrightarrow[\;R = C_8F_{17}\;]{\text{Vis, } Ru(bpy)_3{}^{2+}, RhB} \quad R\diagup\diagdown^{R^1}_{I} \tag{5.10}$$

The postulated mechanism of this atom transfer radical addition (ATRA) of haloalkanes onto alkenes represents an archetype of a catalytic cycle induced by a photochemical SET step producing a photocatalyst radical cation, iodide and the alkyl radical **a** (Scheme 5.7) [12]. The latter undergoes a $C(sp^3) - C(sp^2)$ coupling to the new radical **b** which in turn undergoes a thermal SET to the photocatalyst radical cation producing the photocatalyst ground state and intermediate carbonium ion **c**. Nucleophilic attack of iodide leads to the addition product **d**.

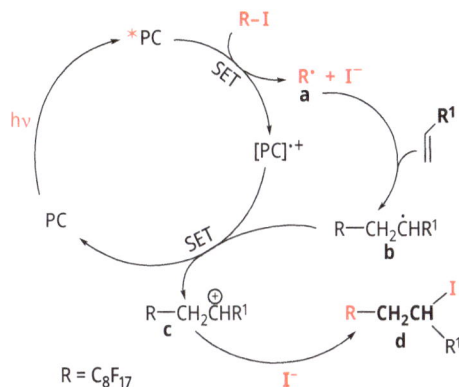

Scheme 5.7: Atom transfer radical addition (ATRA) of haloalkanes onto alkenes. PC = photocatalyst.

5.1.3.2 Photocatalysis combined with thermal organometallic catalysis

C–C coupling reactions involving $C(sp^3)$ carbon atoms are challenging in synthetic chemistry. Recently, the combination of photoredox and thermal organometallic catalysis enabled a simple and efficient method for such a type of alkyl group C–H activation (Equation 5.11) [13].

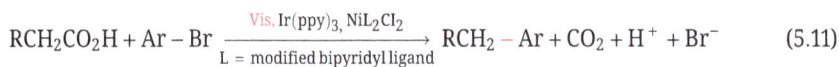

$$RCH_2CO_2H + Ar - Br \xrightarrow[\text{L = modified bipyridyl ligand}]{\text{Vis, Ir(ppy)}_3,\ \text{NiL}_2\text{Cl}_2} RCH_2 - Ar + CO_2 + H^+ + Br^- \qquad (5.11)$$

The photocatalytic cycle (**A** in Scheme 5.8) starts with oxidative decarboxylation of the carboxylic acid affording an alkyl radical **a** and an iridium(II) intermediate. In the thermal cycle **B**, the aryl halide undergoes an oxidative addition to the *in situ* generated nickel(0) catalyst forming the nickel(II) complex **b**. Subsequent addition of radical **a** affords the nickel(III) complex **c**, which undergoes reductive elimination to the final product and a bromonickel(I) intermediate. The latter is reduced back to the starting nickel(0) complex by the iridium(II) species. Thus one metal complex serves as the photocatalyst and another as the thermal catalyst.

An analogous mechanistic scheme applies for C–N cross-coupling reactions between aryl bromides and amines such as pyrollidine [14]. Since the resulting products are of basic interest in pharmaceutical research, attempts at industrial production continue. By exploiting high intensity LED lamps, impressive scaling-up was recently reported (see textbox) [15].

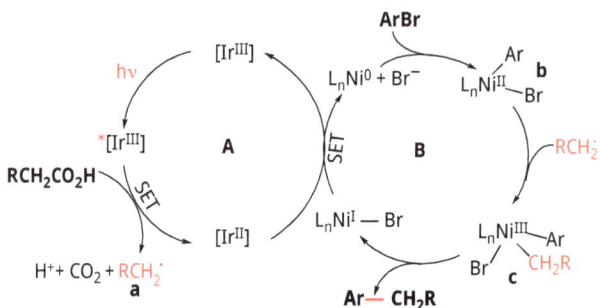

Scheme 5.8: Coupling of photocatalysis (A) with thermal catalysis (B).

Photoredox Catalysis on a Kilogram Scale

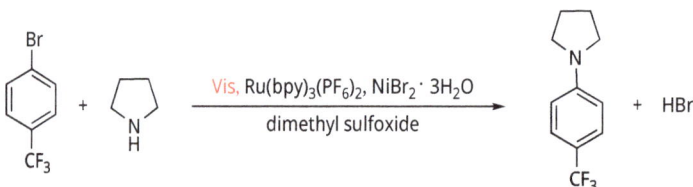

- 1,4-Diazabicyclo(2.2.2)octane for HBr removal
- 15 LEDs of size 5 cm^2 mounted on aluminium plate
- Total power output at 440 nm of 390 W ~ 30 Einstein/h
- Flow reactor of 0.7 L volume, flow rate of 0.1 L/min, tube diameter of 8 mm
- 1.13 kg in 130 min at 90% conversion ~500 g/h.

5.1.4 Hydrogen atom transfer (HAT) photocatalysis

Selective functionalisation of C(sp^3)–H bonds in the absence of neighbouring activating groups can still be considered as the Holy Grail of chemistry. Currently homolytic C–H cleavage through concerted electron and proton transfer (HAT) is gaining increasing interest [16]. Classical hydrogen abstractors are aromatic ketones, dyes and polyoxometalates. According to the basic catalytic cycle the excited photocatalyst abstracts a hydrogen atom from the substrate generating intermediates **a** and **b** (Scheme 5.9). The latter is converted to PC by the hydrogen acceptor Y. The benzophenone photocatalysed addition of isopropanol to maleic acid was the very first example for such a photochemical HAT mechanism [17].

In this case isopropanol acts as hydrogen donor (Scheme 5.10). In the primary step the benzophenone excited state abstracts a hydrogen atom from the alcohol generating Ph$_2$C(OH)$^•$ and Me$_2$C(OH)$^•$ radicals. The latter adds to the maleic acid double

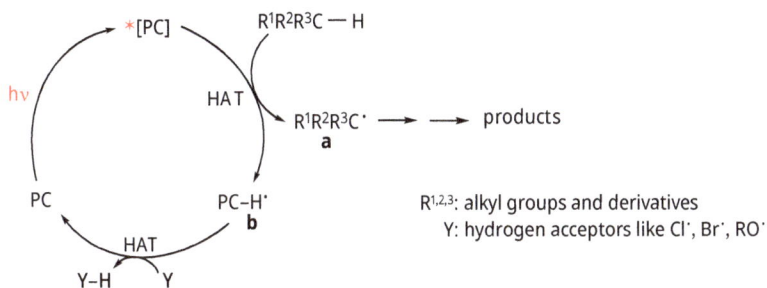

Scheme 5.9: General catalysis cycle of alkane activation through photocatalysed hydrogen atom transfer. PC = photocatalyst, HAT = Hydrogen Atom Transfer.

bond affording radical **b**. Successive back hydrogen atom transfer from **a** regenerates the photocatalyst and produces intermediate **c** which undergoes spontaneous lactonisation to the final product terebic acid **d**. The reaction has been repeated under concentrated sunlight on a ten gram scale [18].

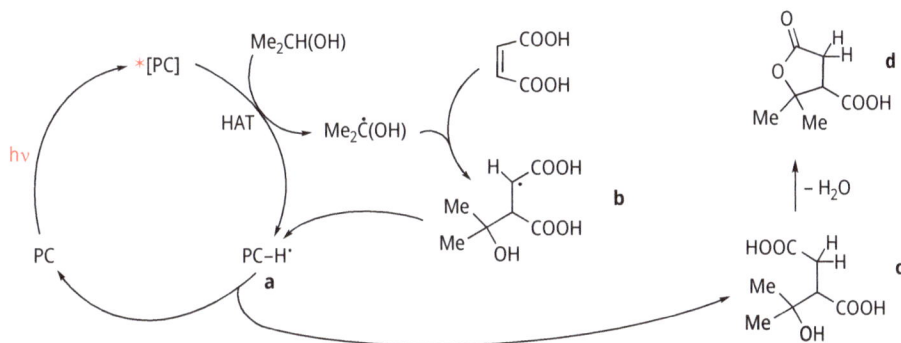

Scheme 5.10: Benzophenone photocatalysed addition of isopropanol to maleic acid.

5.1.5 Asymmetric photocatalysis by photoredox and organocatalysis

In chemical synthesis asymmetric reactions are of basic importance since different enantiomers can have different biological activities.[5] The use of chiral catalysts [19] may introduce enantioselective product formation, thus making troublesome enantio-

---

5 The worst example is the drug thalidomide that was often taken against morning sickness. One of its enantiomers led to birth defects in more than 10,000 babies. Therefore, since the early 1960s the biological action of both enantiomers of pharmaceutical products must be checked.

meric purification obsolete. Weak intermolecular catalyst-substrate interactions can induce enantiospecific bond formation. Whereas thermal catalytic asymmetric processes belong to a highly developed field, asymmetric photocatalysis started flourishing only in the last two decades as acknowledged by the Nobel Prize in 2021 awarded to Benjamin List and David MacMillan.[6] This research area can be considered as the summit of chemical synthesis.

Figure 5.2 depicts how the merging of photoredox catalysis with asymmetric organocatalysis allows convenient access to enantiopure compounds [13c, 20]. The left side resembles Scheme 5.8 with the difference that only cycle **A** is a photoredox cycle while **B** displays the asymmetric catalysis component. Enantiospecific intermolecular interaction between chiral catalyst *Cat** and substrate generates chiral **a**, which undergoes a photo-induced SET from the photocatalyst generating intermediate **b**. Subsequent bond rearrangement leads to the product radical cation **c** which forms the product by an SET from the photocatalyst radical anion. To increase the efficiency of chiral preorganisation, a charged, achiral photocatalyst cation was combined with a chiral anion responsible for enantioselective interaction with the substrate as depicted on the right side of Figure 5.2 [21].

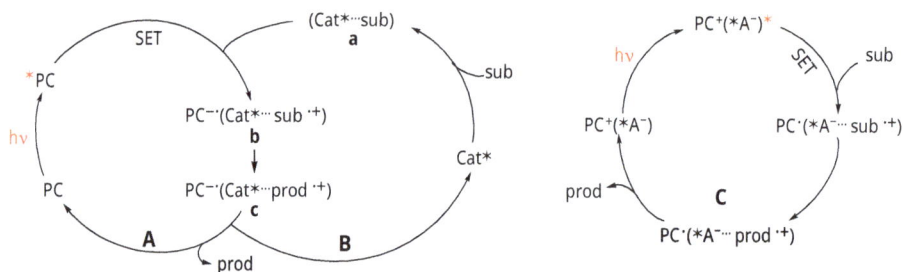

Figure 5.2: Two basic schemes for enantioselective photocatalysis [21].

A recent example for the two-cycle system is the Aza-Paterno-Büchi reaction photocatalysed by a thioxanthone dye inducing a high enantiomeric excess (*ee*) of 94%[7] (Scheme 5.11) [22]. Stereodifferentiation occurs through steric shielding by thioxanthone in an intermediate van der Waals complex. Attack by styrene is therefore highly favoured only from the face opposite to thioxanthone. As result, the azetidine product contains the phenyl group in the position trans to the bicyclic substituent.

As mentioned above (see Scheme 5.7), the introduction of fluorine into organic molecules is of great interest both in basic and applied chemistry. The visible light-

---

6 Benjamin List (born 1968) is a German chemist working at the Max Planck Institut für Kohlenforschung (Max Planck Institute for Coal Research). For David MacMillan see footnote 19 in Chapter 2.
7 The *ee*-value is defined as the difference between the relative abundance of the two enantiomers. Therefore, if a mixture contains 75% of the R enantiomer and 25% S, the enantiomeric excess is 50%.

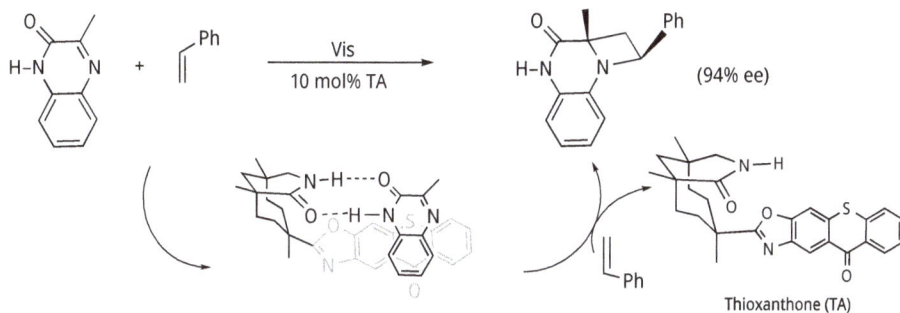

Scheme 5.11: Aza-Paterno-Büchi cycloaddition between a quinoxalinone and styrene.

driven trifluoromethylation of aldehydes illustrates a unique combination of photo-redox catalysis with enantioselective organocatalysis as summarised in Equation 5.12 [20]. Here, tris(bipyridyl)ruthenium(II) is the redox photocatalyst and the chiral organo-catalyst R(R*)NH is a commercially available imidazolidinone derivative. The net reaction consists of an enantioselective substitution of an α-hydrogen atom by a trifluoromethyl group.

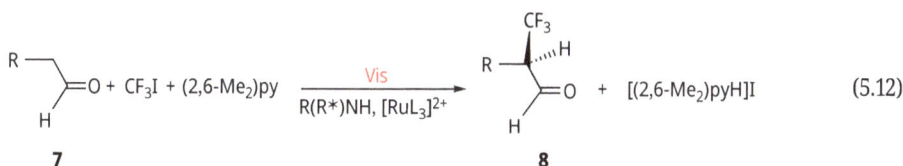

$$(5.12)$$

A mechanistic proposal is summarised in Scheme 5.12. In the initiating step, the ex-cited complex $^*[Ru]^{2+}$ reduces CF_3I to the trifluoromethyl radical and iodide. The radi-cal then starts the cycle by adding to the chiral imine produced from the aldehyde 7 and the organocatalyst. Carbon-centred radicals are good reductants and therefore **a** is easily oxidised by $[Ru]^{3+}$ to the iminium salt **b** and $[Ru]^{2+}$. Hydrolysis generates the product 8 and reforms the chiral organocatalyst.

In the asymmetric syntheses discussed above, stereodifferentiation was achieved with a combination of a photoredox with an asymmetric catalysis cycle. Very recently it was shown that the two cycles can be "short-circuited" in the form of an ion pair consisting of a photocatalyst cation and an enantiopure anion [21]. Scheme 5.13 sum-marises the proposed catalysis cycle. Accordingly, the cation of the photoexcited ion pair undergoes a SET with the arylolefin **9** producing its intermediate radical cation in which the catalyst is a neutral radical. Subsequent stereoselective C–C coupling with the olefin **10** affords the cyclobutyl radical cation **11**$^{*+}$, which undergoes a single electron reduction to product **11** with the regeneration of the catalyst ion pair.

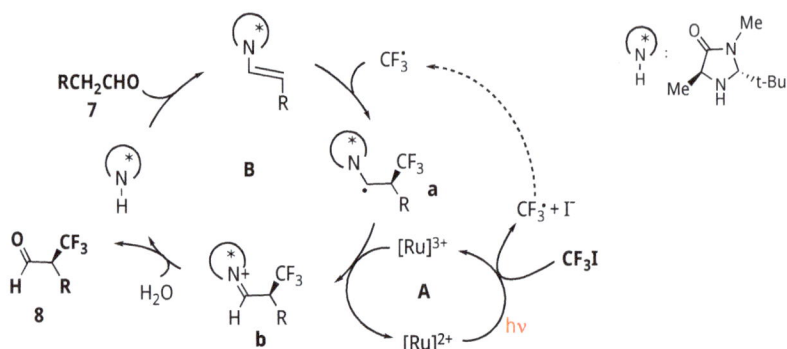

Scheme 5.12: Proposed reaction scheme for the enantioselective trifluoromethylation of aldehydes photocatalysed by ruthenium. [Ru]$^{2+}$ refers to [Ru(bpy)$_3$]$^{2+}$. **A** and **B** symbolise the photoredox and organocatalysis cycle, respectively.

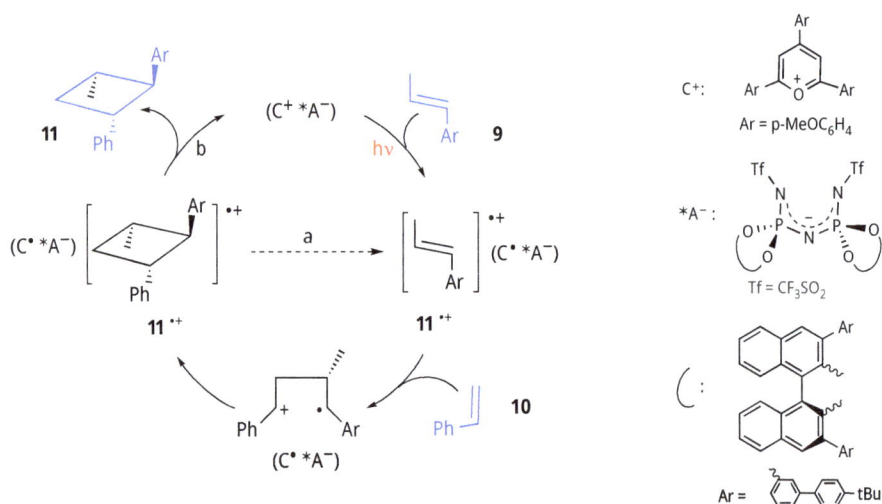

Scheme 5.13: Proposed catalytic cycle of the [2 + 2] cycloaddition between styrene and electron-rich olefins [21].

5.1.6 Photocatalysed reduction and oxidation of water

In the late nineteenth century Jules Verne, one of the first science fiction authors, wrote in his book *The Mysterious Island: when the deposits of coal are exhausted we shall heat and warm ourselves with water. Water will be the coal of the future.* Since the first oil crisis in 1973 this became one of the most popular citations in the field of solar energy conversion science. Obviously, a commercial process of sunshine-driven water splitting to hydrogen and oxygen would be the ideal solution for our energy demanding society. At the present a two-step commercial process consisting of combining photovoltaics with water electrolysis is already in commercial operation. Although a one-step process is feasible by semiconductor photocatalysis, it has not yet reached a commercial scale (see Chapter 5.3). This process resembles the photosynthesis of green plants, the most admirable chemical process on earth, the basis of human life. In photosynthesis the photogenerated charges oxidise water to oxygen and reduce water to NADPH, which in the subsequent Calvin-cycle selectively converts the small amounts of atmospheric carbon dioxide (0.04 vol%) to carbohydrates. Instead of carbohydrates an artificial photocatalyst should produce hydrogen as a clean and recyclable fuel.

At present no *homogeneous* system is able to photocatalyse water splitting, *i.e.* the photochemical cleavage to hydrogen and oxygen. The advantages of homogeneous systems lie in the ability to characterise the complexes in detail, including their molecular structure and electrochemical properties in ground and excited states. The complexes may be tuned systematically by varying substituents. With the aid of time-resolved spectroscopy, the catalytic mechanism may be established in detail. The study of photocatalysts often goes hand-in-hand with electrocatalysts, though they are not reported here [23].

The standard free energy of the endothermic reaction $H_2O_{(l)} \rightarrow H_{2(g)} + \frac{1}{2} O_{2(g)}$ is 237 kJ/mol. At pH 7 the standard reduction potentials ($E^{0'}$) for the reduction and oxidation are –0.41 V and +0.82 V, respectively (Equations 5.13 and 5.14).

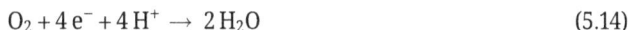

$$2\,H_2O + 2\,e^- \rightarrow H_2 + 2\,OH^- \tag{5.13}$$

$$O_2 + 4\,e^- + 4\,H^+ \rightarrow 2\,H_2O \tag{5.14}$$

Since water absorbs only below 180 nm, *i.e.* in the vacuum-UV (100–200 nm) region, direct photolysis requires vacuum conditions to avoid light absorption by oxygen and nitrogen which start absorbing below 240 nm and 100 nm, respectively. Therefore, water cannot be split by sunlight arriving at the Earth's surface since it contains no radiation in this region (see Chapter 4.4). Instead, photochemical water splitting by UV or Vis light requires sensitisation by electron transfer. As can be seen from Equations 5.13 and 5.14, water reduction is a 2e⁻, 2H⁺ process and water oxidation a 4e⁻, 4H⁺ process. Additionally, water oxidation can generate a variety of unwanted reactive oxygen species (ROS) such as H_2O_2 and O_2^-. These features make both processes challenging. Due to their excellent redox properties, transition metal complexes constitute promising candidates. The best-investigated example is $[Ru(bpy)_3]^{2+}$. Excited state reduction and oxidation potentials of –0.84 V and +0.86 V render both the reduction and oxidation of water thermodynamically feasible. However, only reduction *or* oxidation but not *water splitting* occurs upon irradiating an aqueous solution of the complex with visible light. These are referred to as Hydrogen Evolving Reaction (HER) and Oxygen Evolving Reaction (OER), respectively. Hydrogen or oxygen formation is observable only in the presence of an added reductant or oxidant (Scheme 5.14). In the literature these reactions are in general referred to as "sacrificial water splitting", "sacrificial water reduction" and "sacrificial water oxidation" to emphasise that reductant and oxidant are stoichiometrically consumed.[8]

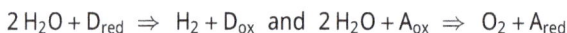

$$2\,H_2O + D_{red} \;\Rightarrow\; H_2 + D_{ox} \quad \text{and} \quad 2\,H_2O + A_{ox} \;\Rightarrow\; O_2 + A_{red}$$

Scheme 5.14: "Sacrificial" water splitting. D_{red} and A_{ox} are the sacrificial reductants and oxidants, respectively.

In water reduction, an **electron relay** in its oxidised form (Ox) and a **thermally active redox catalyst** (□) were originally used for hydrogen evolution (Scheme 5.15). Typical examples are methylviologen (MV^{2+})[9] and colloidal platinum, respectively. Additionally, a fast and irreversibly reacting reductant D_{red} for the oxidised photocatalyst such as triethylamine, triethanolamine or EDTA, is required. However, a single metal complex serves as relay and redox catalyst in more modern systems. In most cases tris(bipyridyl)ruthenium(II) complexes, metal porphyrins or organic dyes are used as photocatalysts [24].

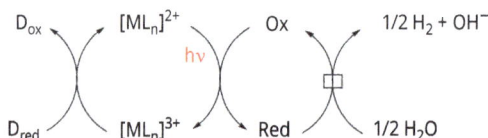

Scheme 5.15: Homogeneous Vis light photoreduction of water by D_{red} if one photon generates one molecule of reductant (see text). □ symbolises the thermally active redox catalyst.

Visible light absorption generates the excited state $^{*}[ML_n]^{2+}$ which reduces the relay Ox to Red (Equation 5.15). In the case of MV^{2+} the blue radical cation $MV^{+\bullet}$ is generated. Even though its reduction potential is more negative than that of water by only 0.03 V, fast hydrogen evolution is observable in the presence of a colloidal platinum catalyst (□). This fast reaction decreases the rate of the competitive, strongly exergonic back-electron-transfer (BET) according to Equation 5.16. BET is also slowed down by the fast and irreversible re-reduction of the oxidised photosensitiser by D_{red}, thus regenerating the photocatalyst (Equation 5.17). In the case of triethylamine, a carbon-centred radical is produced capable of reducing a second MV^{2+} ion (Equation 5.18).

---

8 This is an unnecessary and rather misleading habit. In any redox reaction the reductant and oxidant are consumed, of course.

9 In general added as dichloride: 1,1'-dimethyl-4,4'-dipyridinium-dichloride.

This means that absorption of one photon generates two electrons for the reduction of water ($1H^+/2e^-$ process).[10] The iminium cation is efficiently hydrolysed to diethylamine, acetaldehyde and a proton (Equation 5.19). Finally, water reduction proceeds according to Equation 5.20. Thus, the overall stoichiometry is given by Equation 5.21.

$$[ML_n]^{2+} + MV^{2+} + h\nu \rightarrow [ML_n]^{3+} + MV^{+\bullet} \tag{5.15}$$

$$[ML_n]^{3+} + MV^{+\bullet} \rightarrow [ML_n]^{2+} + MV^{2+} \tag{5.16}$$

$$NEt_2CH_2CH_3 + [ML_n]^{3+} \rightarrow [NEt_2CHCH_3]^{\bullet} + [ML_n]^{2+} + H^+ \tag{5.17}$$

$$[NEt_2CHCH_3]^{\bullet} + MV^{2+} \rightarrow [Et_2N = CHCH_3]^+ + MV^{+\bullet} \tag{5.18}$$

$$[Et_2N = CHCH_3]^+ + H_2O \rightarrow Et_2NH + CH_3CHO + H^+ \tag{5.19}$$

$$2\,H_2O + 2\,MV^{+\bullet} + Pt_{coll} \rightarrow H_2 + 2\,OH^- + 2\,MV^{2+} + Pt_{coll} \tag{5.20}$$

$$H_2O + NEt_2CH_2CH_3 + h\nu \rightarrow H_2 + Et_2NH + CH_3CHO \tag{5.21}$$

One of the most effective completely homogeneous systems consists of an iridium complex as photocatalyst $[Ir(ppy)_2(bpy)]^+$ (ppy represents cyclometalated phenyl pyridine, Figure 5.1), together with a rhodium bipyridine derivative $[Rh(bpy)_3]^{3+}$, as combined relay and catalyst with triethylamine as sacrificial reductant. The sample is irradiated in a THF/H_2O mixture at 460 nm where the iridium complex absorbs but the rhodium complex does not [25]. However, this reaction requires two precious metal salts and emphasis has shifted to Earth-abundant metals instead [23b]. A good example is the use of $[Ru(bpy)_3]^{2+}$ as photocatalyst in conjunction with a cobalt polypyridine complex as electron relay and water reduction catalyst and ascorbate as sacrificial reductant. The mixture is irradiated in water at pH 4 with simulated solar light giving a turnover frequency up to 3200 mol H_2/mol Co hour (Figure 5.3). The system is limited by the photostability of $[Ru(bpy)_3]^{2+}$ [26]. Precious metals have been completely eliminated by replacing $[Ru(bpy)_3]^{2+}$ by quantum dots. Organic photocatalysts have been used successfully in conjunction with nickel catalysts.

Analogous to water reduction, water oxidation requires a thermally active oxidation catalyst ($\square$) and an efficient oxidant A_{ox}. Classical examples are colloidal IrO_2, M_xO_y (M = Rh, Mn) and peroxodisulfate or Co(III) complexes (Scheme 5.16) [27]. Due to favourable kinetic conditions, an electron relay is not usually needed. Therefore, BET according to Equation 5.22 must be much slower than the oxidation of water by the strong oxidant $[ML_n]^{3+}$ generated according to Equation 5.23. In the case of $[CoCl(NH_3)_5]^{2+}$ as oxidant, fast aquation (Equation 5.24) of A_{red} lowers its concentration significantly, thus rendering BET too slow to efficiently compete with water oxidation [28].

---

10 The same type of process may always be observed when a strongly reducing intermediate is produced in a primary oxidation step. In photoelectrochemistry this phenomenon is called *current amplification* or (historically) *current-doubling* (See Chapter 5.1.3).

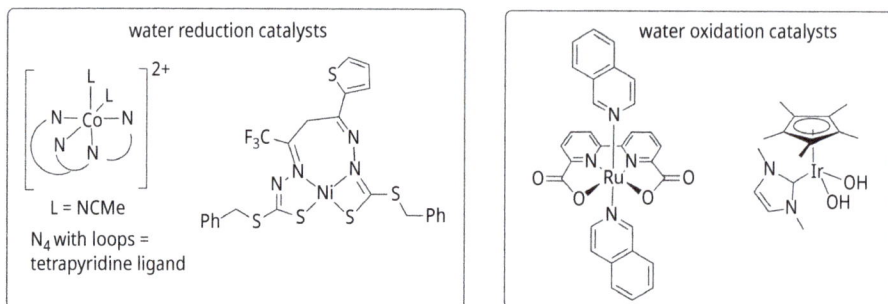

Figure 5.3: Left water reduction catalysts, right water oxidation catalysts.

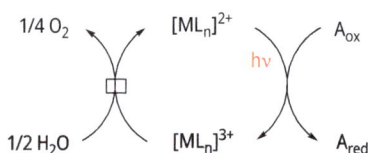

Scheme 5.16: Homogeneous Vis light photooxidation of water by A_{ox}.

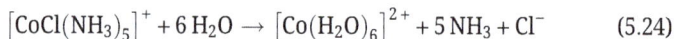

$$[ML_n]^{3+} + A_{red} \rightarrow [ML_n]^{2+} + A_{ox} \tag{5.22}$$

$$[ML_n]^{2+} + [CoCl(NH_3)_5]^{2+} + h\nu \rightarrow [ML_n]^{3+} + [CoCl(NH_3)_5]^{+} \tag{5.23}$$

$$[CoCl(NH_3)_5]^{+} + 6 H_2O \rightarrow [Co(H_2O)_6]^{2+} + 5 NH_3 + Cl^{-} \tag{5.24}$$

In recent years, the emphasis of water oxidation has been on electrocatalysts, but some of these have been investigated as thermal catalysts in conjunction with photochemical methods. Two entirely homogeneous examples are depicted in Figure 5.3: they have proved effective in conjunction with $[Ru(bpy)_3]^{2+}$ as photocatalyst and $S_2O_8^{2-}$ as oxidant at pH 7 with 450 nm irradiation [28].

Astonishing aspects of the formation of oxygen by biological photosynthesis include the ability to accumulate four oxidising equivalents – holes – in the Mn_4CaO_4 cluster of photosystem II (see Chapter 6.1) and the ability to link the transfer of the four holes to the transfer of four protons. Accumulative charge separation has been mimicked by the synthesis of molecules containing organic electron-accepting chromophores, a ruthenium polypyridyl and a dimanganese complex, all covalently linked. However, water splitting has not yet been observed [29].

In summary, the separate reduction and oxidation of water occurs only if the back-electron-transfer between the primary products is slowed down by fast competitive redox reactions, with A_{ox}, D_{red} and H_2O, respectively. We return to this subject in the context of heterogeneous photocatalysis (Chapter 5.3.2) where we also consider complete systems for water splitting.

5.1.7 Photocatalysed reduction of CO_2

The reduction of carbon dioxide offers the potential to recycle this greenhouse gas in conjunction with oxidising another substrate. It has therefore become a focus of research in both homogeneous and heterogeneous catalysis. As can be seen from Table 5.1, 1-electron reduction demands extremely powerful reducing agents whereas multi-electron reduction is much less demanding. However, multielectron reduction has to be coupled to proton transfer. The two most important reactions for homogeneous photocatalysis are reduction to carbon monoxide or to formate – they are both $2e^-$, $2H^+$ processes. They may compete with reduction of protons to form hydrogen. Consequently, selectivity between the different possible products becomes a significant issue. Homogeneous photocatalysts often achieve remarkably high selectivity. If reactions are to be performed in water, there are additional considerations – the presence of carbonate and hydrogen carbonate at equilibrium with dissolved carbon dioxide and the pH-dependent speciation between these three species (see below). As with water splitting, purely homogeneous photocatalysts typically perform the reduction in the presence of a reducing agent. The study of electrocatalysts goes hand-in-hand with that of photocatalysts. More complex devices are required in order to achieve CO_2 reduction coupled to oxidation of water or, for example, waste polymers.

Table 5.1: Some standard potentials for carbon dioxide reduction at pH 7 [30].

Reaction	E^0_{red} / V
$CO_2 + e^- \rightarrow CO_2^{-\cdot}$	−1.90
$CO_2 + 2H^+ + 2e^- \rightarrow HCOOH$	−0.61
$CO_2 + 2H^+ + 2e^- \rightarrow CO + H_2O$	−0.53
$CO_2 + 6H^+ + 6e^- \rightarrow CH_3OH + H_2O$	−0.38
$CO_2 + 8H^+ + 8e^- \rightarrow CH_4 + 2H_2O$	−0.24

The photochemical reduction of CO_2 goes back to the work of Lehn et al. [31] who discovered that a rhenium(I) bipyridine complex, *fac*-Re(bpy)(CO)$_3$Cl (Figure 5.4A), acts as a photocatalyst for the reduction of CO_2 to CO in the presence of triethanolamine (TEOA, Figure 5.4) as sacrificial reductant and DMF as solvent.[11] Unlike the catalysts discussed for water splitting, this rhenium complex acts both as light absorber and as electron transfer catalyst, but it requires near UV radiation. Like [Ru(bpy)$_3$]$^{2+}$, the rhenium complex has a low-lying MLCT excited state. It absorbs at ca. 370 nm forming the ^{1}[MLCT] state and emits at ca. 600 nm from the ^{3}[MLCT] state. Its excited state oxidation potential is ca. 1 V and the complex is therefore capable of reducing CO_2 to any

---

11 Lehn et al. confirmed that CO_2 was the source of CO by ^{13}C labelling.

of CO, formate, CH_3OH and CH_4. Similar behaviour has been observed for charged derivatives such as *fac*-[Re(bpy)(CO)$_3$(NCMe)]$^+$ (Figure 5.4B). One advantage of the catalysts in Figure 5.4 over metal polypyridine catalysts is the presence of carbonyl groups that enable reactions to be followed by IR spectroscopy via the structure-sensitive CO-stretching modes.

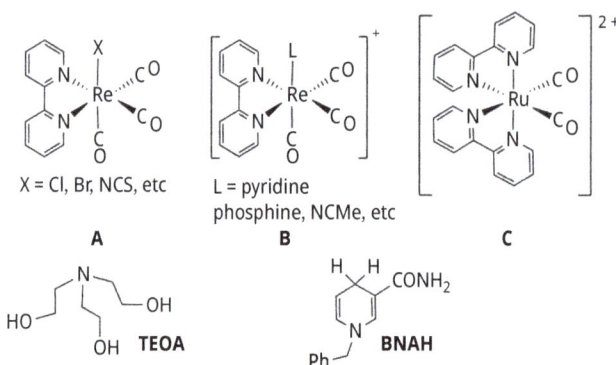

X = Cl, Br, NCS, etc   L = pyridine
                            phosphine, NCMe, etc

      A   **B**   **C**

TEOA   **BNAH**

Figure 5.4: Examples of photocatalysts for CO_2 reduction. Catalysts **A** and **B** do not require a photosensitiser whereas catalyst **C** must be used in conjunction with a sensitiser. Below: examples of "sacrificial" electron donors used for CO_2 reduction triethanolamine (TEOA) and 1-benzyl-1,4-dihydronicotinamide (BNAH).

These rhenium complexes are highly selective for photocatalytic CO production. There is a double role for the rhenium complex as has been shown by Ishitani [32]. Following light absorption, the excited state accepts an electron from the reductant. This one-electron reduced state (OERS) undergoes substitution of the halide ligand (or analogues with MeCN, phosphine etc.) by CO_2. The OERS has a second role which is to supply an electron to the CO_2 complex which can undergo protonation and reduction to form CO and water (Scheme 5.17a). The quantum yield for CO production varies greatly with the substituent, one of the highest is for *fac*-{Re(bpy)(CO)$_3$[P(OEt$_3$)]}$^+$.

There is an added complication: TEOA may become involved in the reaction by coordination as an alkoxide complex. CO_2 can insert into this species and on reduction and protonation yield CO and water: this alternative mechanism is shown in Scheme 5.17b.

Since the rhenium complexes absorb in the near UV, they are unsuitable for solar energy utilisation. This problem has been addressed by coupling them covalently to chromophores that can supply an electron to the rhenium centre. Two approaches are illustrated in Figure 5.5: either to link to a ruthenium bipyridine centre or to link to a zinc porphyrin centre. Irradiation at wavelengths greater than 400 nm or 520 nm, respectively, induces electron transfer to the rhenium centre. These methods act to convert CO_2 to CO with visible radiation. The best performance is ca. 3000 turnovers per mol for the Ru-Re complex with BNAH (Figure 5.4) as donor and ca. 360 turnovers per mol for the Zn(porphyrin)-Re system with TEOA as donor [33]. Supramolecules of

(a) CO$_2$ coordination and protonation

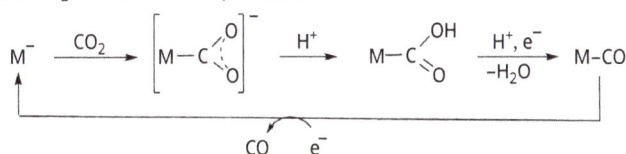

(b) CO$_2$ insertion into metal alkoxide

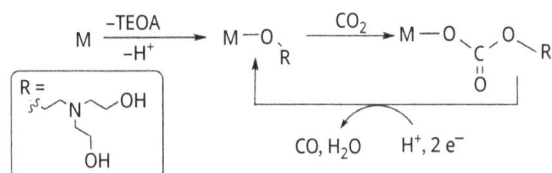

(c) CO$_2$ insertion into metal hydride

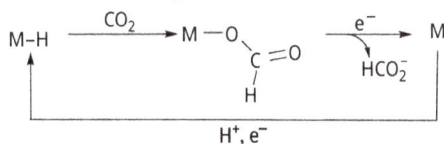

Scheme 5.17: Mechanism of CO$_2$ reduction to form (a) CO via M-CO$_2$, (b) CO via metal alkoxide, (c) formate via metal hydride.

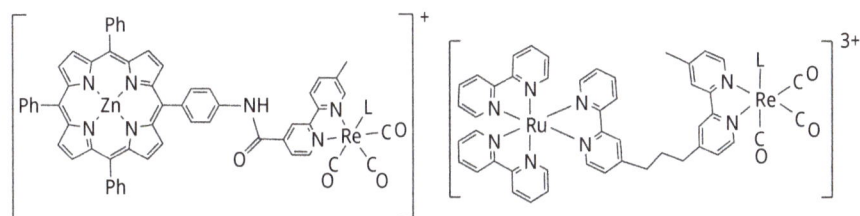

Figure 5.5: Dyads enabling photocatalytic CO$_2$ reduction to CO with visible radiation.

this type that contain a light absorber covalently linked to a catalytic centre are called dyads. Similarly, systems with an electron acceptor, a light absorber and an electron donor are called triads (Chapter 4.3.3).

The photochemical reduction of CO$_2$ to formate also goes back to Lehn's research in the 1980s. The ruthenium(II) complex cis-[Ru(bpy)$_2$(CO)$_2$]$^{2+}$ (Figure 5.4c) is selective for CO$_2$ reduction to formate when used in conjunction with [Ru(bpy)$_3$]$^{2+}$ as photosensitiser in anhydrous TEOA/DMF. However, in aqueous BNAH/DMF, a mixture of CO and formate is generated. In attempts to avoid precious metals, CO$_2$ has been reduced photochemically with either an organic dye or a copper complex as photosensitiser with a manganese analogue of the rhenium catalysts in Figure 5.4, but a mixture of formate and CO is generated [33]. Formic acid is of increasing interest as a hydrogen storage molecule since it can be converted back to CO_2 and H_2 catalytically (CO/H_2 mixtures are known as syn-gas and are used in industry extensively).

5.2 Energy transfer photosensitisation

In the examples discussed above the excited photocatalyst started the reaction cycle through a single electron transfer with the substrate. Only recently, it has been found that the other well-known classical primary reaction of an excited molecule, energy transfer (EnT), plays an increasing role in selective photochemical syntheses. As an example, Scheme 5.18 summarises an intramolecular [2 + 2] cycloaddition photosensitised by the triplet excited state of an iridium(III) complex [34]. Therein triplet-triplet energy transfer generates the dienone triplet state which undergoes C-C bond formation to a 1,4-diradical and the final bicyclic product. These compounds open access to pharmaceuticals for inflammatory diseases.

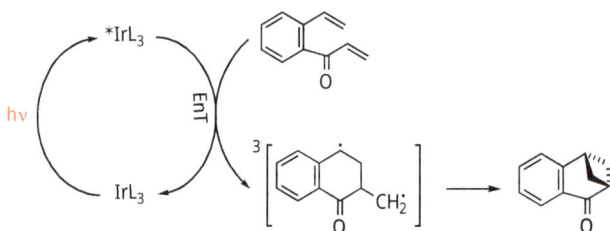

Scheme 5.18: Intramolecular [2 + 2] cycloaddition photosensitised by an iridium(III) complex, L = ppy (deprotonated phenylpyridine).

In another example the excited state of $Ir(ppy)_3$ transfers its excitation energy to a nickel(II) complex that catalyses the coupling of benzoic acid to 4-bromomethylbenzoate to form a new ester [35]. This example serves as a warning that single-electron transfer mechanisms should not be assumed to be correct in all cases.

5.3 Heterogeneous photocatalysis at semiconductor surfaces

Sunlight-induced physical and chemical effects such as sensitisation of electrical conductivity and surface deterioration of zinc oxide containing paints were known already in the beginning of the twentieth century.[12] In 1912 G. Ciamician stated in his

12 For an exhaustive summary of the history of semiconductor photocatalysis see ref. [36].

famous talk "The Photochemistry of the Future" that in the future we may chemically fix solar energy by using suitable photocatalysts: *"glass buildings will rise everywhere; inside of these will take place the photochemical processes that hitherto have been the guarded secret of the plants"*[13] [37].

One of the first semiconductor photocatalysed reactions was the solar photo-reduction of azobenzene to N,N'-diphenylhydrazine in the presence of ZnO suspended in deoxygenated aqueous alcohol [38]. The presence of deoxygenated alcohol was essential since oxygen inhibits the reaction. In the following we first propose a simple classification of semiconductor photocatalysed reactions before discussing water splitting, aerial oxidations of pollutants, nitrogen fixation, carbon dioxide fixation and a few novel organic reactions including atom economic C–C and C–N couplings.

5.3.1 General classification of reactions

The chemical utilisation of solar energy is based on the availability of photocatalysts capable of undergoing visible light-induced electron transfer with appropriate substrates. In the case of *solar energy storage,* the overall reaction must be endergonic, whereas for *solar energy utilisation* it may be also exergonic.

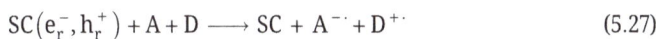

$$SC \xrightarrow{h\nu} SC\left(e_r^-, h_r^+\right) \tag{5.25}$$

$$SC\left(e_r^-, h_r^+\right) \longrightarrow SC + h\nu/heat \tag{5.26}$$

$$SC\left(e_r^-, h_r^+\right) + A + D \longrightarrow SC + A^{-\cdot} + D^{+\cdot} \tag{5.27}$$

A semiconductor photocatalysed reaction may be broken down into three basic reaction steps according to Equations 5.25–5.27 (see Chapter 3.3) where e^-_r, h^+_r represents the reactive electron-hole pair.[14] In the final reaction step the primary intermediates from interfacial electron transfer IFET are converted into an oxidised and reduced product, in perfect analogy with photoelectrochemistry. Almost all semiconductor photocatalysed reactions follow this classification and we proposed to categorise them as *semiconductor type A photocatalysis* (Equation 5.28).[15] In a very few cases the primary products undergo intermolecular bond formation, a reaction type unknown

---

13 Giacomo Ciamician, Professor of Chemistry at the University of Bologna, conducted organic solar photoreactions on the roof of his laboratory.

14 We note that the excited semiconductor, similarly to a molecule (see Chapter 2), may also transfer *energy* to an acceptor like oxygen. However, the observed $^1O_2$ seems to be formed through hole oxidation of superoxide. For a review see ref. [39].

15 This classification has the advantage that it does not imply any mechanism, which may change as function of time because of progress in the field.

in organic photoelectrochemistry. We name it *semiconductor type B photocatalysis* (Equations 5.29, 5.30). Scheme 5.19 summarises the two reaction modes pictorially.

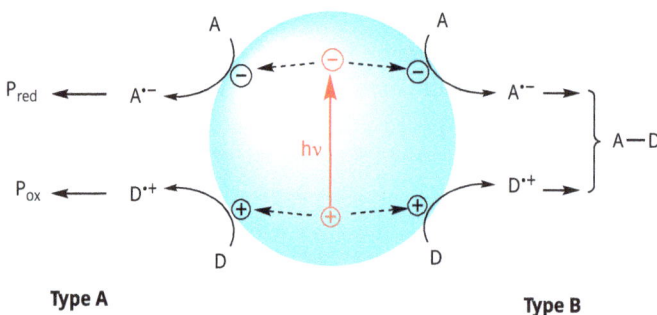

Type A   **Type B**

Scheme 5.19: Schematic representation of type A and type B semiconductor photocatalysis.

In molecular photocatalysis the primary process is a single-electron transfer (SET) between photocatalyst and one substrate, induced by absorption of one photon. In contrast, in semiconductor photocatalysis absorption of one photon induces two one-electron primary processes, one each with two substrates.

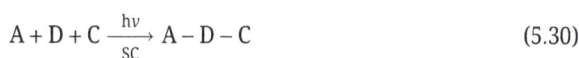

$$A + D \xrightarrow[\text{SC}]{h\nu} B_{red} + C_{ox} \tag{5.28}$$

$$A + D \xrightarrow[\text{SC}]{h\nu} A - D \tag{5.29}$$

$$A + D + C \xrightarrow[\text{SC}]{h\nu} A - D - C \tag{5.30}$$

The thermodynamic feasibility of a semiconductor photocatalysed reaction is estimated by checking the following.

– *The reduction potentials of the substrates must be located within the bandgap of the semiconductor.*[16]
– *The light absorbing species must be determined through comparison of the substrates' electronic absorption spectra with the semiconductor diffuse reflectance spectrum.*
– *For prolonged irradiations, the product light absorption must also be considered to prevent consecutive photoreactions.*

In general, the semiconductor is the light-absorbing species, although in many cases it may be the substrate or a substrate-semiconductor surface complex. We named these two mechanisms direct and indirect semiconductor photocatalysis. In the direct pro-

---

16 In general reduction potentials refer to solvated substrates. In the adsorbed state the value may change by up to 1 V in the case of very strong adsorption [40].

cess light (see Scheme 5.20A) generates the reactive electron-hole pair, followed by IFET reactions with substrates A and D.

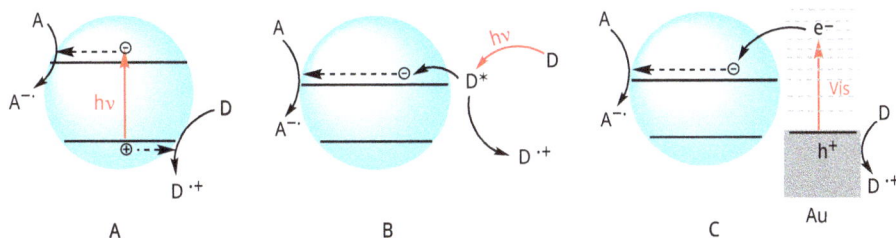

Scheme 5.20: Direct (A) and indirect (B, C) semiconductor photocatalysis.

This standard mechanism is quite often anticipated without having experimental evidence on the nature of the light-absorbing species. A classic example for indirect semiconductor photocatalysis is the aerial oxidation of dyes (Scheme 5.20B). In that case the excited state reduction potential must be equal to or more negative than the conduction band edge potential.[17] The resulting spatial separation of the two charges may partially prevent recombination.[18] In the field of semiconductor photosensitisation, this type of photo-induced electron transfer (PET) is also termed the Sakata-Hashimoto-Hiramoto mechanism [41]. Quite often CT interaction of the donor with the semiconductor surface generates a new absorption band or shoulder at the low-energy bandgap absorption onset. In the case of titania, sulfur dioxide or aromatic 1,2-diols generate yellow and red surface CT complexes, respectively [42]. Excitation of the CT band results in optical electron transfer (OET, Creutz-Brunschwig-Sutin mechanism) [43] as summarised in Scheme 5.21. Thus, the two types of indirect photocatalysis involve different charge generation steps. In the PET mechanism it is a two-step process consisting of light absorption and subsequent electron transfer while in OET it is a one-step process. Light absorption and electron transfer occur instantaneously. When the oxi-

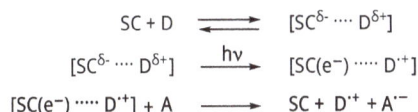

Scheme 5.21: Indirect Photocatalysis via Optical Electron Transfer (OET).

---

17 When this excited state potential is more positive than the valence band edge, an oxidative IFET may occur as suggested for chromate(VI) modified titania [39].
18 Some authors prefer the term "photosensitised photocatalysis" for such a type of reaction. This is not correct, since according to its definition a sensitiser must not be consumed.

dised donor D$^{+\bullet}$ is reduced back to D, the process constitutes the case of electron transfer photosensitisation.

A standard experiment to find out which mechanism may be operating is to measure the optimal reaction rate as a function of the excitation wavelength, the so-called action spectrum (see Figure 1.10).[19] If it compares well with the diffuse reflectance spectrum of the photocatalyst, a *direct* photocatalysis is operating. If it is similar to the spectrum of the substrate or a substrate-surface CT complex, an *indirect* mechanism is present [44].

According to this classification the primary redox products are the same for the direct and indirect mechanisms (Chapter 3.5). This is also true for the case when the sensitiser is not a molecule but a 2–100 nm small noble metal nanoparticle. In the nanoparticles the metal electrons interact with the electrical field of the light wave generating a local surface plasmon resonance (LSPR) resulting in absorption at much longer wavelength than the bulk semiconductor. Such plasmons are longitudinal electronic vibrations parallel to the metal surface. Consequently, metal electrons are excited to high energy states leaving a hole in the bulk solid. If they have a lifetime in the range of picoseconds, they can be transferred to the semiconductor conduction band and undergo a reductive IFET with the acceptor A (Scheme 5.20C). The hole remaining at the metal is neutralised by IFET from the donor D. Its oxidation potential is given by the Fermi level which is located for gold at 0.5 V. Although water oxidation requires a potential of at least 0.8 V, oxygen formation has been reported for TiO$_2$/Au particles.[20] Thus, the overall process belongs to the class of indirect photocatalysis. It is remarkable that the LSPR strongly depends on the size and shape of the metal particle and on the solvent refraction index. Spheres of gold, silver and copper absorb at 530, 400 and 580 nm, respectively. Nanorods exhibit two absorption bands, one each along long and short axes. The morphology of the metal-semiconductor contact may also influence light absorption and efficiency of photocatalysis. These nanoparticles are generally deposited in amounts of 0.5–5.0 weight% on inorganic oxides like titania by photoreduction of corresponding metal salts (see Chapter 3.5) [46].

In summing up the above discussions, the following recommendations emerge for the planning of a semiconductor photocatalysed reaction (see textbox).

---

19 An *action spectrum* is defined as the plot of a chemical or physical response as function of the wavelength of incident photons (see Chapter 1.7).

20 It was assumed that the TiO$_2$/Au contact generates an extremely strong electrical field upon excitation making the hole a stronger oxidant [45].

A Code for Practical Semiconductor Photocatalysis
- Find the redox potentials of the substrates
- Estimate the position of reactive electron-hole pairs from quasi-Fermi and bandgap energies (see Chapter 3.5.4)
- Estimate the thermodynamic feasibility of the IFET reactions
- Compare the absorption spectra of all reaction components to identify the light absorbing species and to avoid absorption of the exciting radiation by the product.

5.3.2 Type A reactions

5.3.2.1 Water splitting

Photocatalytic water splitting and its historic development are summarised in a great number of publications, which can be found in recent overviews [47]. Since water absorbs only at wavelengths shorter than 180 nm, solar water splitting requires the presence of a visible light absorbing semiconductor capable of generating reactive-electron hole pairs of appropriate redox potentials (Equations 5.13, 5.14, values shown apply for pH 7). Accordingly, a minimum bandgap of 1.23 eV and a quasi-Fermi level of at least −0.42 V are required (Figure 5.6).[21] This consideration is made assuming zero overpotential[22] and zero reorganisation energies of the IFET reactions. In reality, the latter are in the range of 0.4 eV resulting in a minimum bandgap of about 2.0 eV.[23,24] Several inorganic semiconductors fulfil that requirement (see Figure 3.16). In 1971 Fujishima and Honda reported the first *solar-electrochemical* cell for water splitting [50]. It is composed of an array of twenty rectangular (85 × 100 mm) titanium metal plates covered by a rutile layer and a platinum black cathode (Figure 5.7). After exposing it for a full day to Tokyo sunlight, 1.1 litres of hydrogen were produced.[25]

In the late 1970s, Bard et al. prepared a type of "short-circuited" photoelectrochemical cell by depositing the counter electrode in the form of a few weight percent of platinum nanoparticles onto the titania surface (see Chapter 3.5). Platinum functions as catalyst for the hydrogen evolving reaction (HER, Equation 5.13). The oxygen

---

21 In other words, the potentials for water spitting have to be located within the bandgap.
22 The overpotential refers to the potential difference (voltage) between the standard reduction potential and the potential at which the reaction is observed experimentally [48].
23 In photosystem II the overpotential for water oxidation is 0.3 V.
24 We note that both reorganisation energy and overpotential strongly depend on the nature of the semiconductor surface and redox catalysts. A rather low overpotential for oxygen formation of 0.41 V was reported for a cobalt phosphate catalyst [49].
25 Prepared by heating a titanium plate with the Bunsen burner to 1300 °C. Application of a 0.3 V cathodic external voltage to the platinum cathode increased the reaction rate.

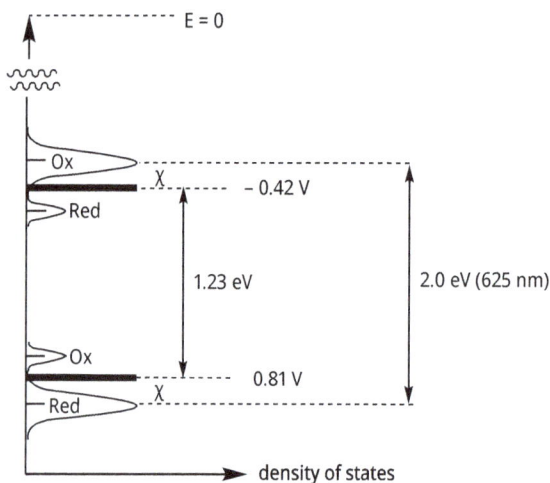

Figure 5.6: Thermodynamic requirements for water splitting (see also Equation 3.7).

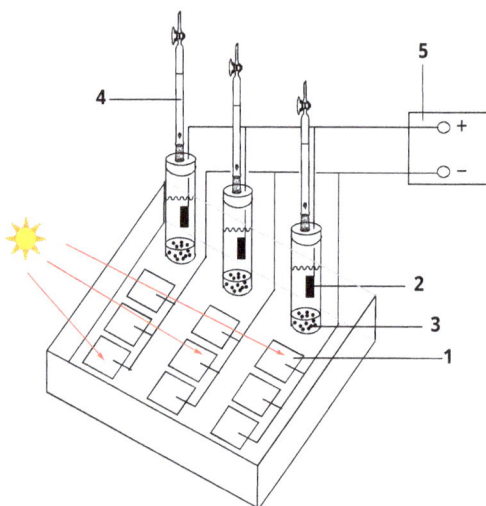

Figure 5.7: Simplified construction scheme of the solar-electrochemical cell. (1) Anatase photoanode dipping into 1 M NaOH, (2) platinum black cathodes immersed into 0.5 M H_2SO_4, (3) agar salt bridges, (4) gas burettes, (5) ammeter. Adapted from ref. [50].

evolution reaction (OER, Equation 5.14) occurs at the unmodified titania surface.[26] By analogy to natural photosynthesis an artificial Z-scheme was proposed. It should consist of two semiconductors able to undergo interparticle electron exchange [51]. Later

---

26 Titania requires UV irradiation and the efficiency of water splitting is very low.

such systems were named *one-step* and *two-step* systems. We feel a classification as *one-* and *two-particle photocatalysis* is more appropriate (Scheme 5.22).[27] Subsequent research followed these ideas trying to understand and control the factors determining the quantum yield of water splitting at one-particle systems.

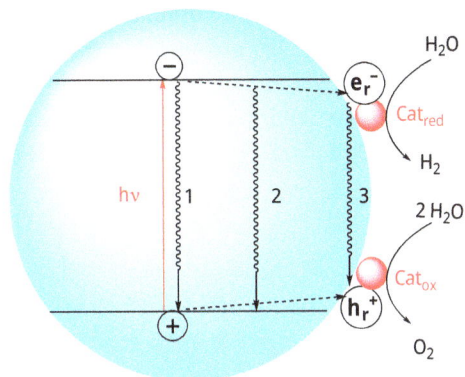

Scheme 5.22: Schematic mechanism of one-particle water splitting. Cat_{ox}: ruthenium and cobalt oxides, Cat_{red}: platinum and rhodium metal. 1–3 symbolise radiationless deactivation; radiative processes are not depicted.

As a first step, the redox catalyst for HER (Cat_{red}, Scheme 5.22) was optimised, followed by adding a catalyst for OER (Cat_{ox}, Scheme 5.22). The principles of semiconductor photocatalysis were applied for an impressively simple preparation of these multifunctional heterogeneous materials (see textbox) [52a, 53a]. It turned out that the two photodeposited redox catalysts are located about 150 nm apart at different crystal facets (Figure 5.8). This concept of crystal facet engineering has been reported repeatedly [45].

Sequential, site-selective functionalisation of strontium titanate

In the first step $SrTiO_3:Al^{3+}$ powder (1.0 atom% Al_2O_3 relative to Ti, d = 200–300 nm) is calcined, followed by deposition of rhodium metal through irradiation of an aqueous suspension containing $RhCl_3$ (0.1 wt% Rh) with a 300 W xenon lamp. Subsequent addition of K_2CrO_4 solution (0.05 wt% Cr) and further irradiation results in deposition of Cr_2O_3 at the rhodium particles. In the final step dissolved $Co(NO_3)_2$ (0.05 wt% Co) is added resulting in the oxidative deposition of CoO(OH) onto the surface that is free of Rh/Cr_2O_3.

The risk of an explosion during the water splitting experiments can be minimised by keeping the H_2 and O_2 concentrations in the sample gas space below 4%.

---

27 The reason for this definition is that in chemistry, the word "step" in general is used for *one process* out of a series of *successive reaction steps* constituting the overall reaction.

Recently it was found that the material $SrTiO_3:Al^{3+}/Rh/Cr_2O_3/CoO(OH)$ exhibits almost quantitative utilisation of the absorbed light energy. Doping by aluminium removes charge-recombination centres, rhodium and cobalt [49, 53] accelerate HER and OER, while chromium prevents the back reaction to water. There is good evidence that the different crystal facets have different Fermi levels resulting in a strong electrical field that slows down electron-hole recombination. The latter is further diminished by the two highly active redox catalysts Cat_{red} and Cat_{ox}. No other light-to-energy system has reached such a high catalytic activity: the quantum yield of hydrogen formation at 360 nm is 0.96 [54]. An extension of that approach resulted in a system active with visible light: $BaTaO_2N:Mg$ modified with $(Na)Rh/Cr_2O_3$ and IrO_2 as co-catalysts, but the quantum yield at 420 nm is only about 0.08 [55].

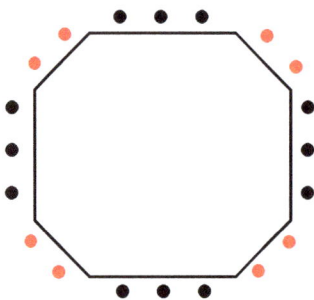

Figure 5.8: Sketch of surface structure of a $SrTiO_3:Al^{3+}$ photocatalyst loaded with Rh/Cr_2O_3 (black dots) and CoO(OH) (red dots, 300–500 nm size). The two co-catalysts are smaller than 5 nm. Redrawn from ref. [53a].

While photocatalyst systems for suspensions are appropriate for basic research, they are less suited for a commercial process. Immobilisation of the semiconductor particles onto a planar panel seems to be a more realistic step towards practical application. Recently, an impressive 100 m^2 array of flat panels was constructed on glass covered by modified $SrTiO_3:Al^{3+}$ particles. Exposure to solar light on a September day in the Tokyo area gave rise to a hydrogen evolution rate of about three litres per minute. Separation of the H_2/O_2 product gas was performed with a commercial polymer membrane [56]. In an extension of this system the photocatalyst was combined with a solar vapour generator consisting of a porous carbon grid. Solar-thermal heating of the latter produces steam which is split at the photocatalyst located above the surface. According to this compartmentalisation seawater and waste streams can also be used [57].

In *two-particle* systems, the analogy with natural photosynthesis is even stronger than in the *one-particle* approach. By analogy to PS I and PS II of the photosynthetic membrane, the semiconductors SC-1 and SC-2 must have bandgap positions [58] appropriate for the HER and OER processes. The latter can successfully compete with charge-recombination only if a fast interparticle ET from SC-1 to SC-2 along the conductive sheet is possible (Scheme 5.23) [56]. Good visible light water splitting activity was reported for the photocatalyst consisting of a heterojunction between red phosphorus and 2.5% WO_3 [58].

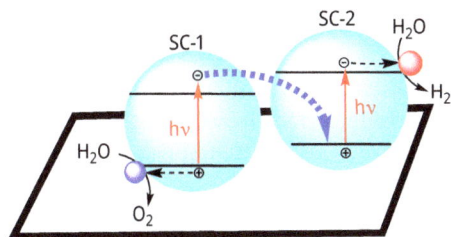

Scheme 5.23: Two-particle photocatalyst on a conductive *reduced graphene oxide* sheet. SC-1 = $BiVO_4/CoO_x$, SC-2 = $CuGaS_2/Pt$. Redrawn after ref. [59].

Reduction and oxidation of water

To find the most active redox catalysts for both HER and OER, the two reactions are conducted separately in the presence of a reducing or oxidizing agent such as sulfites, alcohols, amines or silver nitrate, and potassium peroxodisulfate, respectively.[28] Since the amounts of products are usually on a micro- or millimole scale, the experiments are performed in D_2O or $H_2{}^{18}O$ to prove the origin of the evolved gases and the formation of D_2 or $^{36}O_2$ is usually taken as proof for overall water reduction and oxidation, respectively. However, this is a necessary but not sufficient criterion since water may be reformed during the reaction path (see also Section 5.1.6). While at early reaction stages D_2 is formed, prolonged irradiation also generates HD and H as summarised for ethanol as reductant in Equations 5.31–5.34. Although water is consumed according to Equation 5.34, it is reformed through reaction steps according to Equations 5.31–5.34 ($2\,H^+ + 2\,OD^-$) and the overall stoichiometry corresponds to Equation 5.35 which is the dehydrogenation of ethanol and not reduction of water.

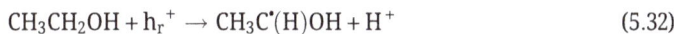

$$SC + h\nu \rightarrow SC + e_r^- + h_r^+ \tag{5.31}$$

$$CH_3CH_2OH + h_r^+ \rightarrow CH_3C^{\bullet}(H)OH + H^+ \tag{5.32}$$

$$CH_3C^{\bullet}(H)OH \rightarrow CH_3CHO + e_r^- + H^+ \tag{5.33}$$

$$\underline{2\,D_2O + 2\,e_r^- \rightarrow 2\,D_2 + 2\,OD^-} \tag{5.34}$$

$$CH_3CH_2OH \rightarrow 2\,H_2 + CH_3CHO \tag{5.35}$$

While two electrons are involved in the HER, four electrons must be released in OER. This multi-electron reaction is kinetically very demanding and, like in photosynthesis, is still a topic of intense mechanistic studies [60]. In the separately conducted OER, relatively strong oxidants such as $K_2S_2O_8$, $(NH_4)_2Ce(NO_3)_6$, $AgNO_3$ having standard potentials of 2.0, 1.7 and 0.8 V, respectively, are therefore required.

Another approach is to attach molecular catalysts to semiconductor particles. The molecular catalysts are usually modified with carboxylic acid or phosphonic acid

---

28 The reagents are termed "sacrificial" reductants or oxidants. It would be much clearer just to name these reactions "reduction" or "oxidation" of water.

groups that react with the semiconductor surface. The semiconductor acts as the light absorber and transfers a hole or electron to the catalyst extremely rapidly. It is also possible to combine photochemistry and electrochemistry by attaching a photosensitiser and a catalyst to an inert transparent electrode surface (see Chapter 3.3). These are important steps toward development of practical devices. With a water oxidation catalyst attached to a transparent electrode as a photoanode and a conventional photocathode for proton reduction, it becomes possible to split water into hydrogen and oxygen. The same principles can be used to make a complete dye-sensitised photo-electrochemical device for water splitting consisting of a photoanode with a molecular oxygen-evolving catalyst attached to one semiconductor and a photocathode with a hydrogen evolving catalyst attached to another semiconductor. Zhang and Sun reviewed such devices and summarise criteria for assessing their effectiveness [23a]. Further information may also be found in two critical summaries of photochemical water splitting [47a, 53b].

5.3.2.2 Aerial oxidation reactions

We mentioned earlier that the oxidation of water to oxygen is the more demanding step in water splitting since a four-electron IFET is involved. In the presence of impurities which are more easily oxidised via one- or two-electron processes, water oxidation is inhibited and the organic impurity RCH_3 is completely oxidised (Scheme 5.24). It is recalled that reactive holes in almost all oxidic semiconductors have very high oxidation potentials in the range of about 2–3 V (see Figure 3.17). In the presence of air, the reactive electrons reduce oxygen to the superoxide radical (Figure 4.12).[29]

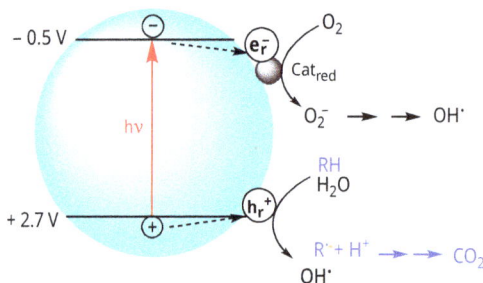

Scheme 5.24: Water splitting and aerobic photooxidation of RH at titania. RH: Organic molecule with functional groups such as OH, CO_2H, NH_2 and SH. In many cases the redox catalyst Cat_{red} is not essential but usually it enhances the degradation rate.

---

29 We prefer to write O_2^- instead of $O_2^{-\bullet}$ since the latter formula suggests that the unpaired electron is localised, i.e. the anion should have a strong radical character. This is not the case, since the electron is delocalised over both atoms in an MO of π^* O-O character. For the chemical reactivity of superoxide see ref. [61].

$$O_2 + e_r^- \longrightarrow O_2^{\cdot -} \tag{5.36}$$

$$O_2^{\cdot -} + H_2O \longrightarrow HO_2^{\cdot} + OH^- \tag{5.37}$$

$$HO_2^{\cdot} + RCH_3 \longrightarrow RCH_2^{\cdot} + H_2O_2 \tag{5.38}$$

$$H_2O_2 + e_r^- + H^+ \longrightarrow OH^{\cdot} + H_2O \tag{5.39}$$

According to the standard potential at pH 7 of $E°(O_2(aq)/O_2^{\cdot -}) = -0.18$ V [62] this IFET (Equation 5.36) has a significantly higher driving force than water reduction. Superoxide is protonated to the hydroperoxyl radical (Equation 5.37) which efficiently abstracts a hydrogen atom from RCH_3 generating an alkyl radical and hydrogen peroxide (Equation 5.38). The latter can be easily reduced by another electron generating an $OH^{\cdot}$ radical (Equation 5.39). Thus, in the presence of oxygen the reductive primary step generates the hydroxyl radical, a strong oxidant. $OH^{\cdot}$ radicals may abstract hydrogen from the substrate or add to double bonds present in the group R. Intermediate $OH^{\cdot}$ and $HO_2^{\cdot}$ radicals may escape to the gas phase and enable remote oxidations [63].

When TiO_2 is replaced by WO_3 the activation of oxygen through a one-electron IFET is not possible since the quasi-Fermi level lies below the $O_2/O_2^{\cdot -}$ potential (Scheme 5.25). However, grafting with copper(II) generates a surface state promoting the two-electron IFET to hydrogen peroxide.[30] Similar results are known for platinum-loaded tungsten oxide [65].

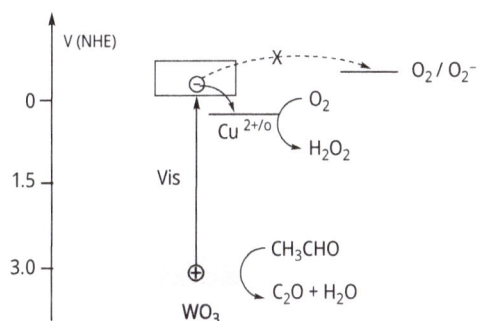

Scheme 5.25: Cu(II) on the WO_3 surface generates a thermodynamically favourable $2e^-$-reduction of oxygen.

In the first oxidative reaction step a further alkyl radical is produced through a dissociative IFET (Equation 5.40). Addition of oxygen generates a peroxyl radical (Equation 5.41) which in turn abstracts hydrogen forming alkylhydroperoxide and a further substrate radical (Equation 5.42). Dimerisation and O–O cleavage lead to the first sta-

---

30 A suspension of WO_3 powder is impregnated with 0.05% of $CuCl_2$ and heated to 260 °C followed by calcination at 650 °C [64].

ble intermediates (Equations 5.43 and 5.44). They are finally completely oxidised to carbon dioxide and water. Potentials of 1.90 V (OH$^{\bullet}$/OH$^-$),[31] 1.29 V (H$_2$O$_2$/H$_2$O) and 0.94 V (O$_2^-$/H$_2$O$_2$) illustrate the oxidising power of some key intermediates.[32] It is therefore not surprising that almost any pollutant can be oxidised, especially at oxidic semiconductors.

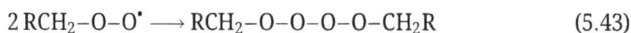

$$RCH_3 + h_r^+ \longrightarrow RCH_2^{\bullet} + H^+ \tag{5.40}$$

$$RCH_2^{\bullet} + O_2 \longrightarrow RCH_2-O-O^{\bullet} \tag{5.41}$$

$$RCH_2-O-O^{\bullet} + RCH_3 \longrightarrow RCH_2-O-OH + RCH_2^{\bullet} \tag{5.42}$$

$$2\,RCH_2-O-O^{\bullet} \longrightarrow RCH_2-O-O-O-O-CH_2R \tag{5.43}$$

$$RCH_2-O-O-O-O-CH_2R \longrightarrow RCH_2OH + RCHO + O_2 \tag{5.44}$$

In the *absence* of oxygen, other acceptors such as organic chlorides and bromides compounds can also be reduced [68]. When their reduction potential is equal or less negative than that of the reactive electron, a dissociative IFET according to Equation 5.45 generates a radical, in the case of CBr$_4$ tribromomethyl. If cyclohexane is also present, it is oxidised by the reactive hole via proton-coupled hole transfer affording a cyclohexyl radical which in turn generates a new alkyl radical (Equations 5.46, 5.47).

$$CBr_4 + e_r^- \rightarrow CBr_3^{\bullet} + Br^- \tag{5.45}$$

$$C_6H_{12} + h_r^+ \rightarrow C_6H_{11}^{\bullet} + H^+ \tag{5.46}$$

$$C_6H_{11}^{\bullet} + CBr_4 \rightarrow C_6H_{11}Br + CBr_3^{\bullet} \tag{5.47}$$

Alkyl radicals, can induce a radical chain reaction. Termination occurs by C–C coupling between tribromomethyl and cyclohexyl radicals. Under appropriate experimental conditions the reaction may proceed further when the light is turned off. It is then called a *photoinitiated* (photo-induced) *chain reaction,* started by a semiconductor.

The literature covers the exhaustive aerial photo-oxidations of pollutants including cyanide, NO$_x$, thiols, phenol derivatives, pharmaceuticals, dyes and polymers. Based on the high oxidation potentials of intermediary oxygen radicals, semiconductor surfaces also exert antibacterial activity [69]. Special attention has been given to visible light active photocatalysts such as doped or modified titania. The latter are prepared by using titania OH groups as a ligand in a coordination complex (see Chapter 3.5) [70]. It was proposed that the slightly yellow powder 4.0% PtCl$_4$–O–TiO$_2$ generates a Pt(III) centre and an adsorbed chlorine atom upon daylight absorption (Scheme 5.26),[33] which oxi-

---

31 At pH = 0 the potential is 2.6 V, see ref. [66].

32 Unless mentioned otherwise, all electrochemical potentials quoted here apply to pH 7 relative to NHE as defined by ref. [67].

33 This is evidenced by the formation of chlorophenols during oxidation phenol degradation.

dises D with reformation of chloride (Scheme 5.26, process B). In agreement with this proposal unmodified titania does not photocatalyse exhaustive oxidation of the chlorinated herbicide atrazine but modified titania does.[34]

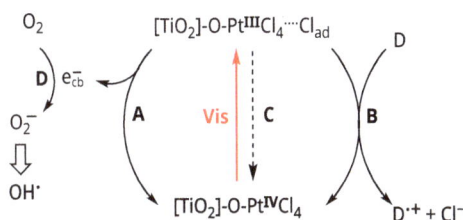

O_2 [TiO$_2$]-O-Pt^{III}Cl$_4$····Cl$_{ad}$ D

D) e$_{cb}^-$

O_2^- A   Vis   C   B

⇩

OH$^•$ [TiO$_2$]-O-Pt^{IV}Cl$_4$ D$^{•+}$ + Cl$^-$

Scheme 5.26: Proposed photocatalysis mechanism at a transition metal complex with titania as a ligand.

Assuming that Pt(III) injects an electron into the titania conduction band, Pt(IV) is reformed and oxygen reduced to superoxide (Scheme 5.26, processes A and D). The latter is final converted to an OH$^•$ radical (see Equations 5.35–5.39). Charge recombination competes with these reaction steps (process C) [71].

The unique properties of photoexcited semiconductor surfaces to generate reactive oxygen species (ROS) offer a variety of commercial applications (see Chapter 5.1.6).[35] Classical air pollutants such as NH_3, H_2S, SO_3, NO_x and HCHO are completely oxidised to sulfate, nitrate and carbon dioxide. A significant example is the sail-like titania-coated facade of the church "Dives in Misericordia" in Rome: by solar light to purity – science meets religion (Figure 5.9). In more mundane applications photocatalysts can be deposited onto glass, asphalt, concrete and tiles. The generated ROS even induce bactericidal and antiviral properties (see Chapter 5.3.4).

5.3.2.3 Nitrogen fixation

Nitrogen fixation is the second most important chemical process in nature after photosynthesis.[36] Unlike photosynthesis, this reduction of aerial nitrogen to ammonia is a dark reaction. It is catalysed by the molybdenum and iron containing enzyme nitrogenase in the presence of adenosine triphosphate, generated by photosynthesis according to Equation 5.48 [72]. Unlike the Haber-Bosch process, nitrogenase catalyses the reductive reaction of N_2 with protons and does not consume H_2 but generates it.

---

34 Intermediate chlorine atoms are also invoked in the visible-light C–H activation of hydrocarbons by surface-chlorinated BiOBr/TiO$_2$.

35 This property is inhibited in titania powders present in everyday products through surface inactivation by alumina.

36 The Haber-Bosch process, $N_2 + 3 H_2 \rightarrow 2 NH_3 + 22.1$ kcal, the most important process of chemical industry is conducted at 400 °C and 200 bar of pressure.

Figure 5.9: Solar photocatalysis in action: Church "Dives in Misericordia", Rome, architect Richard Meier. Reproduced with permission © Frener & Reifer.

$$N_2 + 8\,H^+ + 8\,e^- + 16\,MgATP \;\longrightarrow\; 2\,NH_3 + H_2 + 16\,MgADP + 16\,P_i \tag{5.48}$$

G. N. Schrauzer reported in 1977 that titania loaded with 0.2 wt% of Fe_2O_3 photocatalyses formation of about 10 micromoles of ammonia when irradiated with UV light [73]. Experiments with $^{15}N_2$ proved that ammonia originated from dinitrogen. Water acted as reducing agent leading in the first reaction step to the two-electron reduction to diimine (1,2-diazene) (Equations 5.49, 5.50) Since diimine is a strong reductant it should not be further reduced but may rather disproportionate to hydrazine and dinitrogen (Equation 5.51). The former is finally reduced to ammonia (Equation 5.52).

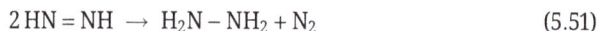

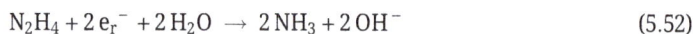

$$2\,N_2 + 4\,H_2O + 4\,e^- \;\longrightarrow\; 2\,HN = NH + 4\,OH^- \tag{5.49}$$

$$2\,H_2O \;\longrightarrow\; O_2 + 4\,H^+ + 4\,e^- \tag{5.50}$$

$$2\,HN = NH \;\longrightarrow\; H_2N - NH_2 + N_2 \tag{5.51}$$

$$N_2H_4 + 2\,e_r^- + 2\,H_2O \;\longrightarrow\; 2\,NH_3 + 2\,OH^- \tag{5.52}$$

To obtain some information on the role of iron(III) in the photocatalytic action, thin layers of iron titanates (about 300 nm thickness) were prepared on glass and irradiated with visible light. It turned out that a thin film of $Fe_2Ti_2O_7$ on glass photocatalyses visible light reduction to ammonia by ethanol, followed by its oxidation to nitrate through traces of air [74]. Product amounts were in the range of 20–30 micromoles and the reaction proceeded also with aerial nitrogen. Detailed work led to the mechanism depicted in Scheme 5.27. The first step is a two-electron reduction of N_2 to 1,2-diazene. While only one reactive electron is generated in the conduction band through absorption of one photon, the second electron originates from the hydroxyethyl radical, which injects an electron into the conduction band. Accordingly, reducing agents which do not have this photoredox amplification property (see Chapter 2.3.2.4),[37] do not enable the overall reaction. Disproportionation of diazene to hydrazine and a second photoreduction step afford ammonia which is oxidised in the presence of $Fe_2Ti_2O_7$ in the dark to nitrate, the final product. In summary, the semiconductor thin film exhibits dual catalytic activity, both in the presence and absence of light.

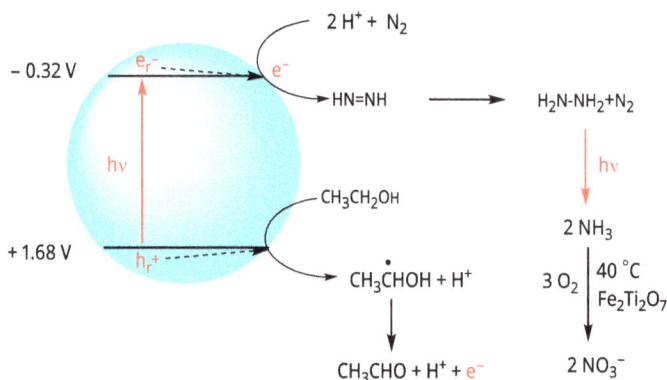

Scheme 5.27: Dual visible light photofixation of aerial nitrogen to ammonia and nitrate.

In the last ten years photocatalytic nitrogen fixation experienced a tremendous new interest [75]. Although considerable increases in photocatalytic activities expressed as quantum yields or chemical yields, based on turnover number per gram of photocatalyst, were reported, a reliable quantitative comparison is not possible as explained in Chapter 3.4. Even when assuming the reported values may be comparable, conclusions about the nature of the rate determining step can very rarely be drawn. This is because measurement of physical properties such as photocurrent and emission spectra are often not conducted under the experimental conditions of the photocatalytic reaction. Quite often emitting and reacting surface charges may not be identical (see Chapter 3.4).

   A central topic of present research is to decrease charge recombination by the rational design of heterojunction photocatalysts [76]. A good example is the organic polymer carbon nitride (C_3N_4, see Scheme 3.5) covalently attached to another inorganic or organic semiconductor. C_3N_4 nanotubes decorated with 20 nm $NaYF_4$:Yb,Tm nanoparticles, enables ammonia formation to occur even with NIR light using the upconversion effect [77]. The high photocatalytic activity of a heterojunction between protonated C_3N_4 and reduced graphene oxide seems to originate from efficient suppression of charge recombination [78]. Surface defects often control charge recombination [79] in a similar way to heterogeneous catalysis in the dark. Metal organic

---

37 Amines, formic acid, oxalates and humic acids also possess this property.

frameworks (MOFs)[38] may offer new design possibilities as one part of a heterojunction [80].

All this work is focused on ammonia formation. Besides its use as a fertiliser, ammonia is the source of nitrogen in all organic nitrogen compounds. The key step in corresponding syntheses is C–N coupling. So, a simple question arises, namely if C–N bond formation may be accomplished directly from N_2 and organocarbon compounds without reduction of N_2 to ammonia. A very rare example is known from coordination chemistry as summarised in Equation 5.53. Irradiation of the tungsten(0) complex (L = 1,2 bis(diphenylphosphino)ethane) in the presence of chlorocyclohexane (R = cyclohexyl) affords N_2 and coordinated RCl. Subsequently, an inner-sphere electron transfer generates W(I) and the intermediate alkylhalogenide radical anion which decomposes to chloride and a cyclohexyl radical. The latter undergoes radical C–N bond formation with the N_2 ligand generating a cyclohexyl diazenido ligand (Equation 5.53) [81]. As organic radicals are key intermediates in many semiconductor

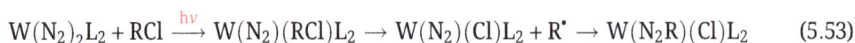

$$W(N_2)_2L_2 + RCl \xrightarrow{h\nu} W(N_2)(RCl)L_2 \rightarrow W(N_2)(Cl)L_2 + R^{\bullet} \rightarrow W(N_2R)(Cl)L_2 \qquad (5.53)$$

photocatalysed reactions, dialkylation to 1,2-diazenes (R–N=N–R) intermediates in the presence of N_2 seems not unlikely. It is known that transition metal catalysts can activate cis-1,2-diazenes to undergo further C–N bond formation [82]. Future research along these lines seems recommendable. In the absence of light, C–N bond formation requires the use of strong reductants such as lithium or sodium for the initial N_2 activation step [83].

5.3.2.4 Carbon dioxide fixation

In the search for renewable fuels and carbon feedstocks for the chemical industry carbon dioxide fixation has become an important topic in basic research (see Chapter 5.1.7). Solar reduction of CO_2 and water to methanol would be a promising counterpart to solar hydrogen production from water [84]. Methanol is much easier to handle than hydrogen and, like the latter, the starting materials are reformed upon use as a fuel. The mechanistic Scheme 5.28 displays some typical photocatalysts and reductants reported in the literature [85]. C1 products like formic acid, carbon monoxide, methanol and even methane dominate the product spectrum while C2 compounds such as oxalate, ethanol and acetate are very rare [86]. There is so much research that reviews report separately on photocatalytic reduction to formic acid [87], reduction in the presence of quantum dots [88], in combination with water oxidation [89], and with metal-organic frameworks [90].

Comparison of the conduction band edges of various semiconductors (see Figure 3.17) with some standard potentials for CO_2 reduction reveals that one-electron re-

---

38 MOFs are organic-inorganic hybrid crystalline porous materials that consist of an array of metal ions surrounded by organic "linker" molecules.

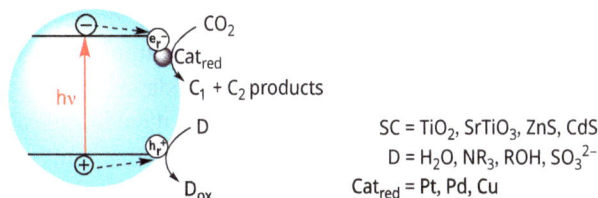

Scheme 5.28: Semiconductor photocatalysed carbon dioxide reduction.

duction requires at least −1.9 V, a potential reached perhaps by zinc sulfide. However, a proton coupled two-electron process requires only −0.61 V (Table 5.1). Such a reaction should be favoured for reducing agents with redox amplification properties. It is likely that oxalate is not produced via dimerisation of an intermediary carbon dioxide radical anion but by hole oxidation of the two-electron reduction product formate.

An alternative could be reduction to methane which only requires a potential of 0.24 V (Table 5.1); however, it is kinetically highly unfavourable requiring the transfer of eight electrons.

As for water splitting, the CO_2 reduction catalysts in Figure 5.4 have been attached to semiconductor particles and electrodes for photoelectrochemistry [23a, 33]. It is also possible to attach enzymes that are specific to CO_2 reduction such as tungsten formate dehydrogenase that reduces CO_2 to formate [91].

For estimates of the thermodynamics, the pH value determines the species present in solution (Equations 5.54, 5.55). Furthermore physi- and chemisorption with surface OH and SH groups may strongly influence the IFET.[39] Depending on the pH value, carbon dioxide is also present in aqueous solution as hydrogen carbonate and carbonate. Thus, in addition to physisorption, chemisorption may also occur at an oxidic or sulfidic semiconductor surface by condensation of hydrogen carbonate. Details of the interfacial electron transfer steps and general mechanisms were recently summarised [85b, 93].

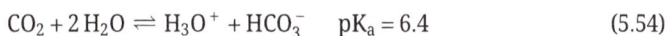

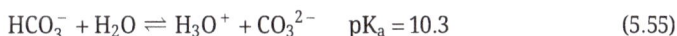

$$CO_2 + 2\,H_2O \rightleftharpoons H_3O^+ + HCO_3^- \quad pK_a = 6.4 \tag{5.54}$$

$$HCO_3^- + H_2O \rightleftharpoons H_3O^+ + CO_3^{2-} \quad pK_a = 10.3 \tag{5.55}$$

In photoelectrochemistry complete cells were constructed in which CO_2 reduction occurs at the photocathode and water is oxidised to oxygen at the photoanode. Designs leading to CO, CO/H_2 and, more remarkably, a mixture of ethanol and 1-propanol have been reported [33, 94]. By using the tungsten enzyme mentioned above, formate has been formed selectively [91]. Reactions with both reduction and oxidation of substrates like this are denoted as artificial photosynthesis. Alternatively, the devices may be called artificial leaves.

---

39 Depending on the detailed structure of the surface-CO_2 complex binding energies of ~17 kJ mol^{-1} for monodentate and ~170 kJ mol^{-1} for bidentate coordination were calculated for the CdS surface [92].

Like photosynthesis, C–C bond formation to generate valuable organic compounds, perhaps in combination with nitrogen fixation, seems to be a promising future research area. Recently it was reported that the allyl radical generated by hole oxidation of acetyl-acetone undergoes C–C coupling with a reductively formed carbon dioxide radical anion [95]. Two very recent papers demonstrate an inspiring C–C coupling that converts CO_2 to acetate using a semiconductor-biohybrid with the bacterium *Sporomusa ovata* [96a,b]. One of them uses CdS as the semiconductor and shows photocatalysed reduction to formate followed by thermal enzyme-catalysed conversion to acetate. The other uses a two-particle photocatalyst (Scheme 5.23) that combines the reduction of CO_2 to acetate with oxidation of water to O_2.

One disadvantage of linking reduction of CO_2 to water oxidation is that it is very difficult to oxidise water, yet the product, O_2, is unwanted! An alternative is to couple CO_2 reduction to oxidation of a waste material. As an example, a device has been designed with a cobalt porphyrin-modified cathode to reduce CO_2 and a photocathode that converts polyethylene terephthalate (PET) to glycollate, $HOCH_2CO_2^-$ [96c]. This cell is envisaged as a means of recycling PET waste.

5.3.2.5 C–C and C–N coupling

Semiconductor photocatalysed organic transformations can be classified as type A reactions in almost all cases. Examples are cis-trans isomerisations, valence isomerisations, substitution and cycloaddition reactions, and reductions. Further examples are oxidative C–N cleavage, intra- and intermolecular C–C and C–N couplings.[40] The situation in which the primary products undergo intermolecular bond formation is also of interest (Scheme 5.29). In the proton-coupled reductive and oxidative IFETs the radicals **12a** and R· are generated. Successive radical C–C heterocoupling affords the product **14**. The overall reaction is a decarboxylative hydroalkylation of an alkene by a carboxylic acid [97].

In all the cases summarised above, two or more well-known products were formed and usually identified spectroscopically but not isolated. One reason is that the photocatalysts are deactivated through photocorrosion on prolonged irradiation. This is especially true for colloidal metal sulfides of zinc and cadmium [98]. In the next section, we discuss a C–C coupling reaction that represented the first synthesis of a previously unknown compound on a preparative scale. Unlike homogeneous photoredox catalysis, product isolation is simple since the photocatalyst powder can easily be removed by filtration.

---

40 For a summary see ref. [82].

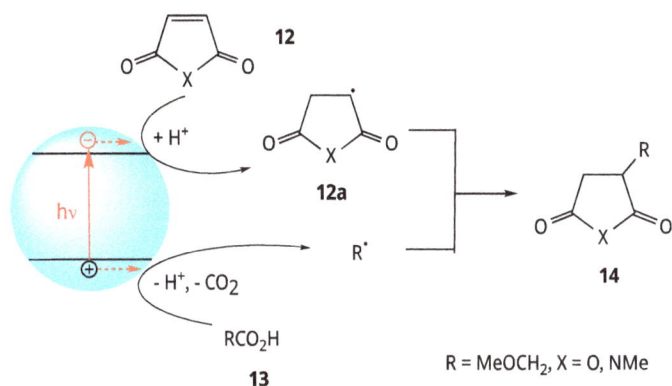

Scheme 5.29: Decarboxylative hydroalkylation of an electrophilic alkene.

Dehydrodimerisation of cyclic olefins

Irradiation of ZnS powder or platinised CdS (Pt/CdS) suspended in an aqueous 2,5-dihydrofuran (2,5-DHF) suspension with UV or visible light, respectively, affords a few litres of hydrogen and gram-amounts of novel dehydrodimers in isolated yields of 60% (Scheme 5.30). Analogous products are obtained from 3,4-dihydropyran, 3-methyl-2,3-dihydropyran and cyclohexene in isolated yields of 30–60%. Initially evolved hydrogen contains about 90% of D_2 when D_2O is employed, but drops to 40% after evolution of one litre of gas, whereas the sum of HD and H_2 increases from 10% to 60%. Surprisingly, no water is consumed and no hydrogen or dehydrodimers are formed in the absence of water. Colloidal zinc or cadmium sulfide and high purity single crystals do not catalyse the reaction [99].

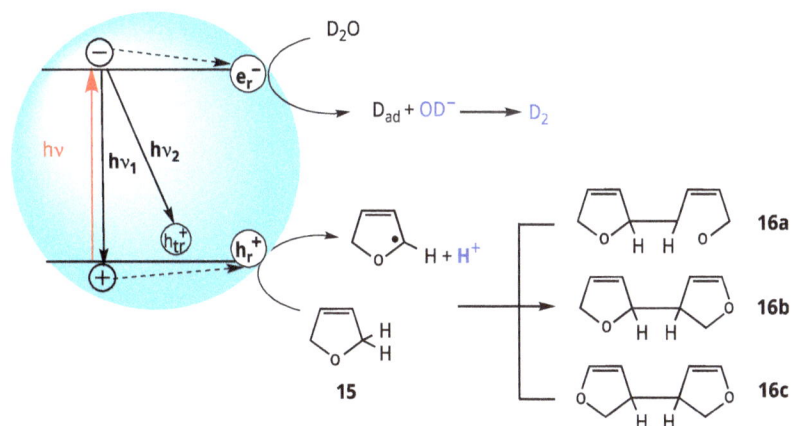

Scheme 5.30: Anaerobic dehydrodimerisation of 2,5-dihydrofuran in D_2O. Band edges at pH 7 are positioned at –1.8 V and +1.8 V or –0.9 V and +1.5 V for ZnS and CdS, respectively.

The mechanistic proposal is depicted within the scheme of semiconductor photocatalysis Type A since reduced (H_2) and oxidised products (**16**) are formed (Scheme 5.30). The light-generated electron-hole pair has a lifetime of 0.1–20 nanoseconds and may recombine or be trapped at emitting (e_{tr}^-, h_{tr}^+) and reacting (e_r^-, h_r^+) surface sites. At the latter, proton-coupled IFET affords allyl radicals (Equation 3.12) that in principle could disproportionate, add to double bonds, undergo electron transfer, or dimerise, as known from their chemistry in homogeneous solution. Surprisingly, according to a complete material balance, dimerisation to regioisomers **16a–c** is observed to about 90%. This high chemoselectivity suggests that C–C coupling does not occur between fully solvated radicals but in the H_2O–2,5-DHF–surface layer (see Figure 3.16). Competition experiments with tetrahydrofuran (THF) support this possibility. Whereas THF reacts only ten times more slowly than 2,5-DHF, THF dehydrodimers or cross-products are not observable when THF is present in tenfold excess. Thus, the solid–liquid interface induces an unexpected chemoselectivity for radical C–C coupling. Adsorption studies reveal that at low concentrations 2,5-DHF is adsorbed in a mixed water-2,5-DHF- surface monolayer. Each hydrated zinc site is occupied by one 2,5-DHF molecule through hydrogen bonding (see Figure 3.16). From Scheme 5.30 it follows that although water is reduced it is reformed in the reductive and oxidative IFETs (see Equations 3.15–3.18). Accordingly, formation of D_2 from D_2O in "sacrificial" hydrogen producing systems is a necessary but not sufficient criterion for "permanent" water reduction.

As discussed in Chapter 3.1.4, emission spectroscopy of zinc sulfide powder suspended in water reveals the presence of two broad bands originating from band-to-band and band-to-surface state emissions at 366 nm ($h\nu_1$ in Scheme 5.30) at 430 nm, ($h\nu_2$ in Scheme 5.30). Addition of zinc or cadmium sulfate or the substrate 2,5-DHF does not significantly alter the two emission bands (Figures 3.5, 3.6), but, unlike emission, product formation is inhibited strongly when cadmium or zinc salts are added. *Therefore, emitting and reacting electron-hole pairs are different.* A plot of relative rates vs. inhibitor concentration (Stern-Volmer plot, see Figure 2.10) gives a straight line only if the concentration of *adsorbed scavenger ions* is plotted on the abscissa and not the concentration in solution. This suggests that the scavenger ions move in a pseudo-homogeneous water-2,5-DHF-surface layer [98].[41] Analogous mechanisms were reported for the visible light homo- and hetero-coupling of benzyl alcohols and benzyl amines at cadmium sulfide [100].

5.3.2.6 Indirect photocatalysis
Most authors generally anticipate a direct photocatalysis mechanism for their reactions although the direct mechanism is often claimed without reliable experimental evidence. Quite commonly, however, substrates or their surface-CT complexes absorb

---

41 The Stern-Volmer equation depends on the assumption of diffusion-controlled processes.

the light in preference to the semiconductor. This is especially true for the visible light degradation of dyes.[42] When the excited state potential of RH is equal or more negative than the conduction band edge, an IFET generates a reactive electron followed by oxygen reduction (Scheme 5.31). Thus, the primary products are the same as in aerial oxidations via a direct photocatalysis mechanism (see Scheme 5.20 and Chapter 5.3.2), but in the indirect case the semiconductor acts as an electron relay, channelling the electron to the oxygen adsorption site and therefore reducing the energy-wasting back-electron transfer. Finally, the primary product R˙ is completely oxidised (see Equations 5.40–5.44).

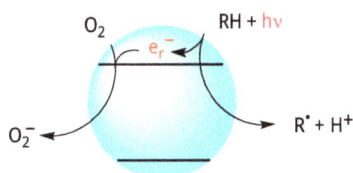

Scheme 5.31: Primary processes in the indirect photocatalysis of an aerial oxidation reaction.

5.3.3 Type B reactions

In recent synthetic organic chemistry, homogeneous photocatalysis has become an important topic (see Chapter 5.1) [101]. In contrast, cleavage and exhaustive aerial oxidations have been at the centre of research in heterogeneous photocatalysis at semiconductor surfaces. After understanding the basic mechanism of the dehydro-dimerisation described above, it turned out that semiconductor photocatalysis may also enable synthesis of novel organic compounds [82].

5.3.3.1 Insertion of 1,2-diazenes and imines into allylic C-H bonds

Novel addition products are obtained on replacement of water as the electron acceptor in the dehydrodimerisation by unsaturated compounds such as 1,2-diazenes **17** or imines **22** (Schemes 5.32 and 5.33). Whereas in the oxidative IFET the same allyl radical is formed as in the dehydrodimerisation, the reductive IFET generates a hydrazyl radical from diazenes. These proton-coupled electron-transfer steps require a protic solvent such as methanol. No product formation is observed in acetonitrile. Chemoselective C–N heterocoupling generates the addition products **19**, isolated in moderate yields of about 40%. Their formation is a $1e^-/1h^+$ process, while the by-products **20** and **21** are generated by a $2e^-/2h^+$ process.

The general applicability of this linear addition reaction is demonstrated by the insertion of ketimines or aldimines into allylic C–H bonds forming novel homoallyl-

---

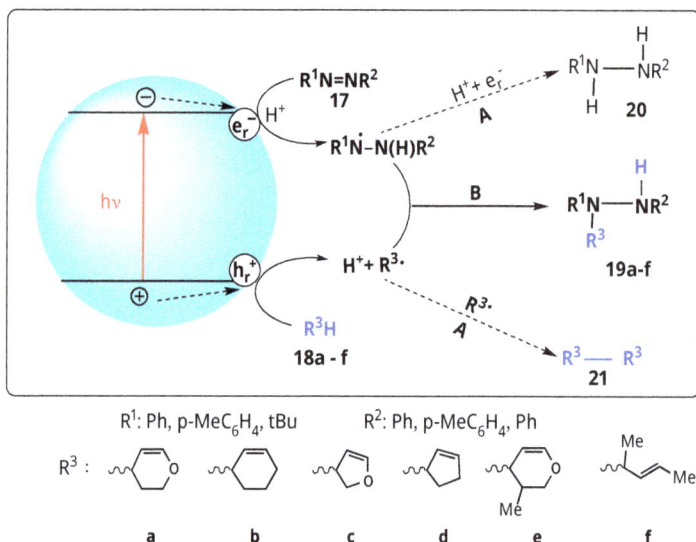

Scheme 5.32: CdS or ZnS photocatalysed insertion of 1,2-diazenes into allylic C-H bonds.

amines **23** in isolated yields of 40–70% (Scheme 5.33). In the case of X = COOR these products may be hydrolysed to unsaturated amino acids of pharmacological interest [102]. While the oxidative IFET is the same as in the dehydrodimerisation, the reductive, proton-coupled IFET affords the α-aminobenzyl radical ArC•(X)–N(Ar)(H).

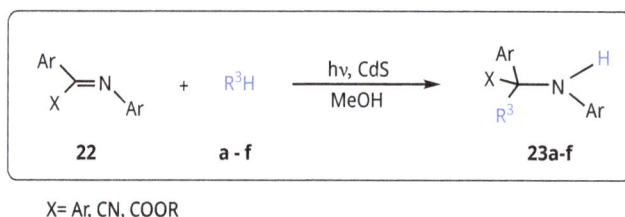

X= Ar, CN, COOR

Scheme 5.33: CdS photocatalysed insertion of ketimines into allylic C-H bonds.

The textbox exemplifies the procedure for the insertion of N-phenylbenzophenone imine into the allylic C-H bond of α-pinene.[43]

---

[43] In these reactions the photocatalyst must be covered by at least one layer of water, otherwise the protonated imine must be used. It is also important to prepare CdS under inert gas.

Semiconductor Photocatalysis Type B – simple and atom-economic
Insertion of C=N into an allylic C–H bond

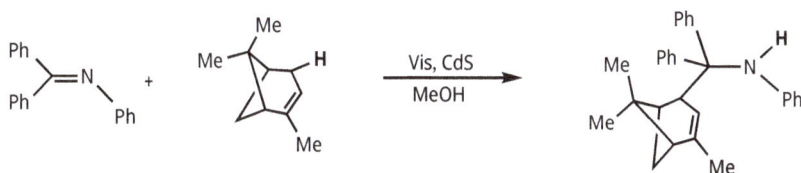

1.55 g of N-phenylbenzophenone imine and 36 g of α-pinene are added to a suspension of 0.30 g of CdS in 200 mL of MeOH. After irradiating for 20 h with a tungsten halogen lamp, removing the solvent and recrystallizing the residue from n-heptane 1.64 g (72%) of the colourless product is obtained.

5.3.3.2 Insertion of oxygen and sulfur dioxide into alkanes

Functionalisation of alkanes is one of the great challenges in chemistry [103]. The sulfoxidation of liquid alkanes by sulfur dioxide and oxygen through UV irradiation is a rare example of an industrial photoreaction (Chapter 4.2.6). In the process, C_{16-20} alkanes are converted into linear sulfonic acids that are excellent biodegradable surfactants.

Surprisingly, titania photocatalyses sulfoxidation by visible light ($\lambda \geq 400$ nm) [42a]. The reason is that SO_2 forms a yellow surface CT complex, exhibiting a broad absorption at 410–420 nm. Visible excitation according to Scheme 5.22 generates a reactive electron $TiO_2(e_r^-)$ and an adsorbed sulfur dioxide radical cation (Scheme 5.34). The latter oxidises the alkane in a proton-coupled step to an alkyl radical (Scheme 5.34, reaction path 1). Successive C–S bond formation induces sulfonic acid production according to Scheme 4.12. Alkyl radicals may also be formed through hydrogen abstraction by hydroperoxyl radicals, obtainable via the reduction of O_2 by $TiO_2(e_r^-)$ and subsequent protonation by water or surface OH groups (Scheme 5.34, reaction path 2).

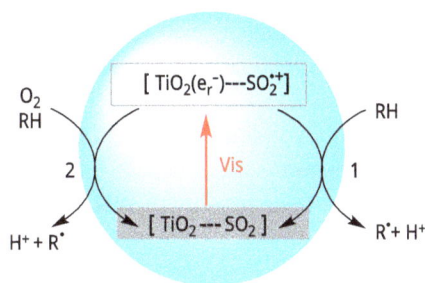

Scheme 5.34: Proposed mechanism for visible light generation of alkyl radicals, R = *n*-heptyl, 1-adamantyl via sulfoxidation with titania photocatalysis.

This visible-light induced C(sp^3)–H activation can be classified as *semiconductor photocatalysis type B*, extending the two-substrate addition to a three-substrate addition scheme (Equation 5.30).

5.3.4 Environmental aspects

In the chapters above we showed that semiconductor photocatalysis is based on the generation of strongly oxidising and reducing surface sites. It was mentioned briefly that the detailed surface properties such as the amount of hydration and hydroxylation may exert a strong influence. In the following, we give a short overview of practical aspects relevant in everyday life [104]. A recent book summarises the present state of commercial applications [105], and there is even a small museum at Tokyo University of Science [106].

5.3.4.1 Amphiphilic properties of titania

Surface wettability of solids plays an important role in industrial and biological processes. The most important example is titania, since in addition to its photoredox activity, it may also have superhydrophobic or superhydrophilic properties[44] [107]. When thin films deposited on glass are exposed to daylight they exhibit an antifogging effect [53b]. In the dark the superhydrophilicity disappears but is renewed upon irradiation. Detailed studies on the influence of light intensity, wavelength dependence, temperature and surface acidity prove the presence of a true photochemical process [108]. The photocatalytic activity of such superhydrophilic layers on glass was also studied [109].

5.3.4.2 Detoxification of air and water

In Chapter 5.3.2 we explained why excited titania and other oxide semiconductors enable complete oxidation of air and water pollutants under ambient aerobic conditions. The final products are carbon dioxide, sulfate, nitrogen, nitrates [110] and water. For air-cleaning a number of commercial products such as tiles, roof tiles, road pavements and wall paints are already on the market.

Cleaning of water is still in the research stage, although almost any pollutant is completely oxidized [111]. Blooming of marine and freshwater algae is also inhibited [112]. An informative summary of ISO[45] standard tests for degradation reactions is available in a review [113].

---

44 Superhydrophilicity refers to the phenomenon of excess hydrophilicity, or attraction to water; in superhydrophilic materials, the contact angle of water is equal to zero degrees. The opposite applies for superhydrophobicity.
45 ISO stands for the International Organization for Standardization.

5.3.4.3 Antiviral and antibacterial effects

The reactive oxygen compounds generated by irradiation of titania and other semiconductor surfaces are active in killing bacteria and viruses. This effect often inactivates bacteria that are resistant to UV irradiation [114]. It works also when the bacterium is separated from the semiconductor by a 50 mm thick porous membrane, due to the *remote oxidation* discussed in Chapter 5.3.2 [63]. Oxidative damage starts at the cell wall and continues in the intracellular area eventually inducing cell death [115]. Even viruses like avian influenza virus [116], HIV virus and cancer cells [117] were deactivated. A $Pt(IV)Cl_4$-grafted titania exhibits activity in photodynamic therapy of mouse melanoma (see Chapter 6) [118].

The lethal effect depends on details of photocatalyst preparation [119]. Microwave synthesis affords a carbon-doped anatase-brookite powder inactivating *Staphylococcus aureus* with visible light [69]. Smaller titania particles cause quicker intracellular damage [115]. Titania-covered cordierite[46] modified by copper(II) improves its antiviral activity in air-cleaners [116].

As practical applications of the photocatalytic "self-cleaning" effect, bactericidal textiles, stainless steel, gypsum-based composites as paints on indoor walls, titania-coated silicone catheters and photocatalytically active building materials have all been mentioned [105]. Solar photocatalytic inactivation of pathogens present in fish aquaculture was also reported [120]. As in abiotic photocatalysis, quantitative comparison of killing efficiencies is almost meaningless due to differing experimental conditions [121].

5.3.4.4 Abiotic photocatalysis

The existence of semiconducting minerals is well known and therefore solar photocatalytic processes, termed geophotocatalysis,[47] may occur at their surfaces. To estimate their thermodynamic feasibility, the flat-band potentials were calculated according to Equation 5.56 [123]. The flat-band potential is based on the electronegativity of the constituent atoms and bandgap. $E_{e,NHE}$ is the energy of a free electron relative to NHE (4.5 eV),[48] and χ corresponds to the geometric mean of the atoms' electronegativities. These values assume an ideal stoichiometric composition of the mineral [124], which is rarely justified.

$$E_{fb} = E_{e,NHE} - \chi + 0.5\,E_g \tag{5.56}$$

Nevertheless, the values in Table 5.2 allow approximate estimates of the thermodynamic feasibility of a desired reaction. According to these values, aerial oxidation re-

---

46 Cordierite is a magnesium iron aluminium cyclosilicate.
47 See also ref. [122].
48 Corresponding to −4.5 V relative to NHE on the potential scale.

actions are especially favoured via both reductive and oxidative primary pathways (see Chapter 5.3.2). Such processes may enable solar self-cleaning of natural waters. Examples are the solar degradation of polyphenols by the mineral birnessite, a manganese (II,III) oxide with small amounts of Na, K and Ca, the degradation of methyl orange by rutile and sphalerite [125], and water oxidation by cerium(IV) nitrate in the presence of hollandite, a barium-manganese manganite [126].

Table 5.2: Approximate properties of semiconducting minerals.[a]

Mineral	Formula	E_g (eV)	λ_g (nm)	E_c (V)	E_v (V)
Sphalerite	ZnS	3.90	319	−1.58	2.32
Tausonite	SrTiO$_3$	3.40	366	−1.26	2.14
Anatase	TiO$_2$	3.20	390	−0.46	2.74
Rutile	TiO$_2$	3.00	414	−0.36	2.64
Ilmenite	FeTiO$_3$	2.80	444	−0.35	2.45
Pyrite	FeS$_2$	0.95	1.31×10^3	0.03	0.98
Hematite	Fe$_2$O$_3$	2.20	565	0.28	2.48
Magnetite	Fe$_3$O$_4$	0.10	1.24×10^4	1.11	1.21

[a] E_g = band gap, λ_g = band gap converted to wavelength, E_c = energy level of conduction band at pH 7, E_v = energy level of valence band at pH 7, adapted from ref. [124a].

5.3.4.5 Abiotic nitrogen fixation

In 1941 Dhar et al. reported that photoreduction of N_2 to NH_3 takes place at the surface of titania minerals. Natural organic compounds or water were assumed as reducing agents [127]. About 30 years later this result was confirmed by Schrauzer *et al.* using $^{15}N_2$ to prove the origin of ammonia. Small amounts of ^{15}N-containing nitrate and nitrite were also detected [73]. Subsequently, desert sands from various parts of the world were tested (Table 5.3). Obviously, a high rutile content favours a higher ammonia yield.

Table 5.3: Origin of sand samples, rutile content and amount of ammonia formed upon solar irradiation.[a]

Site	Rutile (%)	NH$_3$ (nmol)
Imperial Sand Dunes, Imperial Valley (California)	0.051	59
Death Valley Dune, base material (California)	0.036	38
Kuwait desert (Kuwait)	0.011	28
Jumna River near Allahabad (India)	0.019	25
Desert near Cairo (Egypt)	0.030	24
Tengger Desert (China)	0.010	23
Panamint Spring Area, Death Valley (California)	0.016	20

[a]Adapted from ref. [128].

While ammonia yields displayed in Table 5.3 are rather modest, geophotocatalysis may be a relevant factor in natural nitrogen ecology. Assuming the average TiO_2 amount in sands or sandstones of 0.25% [129], solar ammonia production is estimated as about 10 kg of ammonia per 4000 m^2 of desert sand per year. This is about 10% of N_2 reduced biologically [130]. Although speculative, plants in semiarid areas may depend on this non-enzymatic nitrogen. This speculation is supported by the experimental finding that spinach grows faster when impregnated with anatase nanoparticles [131]. For wheat and rape plantlets only root elongation is increased while the plant biomass remains constant [132].

In Chapter 5.3.2 we mentioned that thin films of $Fe_2Ti_2O_7$ catalyse visible light reduction of N_2 by reducing agents such as humic acid with photoredox amplification properties. Since oxidative weathering of $FeTiO_3$ (ilmenite) minerals may generate $Fe_2Ti_2O_7$ surface films, and since humic acids are ubiquitous, a further route to abiotic nitrogen fixation may exist in nature.

5.3.4.6 Titania in food and personal care products

Due to its brightness, high refractive index and resistance to discolouration, titanium dioxide is primarily used as a white pigment.[49] The total global production is in the range of millions of tons per year. Of this, 70% is used as a paint pigment; the rest is contained in glazes, enamels, plastics, paper, fibres, foods, pharmaceuticals, cosmetics and toothpastes. Recently, more attention has been given to the use of titania as a nanomaterial. Its production in 2010 was about 5000 tons and is expected to increase. As a consequence it may accumulate in the human body and environment [133].

The highest titania contents are found in chewing gums and sweets, toothpastes and sunscreens that contain one to two weight percent. Much smaller amounts are present in shaving creams, shampoos and deodorants. Paints contain about 10% of titania. The typical exposure for an US adult is in the order of 1 mg of titanium per kilogram of body weight per day. Children experience a higher exposure because of the higher titania content of sweets [134].

Potential toxicity effects of nanocrystalline titania on the human and animal body is a newly developing field [135]. The presence of anatase or anatase-rutile powders disturbs the structure of porcine skin powders upon exposure to indoor light only slightly.[50] No effect is observable in the case of the rutile [136]. *In vitro* irradiation experiments with commercial sunscreens revealed formation of small amounts of hydroxyl radicals and singlet oxygen [42b].

---

[49] Due to special surface treatments, commercial products exhibit generally very poor photocatalytic activity.

[50] *Porcine skin* serves as a proven model translatable to human skin.

5.3.5 Summary of semiconductor photocatalysis

In contrast to molecular photosensitisers which enable only a single one-electron transfer with one single substrate in the primary step, photoexcited semiconductors also induce two concerted one-electron transfer reactions with two substrates. This difference arises because light absorption generates electron-hole pairs, trapped at distinct surface sites that undergo interfacial electron transfer reactions with donor and acceptor substrates. The photocatalytic properties result not only from light absorption by the semiconductors but also from absorption by complexes between substrates and semiconductors or by the substrates themselves. The primary products are usually radicals, formed by proton-coupled interfacial electron transfer. They undergo chemo- and stereoselective C–C and C–N bond formation resulting in novel organic syntheses. In this sense, the semiconductor photocatalyst functions like an artificial leaf. The high photoredox activity of simple oxide and sulfide semiconductor powders is exemplified by visible light sulfoxidation of alkanes and by the fixation of aerial dinitrogen. Since several minerals are known to have semiconductor properties, solar photocatalysis may be also relevant for prebiotic and environmental chemistry.

Questions

1. What are the three most important primary reactions of a photoexcited homogeneous transition metal catalyst? Give some examples.
2. What product should be formed when $(1,3\text{-butadiene})Cr(CO)_4$ is irradiated in the presence of H_2?
3. The $[Ru(bpy)_3]^{2+}$ photocatalysed addition of an alkyl iodide across the C=C bond of an alkene is initiated by a single electron transfer generating $[Ru(bpy)_3]^{3+}$, iodide and an alkyl radical. The latter undergoes C–C coupling with the alkene generating another intermediate radical. How is the photocatalyst regenerated and the product formed?
4. In asymmetric photocatalysis two basic mechanisms are discussed. One consists of two coupled catalytic cycles. In the thermal cycle (A) the substrate coordinates enantioselectively to the chiral catalyst. The latter undergoes an electron transfer from the substrate to the excited photocatalyst (cycle B). The substrate radical cation rearranges to the product radical cation, which successively is reduced to the product by the photocatalyst radical anion with reformation of the neutral photocatalyst and chiral catalyst. Formulate the combined catalytic cycles.
5. (a) Formulate the net reaction of water splitting and the corresponding reductive (−0.41 V) and oxidative (+0.82 V) partial reactions at pH 7 including the number of exchanged electrons. (b) Give the equations for the reduction and oxidation of

water to hydrogen and oxygen in the presence of a reducing or oxidising agent ("sacrificial" water splitting).

6. (a) The redox potential $Ru^{2+/3+}$ of $[Ru(bpy)_3]^{2+}$ in the ground state is +1.26 V. Why is the excited state capable of reducing water to hydrogen? The MLCT state of the complex is located at 2.1 eV. (b) Discuss the visible light reduction of water by triethanolamine in the presence of $[Ru(bpy)_3]^{2+}$, methyl viologen and colloidal platinum. Explain the function of each component. (c) How may water be reduced without use of an electron relay or colloidal metal?

7. (a) Discuss the general mechanistic schemes of type A and type B semiconductor photocatalysis. (b) How can the thermodynamic feasibility of such a reaction be estimated? (c) What type is analogous to photoelectrochemistry?

8. At pH 7 the conduction band edge of titania is located at −0.5 V, the bandgap has a value of 3.2 eV. (a) At what wavelength does titania start to absorb light? (b) What is the maximum oxidation potential of the light-generated hole?

9. VIS light irradiation of CdS powder in presence of imines and cyclohexene enables insertion of the imine into the allylic C-H bond of the alkene. Explain the mechanism.

10. Identify four half-reactions in this chapter that require proton-coupled electron transfer and transfer of more than one electron. Why are these reactions more difficult to achieve than one-electron transfer reactions? What steps can be taken to achieve proton-coupled electron transfer systematically?

11. For many years, most research on solar energy conversion focused on water splitting to form H_2 and O_2. Identify alternative substrates for reduction and oxidation and explain why such reactions would be useful.

12. Three design features for a solar energy photocatalyst are: (a) it is not photolabile (homogeneous) or liable to photocorrosion (heterogeneous) under irradiation conditions, (b) it absorbs visible light with substantial absorption coefficient, (c) it has an appropriate excited state redox potential. To what extent do the compounds in Figure 5.1 fulfil these requirements? Why are supramolecular photocatalysts such as those in Figure 5.5 thought to be advantageous?

References

[1]   N. A. Romero, D. A. Nicewicz, *Chemical Reviews* **2016**, *116*, 10075–10166.

[2]   P. Melchiorre, *Chemical Reviews* **2022**, *122*, 1483–1484.

[3]   a) L. Buglioni, F. Raymenants, A. Slattery, S. D. Zondag, T. Noël, *Chemical Reviews* **2021**, *122*, 2752–2906; b) R. C. McAtee, E. J. McClain, C. R. Stephenson, *Trends in Chemistry* **2019**, *1*, 111–125; c) L. Candish, K. D. Collins, G. C. Cook, J. J. Douglas, A. Gomez-Suarez, A. Jolit, S. Keess, *Chemical Reviews* **2022**, *122*, 2907–2980.

[4]   a) R. H. Crabtree, *The Organometallic Chemistry of the Transition Metals*, 7th edition John Wiley & Sons, New Jersey, USA **2019**; b) M. Bochmann, *Organometallics and Catalysis: An Introduction*, Oxford

University Press, USA, **2015**; c) J. Kochi, *Organometallic Mechanisms and Catalysis: The Role of Reactive Intermediates in Organic Processes*, Elsevier, **2012**.

[5]   a) P. M. Hodges, S. A. Jackson, J. Jacke, M. Poliakoff, J. J. Turner, F. W. Grevels, *Journal of the American Chemical Society* **1990**, *112*, 1234–1244; b) D. Chmielewski, F. W. Grevels, J. Jacke, K. Schaffner, *Angewandte Chemie International Edition* **1991**, *30*, 1343–1365.

[6]   B. Heller, B. Sundermann, H. Buschmann, H.-J. Drexler, J. You, U. Holzgrabe, E. Heller, G. Oehme, *Journal of Organic Chemistry* **2002**, *67*, 4414–4422.

[7]   a) Y. Wakatsuki, H. Yamazaki, *Journal of the Chemical Society, Dalton Transactions 1978*, 1278–1282; b) H. Boennemann, *Angewandte Chemie* **1985**, *97*, 264–279.

[8]   D. P. Summers, J. C. Luong, M. S. Wrighton, *Journal of the American Chemical Society* **1981**, *103*, 5238–5241.

[9]   N. W. Hoffman, T. L. Brown, *Inorganic Chemistry* **1978**, *17*, 613–617.

[10] M. Schmalzbauer, M. Marcon, B. Konig, *Angewandte Chemie International Edition* **2021**, *60*, 6270–6292.

[11] a) C. K. Prier, D. A. Rankic, D. W. MacMillan, *Chemical Reviews* **2013**, *113*, 5322–5363; b) M. H. Shaw, J. Twilton, D. W. MacMillan, *Journal of Organic Chemistry* **2016**, *81*, 6898–6926.

[12] a) C. J. Wallentin, J. D. Nguyen, P. Finkbeiner, C. R. Stephenson, *Journal of the American Chemical Society* **2012**, *134*, 8875–8884; b) E. Yoshioka, S. Kohtani, T. Jichu, T. Fukazawa, T. Nagai, A. Kawashima, Y. Takemoto, H. Miyabe, *Journal of Organic Chemistry* **2016**, *81*, 7217–7229.

[13] a) Z. Zuo, D. T. Ahneman, L. Chu, J. A. Terrett, A. G. Doyle, D. W. MacMillan, *Science* **2014**, *345*, 437–440; b) J. Twilton, C. Le, P. Zhang, M. H. Shaw, R. W. Evans, D. W. C. MacMillan, *Nature Reviews Chemistry* **2017**, *1*, 0052; c) A. Y. Chan, I. B. Perry, N. B. Bissonnette, B. F. Buksh, G. A. Edwards, L. I. Frye, O. L. Garry, M. N. Lavagnino, B. X. Li, Y. Liang, E. Mao, A. Millet, J. V. Oakley, N. L. Reed, H. A. Sakai, C. P. Seath, D. W. C. MacMillan, *Chemical Reviews* **2022**, *122*, 1485–1542.

[14] E. B. Corcoran, M. T. Pirnot, S. Lin, S. D. Dreher, D. A. DiRocco, I. W. Davies, S. L. Buchwald, D. W. MacMillan, *Science* **2016**, *353*, 279–283.

[15] F. Lévesque, M. J. Di Maso, K. Narsimhan, M. K. Wismer, J. R. Naber, *Organic Process Research & Development* **2020**, *24*, 2935–2940.

[16] L. Capaldo, D. Ravelli, M. Fagnoni, *Chemical Reviews* **2021**, *122*, 1875–1924.

[17] G. Schenck, G. Koltzenburg, H. Grossmann, *Angewandte Chemie* **1957**, *69*, 177–178.

[18] D. Ravelli, D. Dondi, M. Fagnoni, A. Albini, *Chemical Society Reviews* **2009**, *38*, 1999–2011.

[19] a) M. J. Genzink, J. B. Kidd, W. B. Swords, T. P. Yoon, *Chemical Reviews* **2022**, *122*, 1654–1716; b) W. Yao, E. A. B. Bergamino, M. Y. Ngai, *ChemCatChem* **2022**, *14*, e202101292.

[20] D. A. Nagib, M. E. Scott, D. W. MacMillan, *Journal of the American Chemical Society* **2009**, *131*, 10875–10877.

[21] S. Das, C. Zhu, D. Demirbas, E. Bill, C. K. De, B. List, *Science* **2023**, *379*, 494–499.

[22] X. Li, J. Großkopf, C. Jandl, T. Bach, *Angewandte Chemie International Edition* **2021**, *60*, 2684–2688.

[23] a) B. Zhang, L. Sun, *Chemical Society Reviews* **2019**, *48*, 2216–2264; b) K. E. Dalle, J. Warnan, J. J. Leung, B. Reuillard, I. S. Karmel, E. Reisner, *Chemical Reviews* **2019**, *119*, 2752–2875.

[24] a) J. M. Lehn, J. P. Sauvage, *Nouveau Journal de Chimie* **1977**, *1*, 449–451; b) P. Keller, A. Moradpour, E. Amouyal, H. B. Kagan, *Nouveau Journal de Chimie* **1980**, *4*, 377–384.

[25] E. D. Cline, S. E. Adamson, S. Bernhard, *Inorganic Chemistry* **2008**, *47*, 10378–10388.

[26] R. Khnayzer, V. Thoi, M. Nippe, A. King, J. Jurss, K. El Roz, J. Long, C. Chang, F. Castellano, *Energy & Environmental Science* **2014**, *7*, 1477–1488.

[27] a) N. D. Morris, M. Suzuki, T. E. Mallouk, *Journal of Physical Chemistry A* **2011**, *115*, 547; b) D. Shevela, S. Koroidov, M. M. Najafpour, J. Messinger, P. Kurz, *Chemistry – A European Journal* **2011**, *17*, 5415–5423.

[28] E. B. B. Limburg, S. Bonnet, *ACS Catalysis* **2016**, *6*, 5273–5284.

[29] L. Hammarstrom, *Accounts of Chemical Research* **2015**, *48*, 840–850.

[30] M. Aresta, A. Dibenedetto, *Dalton Transactions 2007*, 2975–2992.

[31] J. Hawecker, J. M. Lehn, R. Ziessel, *Helvetica Chimica Acta* **1986**, *69*, 1990–2012.

[32] Y. Kuramochi, O. Ishitani, H. Ishida, *Coordination Chemistry Reviews* **2018**, *373*, 333–356.

[33] H. Kumagai, Y. Tamaki, O. Ishitani, *Accounts of Chemical Research* **2022**, *55*, 978–990.

[34] J. Zhao, J. L. Brosmer, Q. Tang, Z. Yang, K. Houk, P. L. Diaconescu, O. Kwon, *Journal of the American Chemical Society* **2017**, *139*, 9807–9810.

[35] E. R. Welin, C. Le, D. M. Arias-Rotondo, J. K. McCusker, D. W. MacMillan, *Science* **2017**, *355*, 380–385.

[36] N. Serpone, A. V. Emeline, S. Horikoshi, V. N. Kuznetsov, V. K. Ryabchuk, *Photochemical and Photobiological Sciences* **2012**, *11*, 1121–1150.

[37] G. Ciamician, *Science* **1912**, *36*, 385–394.

[38] I. Fukushima, M. Horio, M. Ohmori, *Kogyo Kagaku Zashi* **1932**, *35*, 398.

[39] M. Buchalska, J. Kuncewicz, E. Swietek, P. Labuz, T. Baran, G. Stochel, W. Macyk, *Coordination Chemistry Reviews* **2013**, *257*, 767–775.

[40] J. F. Rodriguez, J. E. Harris, M. E. Bothwell, T. Mebrahtu, M. P. Soriaga, *Inorganica Chimica Acta* **1988**, *148*, 123–131.

[41] T. Sakata, K. Hashimoto, M. Hiramoto, *Journal of Physical Chemistry* **1990**, *94*, 3040–3045.

[42] a) F. Parrino, A. Ramakrishnan, H. Kisch, *Angewandte Chemie International Edition* **2008**, *47*, 7107–7109; b) W. Macyk, K. Szaciłowski, G. Stochel, M. Buchalska, J. Kuncewicz, P. Łabuz, *Coordination Chemistry Reviews* **2010**, *254*, 2687–2701.

[43] C. Creutz, B. S. Brunschwig, N. Sutin, *Journal of Physical Chemistry B* **2006**, *110*, 25181–25190.

[44] M. V. Dozzi, B. Ohtani, E. Selli, *Physical Chemistry Chemical Physics* **2011**, *13*, 18217–18227.

[45] X. Tao, Y. Zhao, S. Wang, C. Li, R. Li, *Chemical Society Reviews* **2022**, *51*, 3561–3608.

[46] a) J. Abed, N. S. Rajput, A. E. Moutaouakil, M. Jouiad, *Nanomaterials* **2020**, *10*, 2260; b) A. T. E. Fudo, S. Iguchi and H. Kominami, *H. Kominami, Sustainable Energy Fuels* **2021**, *5*, 3303–3311.

[47] a) Y. Ma, L. Lin, T. Takata, T. Hisatomi, K. Domen, *Physical Chemistry Chemical Physics* **2023**, *25*, 6586–6601; b) H. Song, S. Luo, H. Huang, B. Deng, J. Ye, *ACS Energy Letters* **2022**, *7*, 1043–1065.

[48] H. Dau, I. Zaharieva, *Accounts of Chemical Research* **2009**, *42*, 1861–1870.

[49] M. W. Kanan, Y. Surendranath, D. G. Nocera, *Chemical Society Reviews* **2009**, *38*, 109–114.

[50] A. Fujishima, K. Kohayakawa, K. Honda, *Journal of the Electrochemical Society* **1975**, *122*, 1487–1489.

[51] a) A. J. Bard, *Journal of Photochemistry* **1979**, *10*, 59–75; b) A. J. Bard, *Journal of Physical Chemistry* **1982**, *86*, 172–177.

[52] a) Y. Luo, S. Suzuki, Z. Wang, K. Yubuta, J. J. M. Vequizo, A. Yamakata, H. Shiiba, T. Hisatomi, K. Domen, K. Teshima, *ACS Applied Materials & Interfaces* **2019**, *11*, 22264–22271; b) T. Ohno, K. Sarukawa, M. Matsumura, *New Journal of Chemistry* **2002**, *26*, 1167–1170.

[53] a) T. Takata, J. Jiang, Y. Sakata, M. Nakabayashi, N. Shibata, V. Nandal, K. Seki, T. Hisatomi, K. Domen, *Nature* **2020**, *581*, 411–414; b) S. Nishioka, F. E. Osterloh, X. Wang, T. E. Mallouk, K. Maeda, *Nature Reviews Methods Primers* **2023**, *3*, 42.

[54] T. Higashi, K. Seki, Y. Sasaki, Y. Pihosh, V. Nandal, M. Nakabayashi, N. Shibata, K. Domen, *Chemistry* **2023**, *29*, e202204058.

[55] H. Li, J. Xiao, J. J. M. Vequizo, T. Hisatomi, M. Nakabayashi, Z. Pan, N. Shibata, A. Yamakata, T. Takata, K. Domen, *ACS Catalysis* **2022**, *12*, 10179–10185.

[56] H. Nishiyama, T. Yamada, M. Nakabayashi, Y. Maehara, M. Yamaguchi, Y. Kuromiya, Y. Nagatsuma, H. Tokudome, S. Akiyama, T. Watanabe, *Nature* **2021**, *598*, 304–307.

[57] C. Pornrungroj, A. B. Mohamad Annuar, Q. Wang, M. Rahaman, S. Bhattacharjee, V. Andrei, E. Reisner, *Nature Water* **2023**, *1*, 952–960.

[58] S. Jiang, J. Cao, M. Guo, D. Cao, X. Jia, H. Lin, S. Chen, *Applied Surface Science* **2021**, *558*, 149882.

[59] A. Iwase, S. Yoshino, T. Takayama, Y. H. Ng, R. Amal, A. Kudo, *Journal of the American Chemical Society* **2016**, *138*, 10260–10264.

[60] J. Schneider, D. W. Bahnemann, *Journal of Physical Chemistry Letters* **2013**, *4*, 3479–3483.

[61]   D. T. Sawyer, J. S. Valentine, *Accounts of Chemical Research* **1981**, *14*, 393–400.
[62]   P. M. Wood, *Biochemical Journal* **1988**, *253*, 287–289.
[63]   F. Yang, Y. Takahashi, N. Sakai, T. Tatsuma, *Journal of Physical Chemistry C* **2011**, *115*, 18270–18274.
[64]   a) H. Irie, S. Miura, K. Kamiya, K. Hashimoto, *Chemical Physics Letters* **2008**, *457*, 202–205; b) Y. Nosaka, S. Takahashi, H. Sakamoto, A. Y. Nosaka, *Journal of Physical Chemistry C* **2011**, *115*, 21283–21290.
[65]   R. Abe, H. Takami, N. Murakami, B. Ohtani, *Journal of the American Chemical Society* **2008**, *130*, 7780–7781.
[66]   U. K. Klaning, K. Sehested, J. Holcman, *Journal of Physical Chemistry* **1985**, *89*, 760–763.
[67]   P. Wardman, *Journal of Physical and Chemical Reference Data* **1989**, *18*, 1637–1755.
[68]   S. Gershuni, N. Itzhak, J. Rabani, *Langmuir* **1999**, *15*, 1141–1146.
[69]   V. Etacheri, G. Michlits, M. K. Seery, S. J. Hinder, S. C. Pillai, *ACS Applied Materials & Interfaces* **2013**, *5*, 1663–1672.
[70]   W. Macyk, H. Kisch, *Chemistry – A European Journal* **2001**, *7*, 1862–1867.
[71]   G. Burgeth, H. Kisch, *Coordination Chemistry Reviews* **2002**, *230*, 41–47.
[72]   B. M. Hoffman, D. Lukoyanov, Z.-Y. Yang, D. R. Dean, L. C. Seefeldt, *Chemical Reviews* **2014**, *114*, 4041–4062.
[73]   G. N. Schrauzer, T. D. Guth, *Journal of the American Chemical Society* **1977**, *99*, 7189–7193.
[74]   H. Kisch, *European Journal of Inorganic Chemistry* **2020**, 1376–1382.
[75]   a) A. J. Medford, M. C. Hatzell, *ACS Catalysis* **2017**, *7*, 2624–2643; b) B. Sun, S. Lu, Y. Qian, X. Zhang, J. Tian, *Carbon Energy* **2023**, *5*, e305.
[76]   M.-H. Vu, M. Sakar, T.-O. Do, *Catalysts* **2018**, *8*, 621.
[77]   Y. Zhu, X. Zheng, W. Zhang, A. Kheradmand, S. Gu, M. Kobielusz, W. Macyk, H. Li, J. Huang, Y. Jiang, *ACS Applied Materials & Interfaces* **2021**, *13*, 32937–32947.
[78]   S. Hu, W. Zhang, J. Bai, G. Lu, L. Zhang, G. Wu, *RSC Advances* **2016**, *6*, 25695–25702.
[79]   M. Cheng, C. Xiao, Y. Xie, *Journal of Materials Chemistry A* **2019**, *7*, 19616–19633.
[80]   K. Hu, Z. Huang, L. Zeng, Z. Zhang, L. Mei, Z. Chai, W. Shi, *European Journal of Inorganic Chemistry* **2022**, e202101092.
[81]   J. Chatt, R. A. Head, G. J. Leigh, C. J. Pickett, *Journal of the Chemical Society, Dalton Transactions 1978*, 1638–1647.
[82]   H. Kisch, *Accounts of Chemical Research* **2017**, *50*, 1002–1010.
[83]   a) L. J. Wu, Q. Wang, J. Guo, J. Wei, P. Chen, Z. Xi, *Angewandte Chemie International Edition* **2023**, *62*, e202219298; b) S. M. Bhutto, R. X. Hooper, B. Q. Mercado, P. L. Holland, *Journal of the American Chemical Society* **2023**, *145*, 4626–4637.
[84]   G. A. Olah, A. Goeppert, G. S. Prakash, *Beyond Oil and Gas: The Methanol Economy*, 3rd edition, Wiley-VCH, Weinheim, **2018**.
[85]   a) W. K. Fan, M. Tahir, *Energy Conversion and Management* **2022**, *253*, 115180; b) Y. Wang, E. Chen, J. Tang, *ACS Catalysis* **2022**, *12*, 7300–7316; c) J. He, C. Janáky, *ACS Energy Letters* **2020**, *5*, 1996–2014.
[86]   a) E. A. Kozlova, M. N. Lyulyukin, D. V. Kozlov, V. N. Parmon, *Russian Chemical Reviews* **2021**, *90*, 1520; b) G. Zhang, Y. Cheng, M. Beller, F. Chen, *Advanced Synthesis & Catalysis* **2021**, *363*, 1583–1596.
[87]   H. Pan, M. D. Heagy, *Nanomaterials* **2020**, *10*, 2422.
[88]   H. L. Wu, X. B. Li, C. H. Tung, L. Z. Wu, *Advanced Materials* **2019**, *31*, 1900709.
[89]   X. Liu, S. Inagaki, J. Gong, *Angewandte Chemie International Edition* **2016**, *55*, 14924–14950.
[90]   C. I. Ezugwu, S. Liu, C. Li, S. Zhuiykov, S. Roy, F. Verpoort, *Coordination Chemistry Reviews* **2022**, *450*, 214245.
[91]   E. Edwardes Moore, V. Andrei, A. R. Oliveira, A. M. Coito, I. A. Pereira, E. Reisner, *Angewandte Chemie International Edition* **2021**, *60*, 26412–26412.
[92]   H. Fujiwara, H. Hosokawa, K. Musrakoshi, Y. Wada, S. Yanagida, T. Okada, H. Kobayashi, *Journal of Physical Chemistry B* **1997**, *101*, 8270–8278.

[93]   A. Wagner, C. D. Sahm, E. Reisner, *Nature Catalysis* **2020**, *3*, 775–786.

[94]   M. A. Rahaman, V. Wright, D. Lam, E. Pornrungroj, C. Bhattacharjee, S. Pichler, C. M., H. F. Greer, J. J. Baumberg, E. Reisner, *Nature Energy* **2003**, *8*, 629–638.

[95]   A. D. T. Baran, M. Aresta, K. Kruczała, W. Macyk, *ChemPlusChem* **2014**, *79*, 708–715.

[96]   a) Y. He, S. Wang, X. Han, J. Shen, Y. Lu, J. Zhao, C. Shen, L. Qiao, *ACS Applied Materials & Interfaces*, **2022**, *14*, 23364–23374. b) Q. Wang, S. Kalathil, C. Pornrungroj, C. D. Sahm, E. Reisner, *Nature Catalysis*, **2022**, *5*, 633–641. c) S. Bhattacharjee, M. Rahaman, V. Andrei, M. Miller, S. Rodriguez-Jimenez, E. Lam, C. Pornrungroj, E. Reisner, *Nature. Synthesis*, **2023**, *2*, 182–192.

[97]   D. W. Manley, R. T. McBurney, P. Miller, R. F. Howe, S. Rhydderch, J. C. Walton, *Journal of the American Chemical Society* **2012**, *134*, 13580–13583.

[98]   G. Horner, P. Johne, R. Kunneth, G. Twardzik, H. Roth, T. Clark, H. Kisch, *Chemistry: A European Journal* **1999**, *5*, 208–217.

[99]   a) J. Buecheler, N. Zeug, H. Kisch, *Angewandte Chemie International Edition* **1982**, *21*, 783–784; b) N. Zeug, J. Buecheler, H. Kisch, *Journal of the American Chemical Society* **1985**, *107*, 1459–1465.

[100] T. Mitkina, C. Stanglmair, W. Setzer, M. Gruber, H. Kisch, B. Konig, *Organic and Biomolecular Chemistry* **2012**, *10*, 3556–3561.

[101] a) L. Marzo, S. K. Pagire, O. Reiser, B. Konig, *Angewandte Chemie International Edition* **2018**, *57*, 10034–10072; b) S. Reischauer, B. Pieber, *iScience* **2021**, *24*, 102209.

[102] J. Kollonitsch, L. M. Perkins, A. A. Patchett, G. A. Doldouras, S. Marburg, D. E. Duggan, A. L. Maycock, S. D. Aster, *Nature* **1978**, *274*, 906–908.

[103] R. G. Bergman, *Nature* **2007**, *446*, 506.

[104] D. D. Dionysiou, G. Li Puma, J. Ye, J. Schneider, D. Bahnemann, *Photocatalysis: Fundamentals and Perspectives*, Royal Society of Chemistry, Cambridge, UK, **2016**.

[105] A. Fujishima, *Photocatalysis Experimental Methods*, Kitano-Shoten Publishing Co. Ltd, Tokyo, **2022**.

[106] https://www.tus.ac.jp/info/setubi/museum/main/bunkan/sciencedojo_eng.html.

[107] Q. Ye, P. Y. Liu, Z. F. Tang, L. Zhai, *Vacuum* **2007**, *81*, 627–631.

[108] A. V. Emeline, A. V. Rudakova, M. Sakai, T. Murakami, A. Fujishima, *Journal of Physical Chemistry C* **2013**, *117*, 12086–12092.

[109] J. Marugan, D. Hufschmidt, G. Sagawe, V. Selzer, D. Bahnemann, *Water Reserch* **2006**, *40*, 833–839.

[110] E. Puzenat, H. Lachheb, M. Karkmaz, A. Houas, C. Guillard, J. M. Herrmann, *International Journal of Photoenergy* **2003**, *5*, 51–58.

[111] a) A. Fujishima, H. Irie, X. Zhang, D. A. Tryk, *Handbook of Self-cleaning Surfaces and Materials: From Fundamentals to Applications*, Wiley-VCH, Weinheim, **2023** b) H. Kisch, *Semiconductor Photocatalysis*, Wiley-VCH, Weinheim, Germany, **2015**.

[112] V. Rodriguez-Gonzalez, S. O. Alfaro, L. M. Torres-Martinez, S.-H. Cho, S.-W. Lee, *Applied Catalysis B* **2010**, *98*, 229–234.

[113] A. Mills, C. Hill, P. K. J. Robertson, *Journal of Photochemistry and Photobiology A* **2012**, *237*, 7–23.

[114] J. A. Ibanez, M. I. Litter, R. A. Pizarro, *Journal of Photochemistry and Photobiology A: Chemistry* **2003**, *157*, 81–85.

[115] Z. Huang, P. C. Maness, D. M. Blake, E. J. Wolfrum, S. L. Smolinski, W. A. Jacoby, *Journal of Photochemistry and Photobiology A: Chemistry* **2000**, *130*, 163–170.

[116] R. Nakano, H. Ishiguro, Y. Yao, J. Kajioka, A. Fujishima, K. Sunada, M. Minoshima, K. Hashimoto, Y. Kubota, *Photochemical and Photobiological Sciences* **2012**, *11*, 1293–1298.

[117] R. Cai, K. Hashimoto, K. Itoh, Y. Kubota, A. Fujishima, *Bulletin of the Chemical Society of Japan* **1991**, *64*, 1268–1273.

[118] A. Janczyk, A. Wolnicka-Glubisz, K. Urbanska, H. Kisch, G. Stochel, W. Macyk, *Free Radical Biology & Medicine* **2008**, *44*, 1120–1130.

[119] C. M. N. Chan, A. M. C. Ng, M. K. Fung, H. S. Cheng, M. Y. Guo, A. B. Djurisic, F. C. C. Leung, W. K. Chan, *Journal of Experimental Nanoscience* **2013**, *8*, 695–703.

[120] S. D. Khan, R. H. Reed, M. G. Rasul, *BMC Microbiology* **2012**, *12*, 285.

[121] J. Prakash, S. B. N. Krishna, P. Kumar, V. Kumar, K. S. Ghosh, H. C. Swart, S. Bellucci, J. Cho, *Catalysts* **2022**, *12*, 1047.

[122] A. Lu, Y. Li, *Geomicrobiology Journal* **2012**, *29*, 236–243.

[123] M. A. Butler, D. S. Ginley, *Journal of the Electrochemical Society* **1978**, *125*, 228–232.

[124] a) M. A. A. Schoonen, Y. Xu, D. R. Strongin, *Journal of Geochemical Exploration* **1998**, *62*, 201–215; b) Y. Xu, M. A. A. Schoonen, *American Mineralogist* **2000**, *85*, 543–556.

[125] J.-M. Herrmann, J.-L. Mansot, *Entropie* **2000**, *36*, 60–63.

[126] M. M. Najafpour, *Geomicrobiology Journal* **2011**, *28*, 714–718.

[127] N. R. Dhar, E. V. Seshacharyulu, S. K. Mukerji, *Annals of Agriculture* **1941**, *11*, 83–86.

[128] G. N. Schrauzer, N. Strampach, L. N. Hui, M. R. Palmer, J. Salehi, *Proceedings of the National Academy of Sciences of the United States of America* **1983**, *80*, 3873–3876.

[129] F. W. Clarke, *United States Geological Survey Bulletin* **1908**, *330*, 20.

[130] G. N. Schrauzer, in *Energy Efficiency and Renewable Energy Through Nanotechnology* (Ed.: L. Zang), Springer-Verlag, London, **2011**, pp. 601–623.

[131] F. G. C. L. F. Yang, J. Su, X. Wu, L. Zhang, F. Hog, P. Yang, *Biological Trace Element Research* **2007**, *119*, 77–88.

[132] C. Larue, G. Veronesi, A.-M. Flank, S. Surble, N. Herlin-Boime, M. Carriere, *Journal of Toxicology and Environmental Health, Part A* **2012**, *75*, 722–734.

[133] D. Kişla, G. G. Gökmen, G. Evrendilek, T. Akan, T. Vlčko, P. Kulawik, A. R. Jambrak, F. Özoğul, *Trends in Food Science & Technology* **2023**, *135*, 144–172.

[134] A. Weir, P. Westerhoff, L. Fabricius, K. Hristovski, N. von Goetz, *Environmental Science & Technology* **2012**, *46*, 2242–2250.

[135] a) C. McCracken, A. Zane, D. A. Knight, P. K. Dutta, W. J. Waldman, Chem. Res. Toxicol **2013**, *26*, 1514–1525; b) S. C. Joshi, U. Kaushik, *Research Journal of Pharmaceutical, Biological and Chemical Sciences* **2013**, *4*, 1396–1410; c) A. Markowska-Szczupak, M. Endo-Kimura, O. Paszkiewicz, E. Kowalska, *Nanomaterials* **2020**, *10*, 2065.

[136] F. Turci, E. Peira, I. Corazzari, I. Fenoglio, M. Trotta, B. Fubini, *Chemical Research in Toxicology* **2013**, *26*, 1579–1590.

6 Photobiology

6.1 Photosynthesis

The essential interactions between light and molecules or materials were described in the previous chapters. To summarise, when a substance absorbs light, the energy of the light can be used for the following processes:

- Emission of light (luminescence)
- Production of heat
- Change in molecular structure (isomerisation, bond breaking and making)
- Transfer of energy (photosensitisation)
- Transfer of electrons (photosynthesis, photocatalysis)

All of these processes are important in biology [1]. From the time of Earth's formation to about 3 or 3.5 billion years ago, the atmosphere of our planet contained no oxygen (O_2). The first traces were formed around this time through photosynthesis by cyano-bacteria. Larger quantities first appeared a few hundred million years later, when the first green algae absorbed sunlight and enabled the formation of oxygen and carbohy-drates from water and carbon dioxide. Plants with green leaves appeared between 400 and 500 million years ago. With plants as food and oxygen for respiration, the needs of animal life were provided. In the mean time, the annual worldwide contribution of photosynthesis to carbohydrate production grew to become equivalent to a mountain of 300 billion sugar lumps [2]. A third of that came from plant photosynthesis, but the majority came from phytoplankton in the oceans.

Although carbon dioxide represents only 0.04 vol% of the air, it forms the basis of life. The overall chemical reaction of photosynthesis consists of the reduction of car-bon dioxide by water (Equation 6.1). The water is oxidised to O_2 by formation of the O–O bonds while the carbon dioxide forms carbohydrates (glucose, starch) by forma-tion of numerous C–C bonds. Since there is a 500-fold excess of oxygen in the atmo-sphere relative to carbon dioxide and oxygen is vastly more reactive than carbon dioxide, this selective reduction reflects remarkable chemical selectivity.

$$6\,CO_2 + 6\,H_2O + \text{light energy} \;\rightarrow\; C_6H_{12}O_6\,(\text{glucose}) + 6\,O_2 \qquad (6.1)$$

This amazing process occurs in chloroplasts (Figure 6.1) inside the green leaf. Chloro-plasts are compartments ("organelles") that are enclosed by a separate membrane within the leaf cell – up to 100 may be present in a single cell. Chloroplasts contain their own DNA and are thought to have evolved from cyanobacteria that were incorpo-rated into plants. They also contain stacks of membrane structures called "thylakoids".

Photosynthesis can be divided into two parts, the "light" reaction and the "dark" re-action. Water is oxidised to oxygen (Equation 6.2) and nicotinamide-adenine-diphosphate ($NADP^+$) is reduced to the corresponding dihydride (NADPH) (Equation 6.3) in the light

https://doi.org/10.1515/9783111029375-006

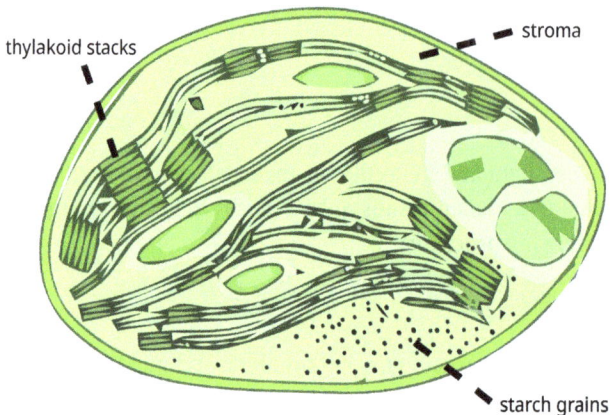

Figure 6.1: Structure of a chloroplast (diameter ~ 4.8 μm). About 300 pigment molecules are located within the thylakoid membrane.

reaction. This reduction is the biological equivalent of formation and storage of hydrogen (H_2). It works by transport of two protons from the outside to the inside (lumen) of the thylakoid membrane. The proton gradient across the membrane provides the energy for the reaction of adenosine diphosphate (ADP) with inorganic phosphate (P_i) to form the biological energy carrier adenosine triphosphate (ATP) catalysed by the enzyme ATP synthase, a motor protein (Equation 6.4).[1] In the dark reaction that occurs on the outside of the thylakoid membrane, carbon dioxide is converted to glucose via a series of enzymatic reactions (Calvin cycle) making use of ATP and NADPH.

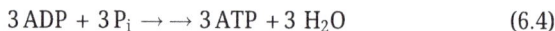

$$2\,H_2O \rightarrow O_2 + 4\,H^+ + 4\,e^- \tag{6.2}$$

$$2\,NADP^+ + 2\,H^+ + 4\,e^- \rightarrow 2\,NADPH \tag{6.3}$$

$$3\,ADP + 3\,P_i \rightarrow\rightarrow 3\,ATP + 3\,H_2O \tag{6.4}$$

The light reaction takes place in the outside part of the membrane of the thylakoid stack. The principal pigments of photosynthesis are chlorophylls which are fully conjugated cyclic organic molecules containing 4 nitrogen atoms coordinated to magnesium (Figure 6.2a).

The absorption maxima of chlorophylls lie in the 650–850 nm region and vary with the substituents. In the thylakoid we find two protein complexes that cross the membrane, designated as Photosystem I (PS I) and Photosystem II (PS II) where the light reactions occur (Scheme 6.1).[2] The arrangement of these centres within the thyla-

---

1 The water in Equation 6.4 stems from the condensation reaction between P-OH groups in ADP and inorganic phosphate P_i ($H_2PO_4^-$).

2 The numbers I and II refer to the sequence in which these protein complexes were identified. Johann Deisenhofer, Robert Huber and Hartmut Michel were awarded the Nobel Prize in 1988 for the

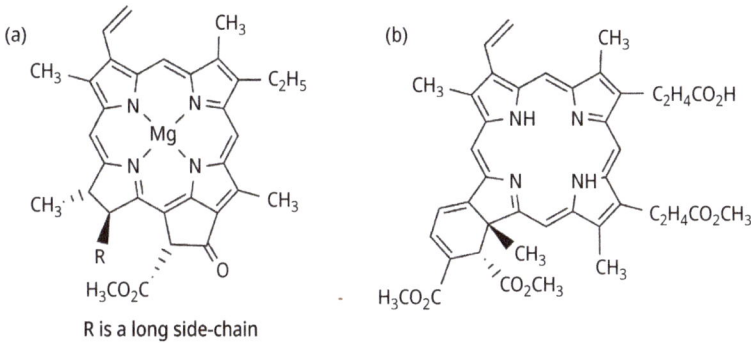

Figure 6.2: (a) Structure of chlorophyll a; (b) Structure of Verteporfin™, a photosensitiser used for treatment of age-related macular degeneration.

koid membrane is shown in Figure 6.3. Surrounding these reaction centres are arrays of pigments that form some of the most beautiful and regular molecular structures in nature. These are the light-harvesting complexes that act like satellite dishes to collect the light and transfer the excitation energy to the reaction centres. Figure 2.12 illustrated the light-harvesting complex from purple bacteria with 9-fold symmetry and three sets of 9 chlorophyll molecules, one set perpendicular to the other two. Sunlight is absorbed principally by the chlorophyll pigments in the antenna and the excitation energy is transferred highly efficiently to the chlorophylls at the reaction centres. Energy transfer (see Chapter 2.2.3) occurs within a few picoseconds unidirectionally with very little energy loss to the reaction centres that are surrounded by the light-harvesting proteins. In the green leaf, both contain pairs of chlorophylls designated P 680 and P 700 for PS I and PS II, respectively, forming excimers with different excited state properties from individual molecules (their absorption maxima lie at 680 and 700 nm, see chapter 2.3.1.2 for excimers).[3] The initial excited state (usually described as an exciton) formed in PS II is extremely short-lived. An ultrafast electron transfer process ensues in which the positive charge is moved in one direction and the negative charge in another direction. The negative charge is transferred initially to pheophytin and a quinone within PS II and then through a series of redox enzymes, plastoquinone (PQ), cytochrome (Cyt), and plastocyanin (PC) to PS I where the charge is neutralised and ATP is formed.[4] The positive charge formed at PS II is used to oxidise water to oxygen (Equation 6.2) at a

---

determination of the three-dimensional structure of PS I. The structure of PS II was not determined until 2004 (James Barber and So Iwata).

3 Further pigments such as carotenoids and xanthophylls (both long-chain polyenes) prevent the formation of reactive oxygen species (ROS) such as $^1O_2$, $OH^\bullet$ and $HO_2^\bullet$, so protecting the chlorophyll against oxidative degradation.

4 Pheophytin is a derivative of a free-base porphyrin. Plastoquinones include a quinone, cytochromes include an iron porphyrin and plastocyanins include a copper-amino acid complex as redox centres.

very unusual Mn_4Ca centre within PS II.[5] The excitation of PSI again results in charge separation and the negative charge is transferred through an iron-sulfur redox protein (ferredoxin Fd) to $NADP^+$ reductase where NADPH is formed (Equation 6.3). As a consequence of the highly efficient charge separation down the chain of redox enzymes, the unwanted back electron transfer (see chapter 5.1.6) that would convert the light energy to useless heat is almost completely suppressed.

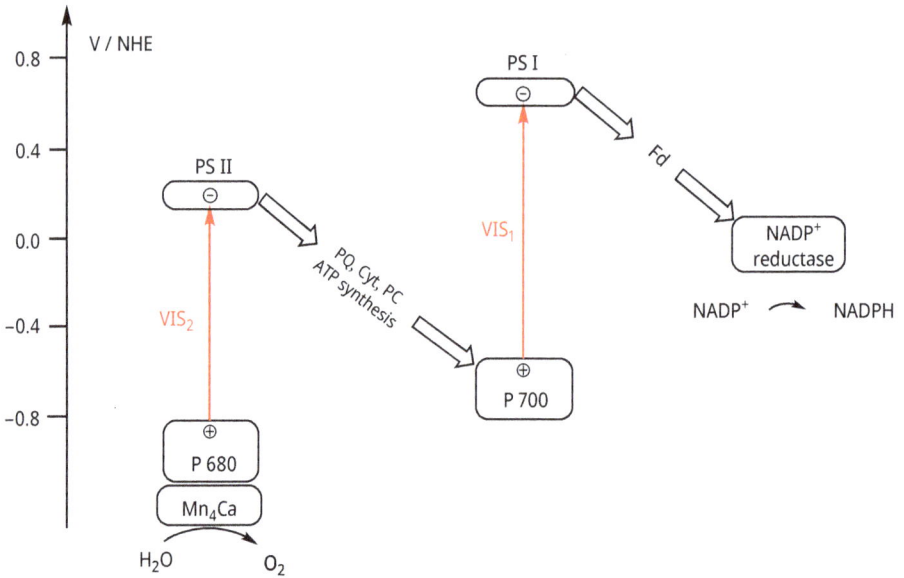

Scheme 6.1: Z-scheme for the primary steps of the light reaction in the reaction centre of photosynthesis.

It's worth mentioning how the rate of photosynthesis, measured by CO_2-utilisation, changes with increasing CO_2 concentrations. Plants are separated into two groups according to the mechanism of CO_2-fixation in the Calvin cycle. Those involving a three-carbon intermediate are called C_3-plants and those with a C_4 intermediate are C_4-plants. The rate increases steeply for C_4-plants (maize, miscanthus grass) but reaches a plateau at about 0.03 vol%. In contrast, C_3-plants (rice, wheat) exhibit a slower but continuous rise, a point that is not often mentioned in discussions of the role of CO_2 in climate change (Figure 6.4). Most plants belong to the C_3-type; C_4-plants represent only about 3%.

Purple bacteria are found in environments where there is no oxygen, both on land and in water. Instead of oxidising water to oxygen, they oxidise different inorganic compounds (anoxygenic photosynthesis). Examples are oxidation of hydrogen

---

5 The redox potential for the oxidation of water to O_2 at pH 7 is 0.82 V. The oxidation requires accumulation of 4 holes (positive charges) at the Mn_4Ca centre. The reverse reduction of O_2 to water is a 4-electron, 4-proton process that occurs in respiration.

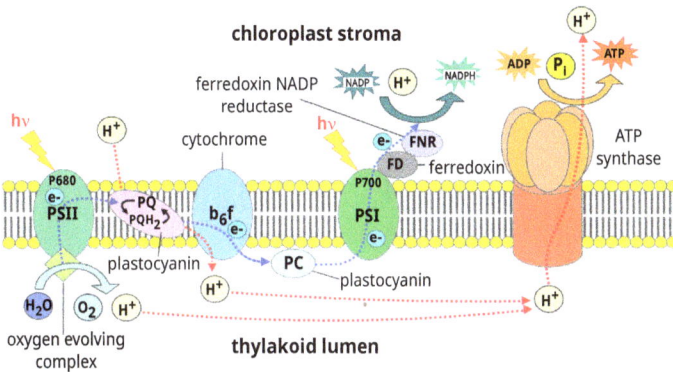

Figure 6.3: Schematic arrangement of the photosynthetic reaction centres and ATP synthase within the thylakoid membrane of a green leaf cell. The light-harvesting proteins are not shown. (from Somepics, CC BY-SA 4.0 <https://creativecommons.org/licenses/by-sa/4.0>, via Wikimedia Commons [3]).

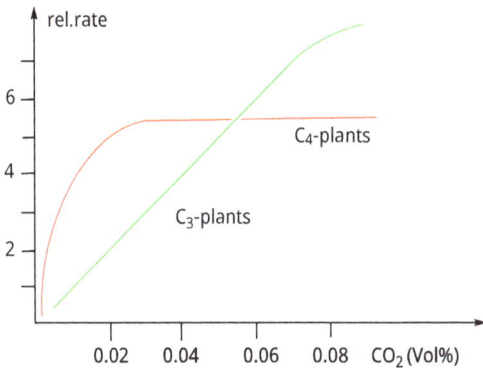

Figure 6.4: Relative rates of photosynthesis as a function of carbon dioxide concentration.

sulfide to sulfur or hydrogen to protons. The light harvesting protein illustrated in Figure 2.12 comes from a purple bacterium.

The authors' generation learnt that animals cannot obtain food directly from a photosynthetic process, but in the last few years this has proved incorrect. The sea slug *Elysia chlorotica* (Figure 6.5) has managed to incorporate chloroplasts from green algae that it eats. The algae continue to photosynthesise, earning it the nickname of "solar-powered sea slug".

Figure 6.5: Plant or animal: the mollusc *Elysia chlorotica* [4].

Glowing drinking water

Most herbicides block electron transport in photosynthesis and therefore prevent the growth of plants. Consequently, the lifetime of the excited pigments and their concentrations increase. The chlorophyll pigments lose their excitation energy radiatively as fluorescence. Water containing the chlorophyll begins to glow! The thylakoid membrane is relatively easy to isolate from the chloroplasts, for example from broad beans, and the fluorescence spectrum can be measured. Indeed, fluorescence is an important method of studying photosynthesis.

The boundary between flora and fauna begins to blur. Will people be able to do something similar in a few million years? In that case our need for carbohydrates would be met without the detour of plants! If we also had incorporated the enzyme nitrogenase from bacteria for fixing nitrogen from air like beans, we could also synthesise the essential amino acids and nucleobases. – bio science fiction?

6.2 Photodynamic therapy (PDT)

In earlier chapters (2.2.2, 2.2.3 and 4.2.5), we discussed the phenomenon of sensitisation in the context of conversion of the triplet ground state of oxygen in the air to its excited singlet state. One example was the use of spinach leaves as a source of chlorophyll for photosensitisation (Chapter 4.2.5). The sensitiser absorbs the light and transfers its energy (94 kJ/mol) to $^3O_2$ to effect the conversion to $^1O_2$: triplet oxygen acts as the quencher of the excited state of chlorophyll. Considering the relatively small energy requirement, it is not surprising that other naturally occurring compounds can take over the sensitiser role of chlorophyll in generating the strongly oxidising singlet oxygen. Bergamot oil is one such example[6] that was already recognised by the ancient Egyptians and Chinese who used it to treat skin cancer; the oil was spread on the affected area and the patient was sent into the sun for a few hours. Singlet oxygen served to destroy the moistened tissue. Nowadays, our treatments use porphyrin derivatives (see Figure 2.16) and related molecules together with artificial light sources [5].

Photodynamic therapy (PDT) is the general name for such treatment and it is used for a wide variety of cancers, not just skin cancer. As we saw in Chapter 2, excitation by light can be followed by energy transfer, electron transfer or proton-coupled electron transfer. It is therefore not surprising that excitation of these pigment molecules in the presence of oxygen leads not just to $^1O_2$ but to other reactive oxygen species (ROS) such as O_2^-, $HO_2^{\bullet}$ and H_2O_2. The ROS attack a variety of unsaturated molecules including lipids, amino acid residues and, of most importance, nucleotides, resulting ultimately in cell death. Treatment makes use of the principle that malignant cells multiply much faster than benign tissue. Another key principle is that red light penetrates deeper into the tissue. It is therefore desirable to use light of long wavelength. Photon upconversion (Chapter 2.2.10) methods are under investigation so that near-infrared light sources can be used. One of the advantages of PDT is the ability to target the treatment precisely to the malignant tissue. Fibre optics can be used to irradiate regions deep inside the body.

The first two drugs to be approved for photodynamic therapy were Photofrin™, a mixture of zinc porphyrin oligomers, and the amino acid δ-aminolevulinic acid, a biosynthetic precursor of porphyrins. This precursor is spread onto the tumour to induce the body's own synthesis of porphyrin derivatives. After 3–4 h, irradiation with red light (~640 nm) can begin. This wavelength can penetrate 3–10 nm which can lead to complete removal of the tumour. One of the problems with use of PDT for skin cancer, however, is that it can lead to general light sensitivity (see below).

A variety of improved photosensitisers (so-called second generation) have been approved for clinical use while others are undergoing clinical trials. Some are based on free-base porphyrins, others on the related macrocycle, phthalocyanine, that has a much higher absorption coefficient in the far-red region. Addition of pendant sulfonate or pyridinium groups has been used to make the drug water soluble. One attractive feature of these pigments, including δ-aminolevulinic acid, is that the same pigment can be used both for diagnostic imaging by fluorescence and for therapeutic use (this combined use is called theranostics) [5a]. These second generation photosensitisers can be excited at longer wavelength than the first generation and cause less prolonged photosensitivity. By 2018, six second generation photosensitisers had been approved for a variety of cancers ranging from oesophogal carcinoma to head and neck cancer. In addition one of the

---

6 Bergamot oil is obtained from the peel of the fruit of bergamot tree (a hybrid between sweet lemon and bitter orange trees). It's a scented and inspiring (according to Voltaire, Goethe and Mozart) component of perfume and Earl Grey tea.

drugs, Verteporfin™ (Figure 6.2b) is used to treat the eye condition macular degeneration [6]. In third generation photosensitisers, the photosensitiser is attached (conjugated) to an antibody or peptide to enable the drug to find its own target in line with Paul Ehrlich's historic "magic bullet" concept. However, it should be emphasised that PDT has proved itself most valuable for the treatment of early-stage skin cancer (actinic keratosis).

Recently, it was found that daylight rather than artificial light can be used. During a long walk, sunlight can act as a healer with the help of a sensitiser synthesised by the body [7]! It should be recalled that UV-radiation can cause skin cancer through sunbathing with inadequate protection, but often the cancer only appears 20–40 years later. Even on a cloudy day the light in the open air is about 10 times more intense (ca. 5000 Lux) than in an office lit by fluorescent lamps. Light begins to have a chrono-biological effect on humans at an intensity of 1000 Lux.[7]

The haem group is obtained by binding an iron(II) ion to the four nitrogen atoms of a naturally occurring porphyrin. In haemoglobin, four haem groups act as prosthetic groups[8] bound to the amino-acid chains of the protein. Oxygen is transported to the tissues in the red cells of the blood of vertebrate organisms by haemoglobin. In some people however, the synthesis of haem stops at the porphyrin stage. As a consequence, little or no haem is formed and therefore too little haemoglobin is present to guarantee a normal supply of oxygen to the tissues. Because of the high concentration of porphyrin, the skin can function as a photosensitiser for oxygen and the patients can suffer very painful burns – this is the same problem as experienced by PDT patients (see above) but lasts indefinitely. Patients who suffer from this inherited disease, porphyria, shun daylight.

Porphyria – Dracula Syndrome

It has been suggested that the British King George III (1738–1820) suffered from porphyria, but this theory has been largely discredited. On the other hand, the symptoms are reminders of vampires: that's why the nickname of the illness is "Dracula Syndrome".[a] The symptoms of porphyria include sensitivity to light, blueish coloration of the teeth and stunted lips and receded gums. Dracula shunned the light, hated garlic and greedily sucked blood from the long, white necks of his victims. His porphyrin-rich skin that produced singlet oxygen aggressively in daylight was to blame. He hated garlic because it contains organic sulfides that remove the small amount of iron available to him. Only foreign blood provided him with haemoglobin necessary to survive – an unusual example of how light and air link poison, healing and a literary Muse through a biogenic photosensitiser.[b,c]

---

7 Cloudless daylight and moonlight have intensities of ca. 10,000 and 0.25 Lux, respectively. 1 Lux corresponds to the light from a standard candle illuminating a vertical surface of area 1 m^2 at a distance of one metre from the light source.

8 A prosthetic group is a molecular component of a protein that is not formed from amino-acid residues and is essential to the action of that protein.

^a *Dracula*, Bram Stoker (1847–1912)
^b *"All things are poison and nothing is without poison – only the dose determines that a thing is not a poison"*
Paracelsus (1493–1541)
^c Christopher Lee as Dracula, 1958, copyright free

Irradiation with near infrared (NIR) light to cause localised heating (PTT – Photothermal Therapy), rather than formation of reactive oxygen species represents an alternative method of treating cancerous tissue. In trial experiments, a solution of sulfur and nitrogen-modified carbon nanoparticles (2–5 nm diameter, carbon dots) is injected into mice and serves as a theranostic method. The particles can be revealed by *photoacoustic spectroscopy* (Chapter 3.1.6) after waiting for them to be taken up by the cancerous tissue. Subsequent irradiation with a laser (wavelength 655 nm) increased the temperature to 70 °C and the malignant tissue was irreversibly destroyed [8].

6.3 Isomerisation: From vision to clocks and optogenetics

The absorption of light has profound chemical consequences as we saw in Chapter 4: it alters the spatial and electronic structure of a molecule and can convert it into a new product – a method of synthesis. The first two effects, in particular, play a major role in biological systems.

6.3.1 Vision

The light-induced change in the spatial structure provides the basis for human and animal vision. The ganglial nerve cells of the retina contain rhodopsin (visual purple) which has an absorption maximum at 500 nm. It consists of seven α-helices[9] (opsin) that cross the cell membrane and a prosthetic group, retinal, that is bound to opsin by an imino carbon-nitrogen double bond. Retinal (related to vitamin A) contains a conjugated chain of five C=C double bonds and terminates with the C=N double bond (Scheme 6.2). Four of the five double bonds are present in the cis configuration. Light absorption leads to the ultrafast (within 2×10^{-13} s) *cis-trans* isomerisation of the double bond marked in red and consequently causes a change in the three-dimensional structure of rhodopsin (see also Chapter 4.2.3). This structural change causes a series of complex changes in associated proteins and alteration of various small molecules that act as signallers. In turn, they stimulate a change in the electrochemical membrane potential and the light stimulus progresses in the form of an electrical signal to the synapses in the visual cortex of the brain. The recovery of cis-retinal and rhodopsin is surprisingly complicated and slow (ca. 15 min),[10] involving a series of intermediates [9].

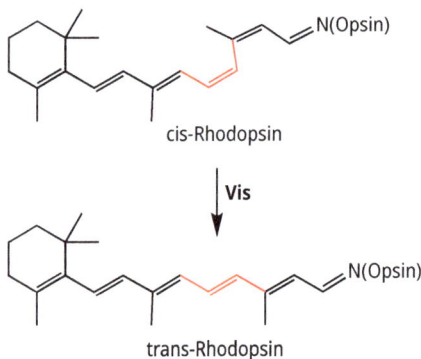

cis-Rhodopsin

Vis

trans-Rhodopsin

Scheme 6.2: The structure of retinal showing its link to the protein opsin and its light-induced *cis-trans* isomerisation.

The human retina contains four types of receptor cells, the rods for low light, and three types of cone for bright light and colour vision with sensitivities peaking at 420, 534 and 564 nm. They all depend on rhodopsin as described above but have different architectures (rods and cones) and different opsins. Many mammal species only have two types

---

9 The α-helix is one of the major types of secondary structure found in proteins in which every C=O group of the peptide bond is linked by a hydrogen bond to the N–H group of a peptide bond four residues further down the chain to form a helical structure.

10 The visual cycle for the conversion of cis-retinal to the trans-isomer and back again was established by George Wald for which he received the 1967 Nobel Prize. The C=N bond is hydrolysed, the resulting aldehyde is converted to an alcohol, esterified, and converted to the cis form, before oxidation and reattachment.

of cone, while most bird species have a fourth cone (tetrachromacy) with maximum sensitivity at 370 nm allowing them to see as far as 330 nm, whereas the human limit is about 380 nm. While males and females of many species, for example tits (genus *Parus*) and starlings (genus *Sturnus*), look identical to the human eye they look different with UV vision. The females search for the males with most strongly UV-reflecting feathers because this feature corresponds to breeding success. Greater UV reflectivity of fruits signals ripening, something that is used by birds and reptiles as they look for food.

Insects such as butterflies and bees also have UV vision. Their compound eyes are also based on rhodopsin but their wavelength sensitivity is different. Both bees and butterflies usually have three different cones with UV, blue and green sensitivity. The UV sensitivity is used to detect patterns in flowers and to communicate with members of the same and other species [10].

In addition to UV vision, many animals can detect the plane of polarisation of linearly polarised light. This sense requires two sets of receptors (ocelli)[11] as observed in the Orchid Bee by X-ray microtomography [11]. These pigments will then behave like a polaroid film in which the electric vector corresponding to the electronic transition lies parallel or perpendicular to the molecular axes. However, details of the detection system are unclear. Animals with the ability to analyse light by polarisation, for example bees, can orient themselves with the aid of the polarisation of the sky, since the pattern changes over the course of the day with the height of the sun. Bees communicate this information with their dance.[12] The UV-portion of the scattered sunlight exhibits the strongest degree of polarisation and the maximum polarisation is at 90° to the sun. Many fish including the green sunfish (*Lepomis cyanellus*) contain polarisation sensitive receptors in their retina (Figure 6.6) [12].

Figure 6.6: The green sunfish orients itself by polarised light [13].

---

11 Insect ocelli are relatively simple eyes that have been assigned various functions not related to pictorial vision.
12 The detection of polarised light by bees and the use of dance to convey information was discovered by Karl von Frisch in 1949. He received the Nobel Prize in 1973.

6.3.2 Internal clock

The retina doesn't just contain rhodopsin as a light detector but also melanopsin with a similar structure;[13] this protein also undergoes light absorption in the visible and UV-A (315–380 nm) regions that once more leads to a cis-trans isomerisation. In contrast to the vision process that sends electrical signals to the brain, melanopsin sends information to the pineal gland. In light conditions, it suppresses the conversion of serotonin into the sleep hormone melatonin.[14] In the dark, this inhibition is relaxed, the concentration of melatonin increases again and reaches three times its initial value in older people and twelve times the value in young people. The phase of deep sleep stimulates the secretion of a growth hormone, the protein somatropin. In winter when the daylight period is short, the melatonin level remains high during the day. As a result, people remain tired and sleep badly: winter depression, winter-blues or, more seriously, SAD-syndrome (seasonal affective disorder). Walks or artificial light (light therapy) are helpful, especially to reduce jetlag [14]. Frequent flyers take melatonin tablets for this purpose, but then our organism throttles its own production of the sleep hormone. Animal experiments indicate that melatonin can prevent growth of tumours [15].

The 24-hour rhythm of our life is controlled by the processes described above. We carry a type of inner clock that determines the secretion of hormones, the body temperature and the metabolism. This clock is controlled by daylight. The pineal gland releases melatonin at dusk and reduces its concentration at daybreak. It functions as a light-controlled hormone – chemical communication through daylight down to a simple *cis-trans* isomerisation of a C=C double bond!

Our internal clock should be controlled by daylight but is easily displaced by artificial light sources such as smartphones, laptops and television since they inhibit the formation of melatonin. The effect of LED-lights in interior spaces can also be significant since they emit 1000 times more blue light (400–500 nm) than incandescent bulbs.

6.3.3 Optogenetics and light-dependent ion channels

Light confers the ability to initiate reactions synchronously and to switch them on and off selectively by choosing the wavelength that matches the absorption. The idea of switching nerves on and off *selectively* with light came from Francis Crick (of double helix fame) in 1979 but was not implemented until Deisseroth and Nagel spotted how to do this in 2003, thereby launching the field of optogenetics. First, we must step back into the field of micro-organisms. Rhodopsin structures are found in the most primitive

---

13 This class of proteins that includes rhodopsin, melanopsin and the melatonin receptor, is called a G-protein-coupled receptor (GPCR).
14 Serotonin and melatonin, like the amino acid tryptophan, are indole derivatives.

micro-organisms, Archaea, where they enable the organisms to open and close ion-channels with light, allowing H^+, K^+, Na^+ or Ca^{2+} to cross the cell membrane. Indeed, the first rhodopsin structure to be determined was that of bacteriorhodopsin from halo-archaea.[15] Unlike rhodopsin that functions by the photochemical conversion of 11-*cis* retinal to all-*trans*-retinal, the photo-isomerisation of bacteriorhodopsin converts all-*trans*-retinal to 13-*cis* retinal. In other words, the isomerisation goes in the opposite direction and generates a different isomer from rhodopsin. A closely related group of rhodopsins called channelrhodopsins control the light orientation, phototaxis, of green algae and have an absorption maximum in the blue region (480 nm) [16]. Like bacterio-rhodopsin, they control the opening of ion channels and convert *trans*-retinal to the 13-*cis* isomer. The key feature is that the relaxation back to the all-*trans* isomer occurs in ca. 10 milliseconds and thus the ion-channel can be opened and shut on a millisecond timescale. It can also be shut by irradiating with green light. Moreover, channelrhodopsins have now been found with different absorption spectra so enabling wavelength control [17].

Optogenetics exploits these channelrhodopsins and bacteriorhodopsin by genetic engineering methods that allow the splicing of their genes into a specific set of nerve cells of experimental animals such as mice or nematode worms. The isolated nerve cell can then be stimulated at will. Alternatively, when a light source (LED or diode laser) is implanted in the relevant region of the brain, the nerve cell can be fired by light pulses, operating on the same timescale as normal nerve signals. The method can be used to map the action of specific groups of nerve cells and to study how to treat abnormalities such as deafness or arrhythmia [16]. Optogenetics provides one of the approaches of "photopharmacology"; numerous other reactions are being exploited that are triggered by light [18].

6.3.4 Germination and growth of plants

The germination of plant seeds is light dependent for numerous plants. Lettuce, fox-glove and tobacco only germinate when they receive light in their planted state whereas other seeds, such as the pigweed family and squashes, germinate only in darkness. The search for the wavelengths of light that are effective in germination led to the discovery that light-germinators are stimulated by red light (R, 660 nm) and in-

---

15 The structure of bacteriorhodopsin was determined by Richard Henderson et al. by cryo-electron microscopy in 1990. Henderson received the Nobel Prize in 2017 for developing this technique. The name bacteriorhodopsin is a misnomer, because it comes from archaea rather than bacteria. There are two domains of single-celled organisms that have no nucleus, archaea and bacteria. Archaea reproduce by fission while some bacteria produce spores; some bacteria are pathogenic but no archaea are pathogenic. Many archaea, but not all, are extremophiles. The mechanism of DNA replication in archaea resembles that in eukaryotes more closely than in bacteria.

hibited by far red light (FR, 730 nm) [19]. The growth promoting effect of R can be stopped by subsequent irradiation with FR, can be stimulated by renewed R, etc. Natural light that is received by the seeds during germination contains both R and FR components in roughly equal proportions. The effect of the R-part of daylight exceeds the FR-part, so explaining the promotion by daylight of light-germinators. The inhibitory effect on dark-germinators is not yet fully explained. As with vision, a light-induced *cis-trans* isomerisation of the prosthetic group of a protein, in this case phytochrome, forms the basis of light-promoted germination (Scheme 6.3).

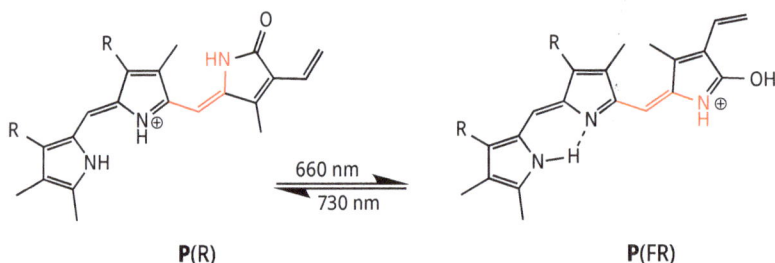

Scheme 6.3: Schematic diagram of the *cis-trans* isomerisation of the prosthetic group of phytochrome.

In addition to germination, the growth of plants also responds to the light spectrum. This process of photomorphogenesis differs significantly from photosynthesis where light is a source of energy [20]. Sunflowers turn toward the light, a phenomenon known as phototropism [21], that may also occur in other organisms such as fungi. Charles Darwin carried out numerous experimental studies on plants and animals and reported that plants respond to blue light in 1881. The light-sensitive protein called cryptochrome contains two chromophores, pterin and flavin adenine dinucleotide (FAD) with absorption maxima at 380 and 450 nm, respectively. The FAD is reduced photochemically in phototropism and photomorphogenesis – see Scheme 6.5 for the molecular structure of the flavin. Cryptochromes also play a role in circadian rhythms of insects and vertebrates as well as magnetoreception of birds (Chapter 6.8).

6.4 Vitamin D – biosynthesis

The preceding sections 6.1–6.3 have shown how simple chemical processes act as the basis of some of the interactions between light and organisms including humans. In these examples, energy transfer and isomerisation induced by absorption of visible light stimulated the effects without chemical change other than change of configuration. In contrast, the formation (biosynthesis) of vitamin D_3 in the skin (commonly denoted vitamin D) involves changes in the molecular structure of the light absorbing molecule.[16] On irradiation with UV-B light (280–315 nm, window glass does not transmit these wavelengths) 7-dehydrocholesterol (7-DHC) undergoes a ring opening of the middle 6-membered ring and synchronous formation and shifting of the double bonds (see Scheme 4.6). This is a conrotatory electrocyclic ring opening of the cyclohexadiene unit that forms the intermediate provitamin D_3. The resulting intermediate undergoes a 1,7-hydrogen shift and rearrangement of the double bonds to form vitamin D_3 (also called cholecalciferol) without participation of light. The vitamin is transferred from the skin, via the blood, to almost all the cells of the body, where it can switch a few thousand genes on and off. It also controls our immune system and induces the formation of an antimicrobial peptide that can protect us from illness. Consequently, a shortage of vitamin D has a wide variety of effects: rickets leading to stunted growth and bow-legs, tiredness, headaches, disturbed sleep, weakened immune system and trembling.

We are dependent on the light-driven synthesis in our own skin since our body can only obtain very little vitamin D_3 from food.[17] In the absence of sunlight, we suffer from deficiency conditions unless we source it in our food. The Inuit did not show signs of vitamin D deficiency despite the long dark winters – their diet included fatty, vitamin-rich fish, whale and seal meat.

The accounts in this chapter demonstrate that numerous functions and indeed our life are controlled by light from the sun: food to eat, oxygen to breathe, synthesis of vitamin D and other hormones to control our internal clock.

6.5 Communication – bioluminescence and fluorescence

Chemical reactions that are exothermic sometimes result in formation of excited states and emission of light. Chapter 4.1 illustrated this phenomenon of chemiluminescence with the oxidation of white phosphorus. Bioluminescence, as the biological analogue is called, is widespread and was already known to Aristotle (384–322 BCE). Robert Boyle demonstrated in the seventeenth century that air is needed for the luminescence of fungi. Glow-worms, the larvae and wingless females of a group of beetles, represent a classic example. Sometimes objects glow that we would not have expected. Even rotting wood can glow, but only when the rotting is caused by the mycelium of the honey fungus that is the real light emitter. There are two types of bioluminescence, primary and secondary. The former is used to describe luminescence that arises directly in the organisms (e.g. glow-worms), the latter when it originates in symbiosis between one organism (e.g. a bacterium) and another (e.g. angler fish or fungi).

---

16 Vitamin D is technically a hormone rather than a vitamin, since vitamins cannot be produced by our body.

17 The optimal blood concentration is ca. 50 nanogram per millilitre.

Darwin: on Phosphorescent Plankton

While sailing a little south of the Plata one very dark night, the sea presented a wonderful and most beautiful spectacle. There was a fresh breeze, and every part of the surface, which during the day was seen as foam, now glowed with a pale light. The vessel drove before her bows two billows of liquid phosphorus, and in her wake she was followed by a milky train.

   Charles Darwin's Journal of Researches into the Natural History and Geology of the countries visited during the Voyage of HMS Beagle Round the World, 1st Edition 1839

6.5.1 Bioluminescence in animals and bacteria

Bioluminescence in the animal world forms a means of communication: it can serve as camouflage, scare the enemy or signal readiness to mate [19]. The quintessential examples of the latter are glow-worms: the females emit pulses of greenish yellow light to attract the males (Figure 6.7). The length and intervals of the light pulses (flashes) of luminescence are species specific. The intervals can be so short that our eyes see continuous light. In some species the males and females fly and show luminescence – these are known as fireflies or, in the USA as lightning bugs. The beetles of the species *Pteroptyx similis* glow with the same rhythm such that a bush or wood can be filled with synchronously flashing lights [22]. In the case of the Japanese species *Hotaria parvula*, both partners emit exactly synchronised light signals that provide characteristic information about their species and sex. As soon as a female answers, an exchange of photons ensues leading to a courtship dance. Some females (of the genus *Photuris*) feign the invitation to a rendezvous – they have other intentions. They flash with the rhythm of a different genus, only to consume the expectant male when he appears: this glowing sex appeal can be deadly!

   The light emission of glow-worms and fireflies originates in most cases from the enzyme-catalysed oxidation of firefly luciferins to oxyluciferin. The enzyme is called firefly luciferase and requires O_2 and ATP.[18] In the first step, a carboxylate group is converted, via an AMP derivative, to a 4-membered ring dioxetanone that undergoes decarboxylation to form oxyluciferin in its electronic excited state. Finally, oxyluciferin decays to its ground state with light emission (Scheme 6.4). The bioluminescence quantum yield is amazingly high: ca. 0.6.[19] Variations in luciferin and luciferase give rise to species-specific differences in the emission spectrum.

---

18 Luciferase and luciferin are generic names that apply to enzymes and their prosthetic groups of many very different organisms. A luciferase catalyses the reaction generating light and the luciferin is an organic compound in a bioluminescent organism that provides the energy for light emission, normally by being oxidised.

19 The bioluminescence yield represents the product of the fluorescence quantum yield, the chemical yield of the light path and the efficiency of populating the excited state of the emitter.

Figure 6.7: Female glow-worm of *Lampyris noctiluca* flashing in search for a partner [23].

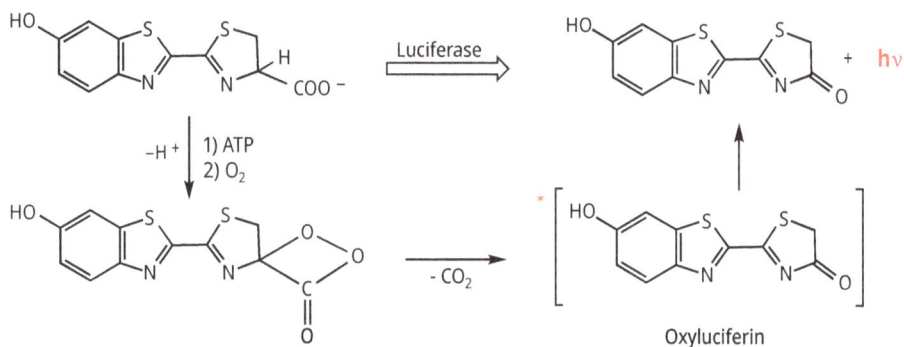

Scheme 6.4: The luciferase reaction of glow-worms.

Since luciferin of glow-worms can be synthesised in the laboratory relatively easily and firefly luciferase is accessible by genetic methods, this greenish yellow (ca. 560 nm) bio-luminescence finds numerous applications. It provides one of the best methods for quantifying the presence of ATP. In diagnostics, it can be used to detect minute traces of ATP that arise from the presence of bacterial contamination. The addition of a buffer solution[20] kills the bacteria and the ATP that is released can be assayed with the lucifer-ase reaction. The luminescence is used in genetics to monitor transfer of genes. The splicing of foreign genes into the genome of another organism in a cell culture is a stan-

---

20 The lethal pH value is achieved with the buffer solution.

dard procedure. The gene that codes for luciferase is expressed (coupled) in the desired gene. On addition of luciferin, the bioluminescence is observed only if the transfer is successful. The luciferase serves as a marker that is easily detected.

The luciferase gene is used in medical research spliced into bacteria, viruses or cells of higher organisms. If, for instance, a rat is infected with salmonella labelled with the luciferase gene, the spread of the infection through the body is even visible by eye! Labelled carcinomas can be made visible optically as they form metastases. If a sensitiser for singlet oxygen is present at the same time, the bioluminescence can even be used to stimulate a PDT-reaction to attack the carcinoma. Genetically modified viruses can be used following the same principles for diagnostics and simultaneous therapeutics (theranostics) – a promising procedure [24].

Bioluminescence is also the basis of animal life in the deep ocean. The midnight zone – depths below 1000 m – is the largest habitat on Earth but is often forgotten. Without the light produced by chemiluminescence, life in complete darkness would not be possible. Only with the aid of light can a partner be attracted or an enemy driven away. Light is therefore the most frequently used information carrier. More than 90% of deep ocean organisms can glow, but for human eyes this can only be perceived indirectly with very sensitive detectors. The light-emitting organs of these creatures are called photophores and may be quite complex as in squid.

Unlike glow-worms, most ocean dwellers glow with blue or blue-green light (ca. 475 nm), corresponding to a different chemical structure of the luminescent substance compared to luciferin. Some crustaceans also have a luciferin, but with a very different prosthetic group. Jellyfish of the species *Aequorea victoria* contain a photoprotein in their photophore called aequorin that is chemiluminescent and another protein that is fluorescent – this is the famous green fluorescent protein (GFP) – more of that later. In life, they glow green from GFP, but isolated aequorin glows blue. Unlike luciferin, the bio-chemiluminescence of aequorin is not the result of an enzymatic oxidation, but the reaction of the protein with calcium ions (see textbox) [19]. The prosthetic group coelenterazine binds to aequorin and reacts very slowly (hours at 5 °C) with oxygen to form a peroxide intermediate. The reaction of the peroxide with two calcium ions triggers structural changes in the protein and conversion to coelenteramide with release of CO_2 and emission of light at 465 nm (Figure 6.8). Importantly, once the intermediate is formed, oxygen is no longer required for chemiluminescence. The structure of aequorin resembles that of calmodulin, a well-known calcium binding protein. In vivo, however, the excitation is transferred to GFP which emits at longer wavelength. Shimomura's discovery[21] of the calcium trigger is reproduced in his own words in the textbox [19a]. This reaction has been used as a biological calcium sensor.

---

[21] Osamu Shimomura, Martin Chalfie und Roger Tsien were awarded the Nobel Prize in Chemistry in 2008 for the discovery and development of green fluorescent protein.

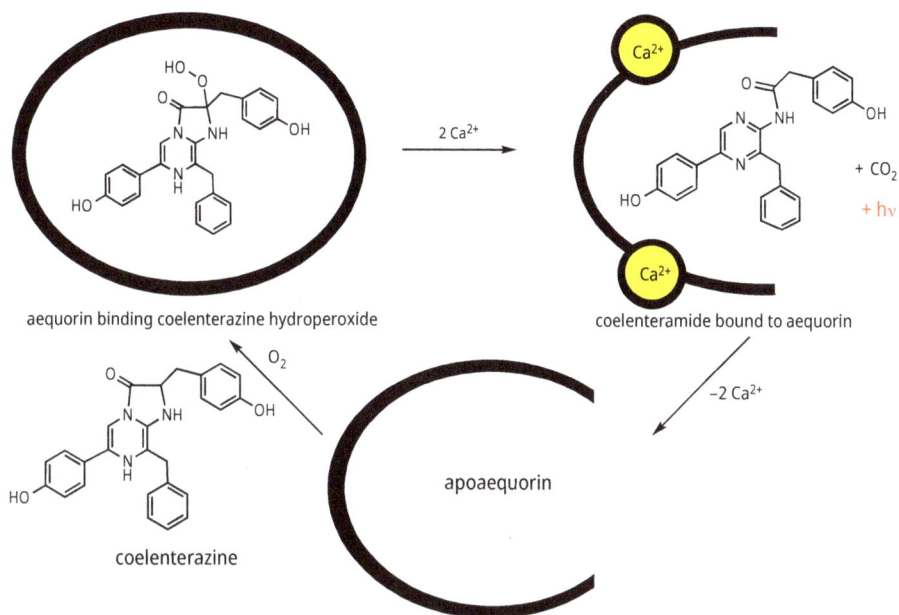

Figure 6.8: Aequorin chemiluminescence triggered activation by oxygen followed by binding of Ca^{2+} to protein. The black figures represent the protein [19].

Osamu Shimomura's discovery of the calcium ion trigger of aequorin

"I tried to reversibly inhibit luminescence [of aequorin] with various kinds of inhibitors of enzymes and proteins. I tried very hard, but nothing worked. I spent the next several days soul-searching, trying to find out something missing in my experiments and in my thought. I thought day and night. I often took a rowboat out to the middle of the bay to avoid interference from people. One afternoon, an idea suddenly struck me on the boat. It was a very simple idea: "The luminescence reaction probably involves a protein. If so, luminescence might be reversibly inhibited at a certain pH". . . I ground the light organs in a pH 4 buffer, then filtered the mixture. The cell-free filtrate was nearly dark, but it regained luminescence when it was neutralized with sodium bicarbonate. . . But a big surprise came the next moment. When I threw the extract into a sink, the inside of the sink lit up with a bright blue flash.. . .Because the composition of seawater is known, I easily found out that Ca^{2+} ions activated the luminescence. The discovery of Ca^{2+} as the activator suggested that the luminescence material could be extracted utilising EDTA, and we devised an extraction method of the luminescent substance."

Every spring, there is a remarkable natural spectacle in the Toyama Bay off the coast of Japan. Firefly squid (*Watasenia scintillans*) live in the ocean at a depth of about 300 m. Their 7 cm long bodies are covered by photophores that glow with blue and green chemiluminescence (Figure 6.9) and act as camouflage against predators (such as salmon). In addition, they possess blue-emitting photophores (470 nm) round the eyes and on the tentacles. The latter can be directed to dazzle attackers or prey like headlights. In spring, the squid come to Toyama Bay to pair and lay their eggs. Then even the beach is covered in blue flashes as thousands of these squids are washed up onto the shore. The local people wait there with nets and headlamps to gather these delicacies. The light-emitter is a coelenterazine derivative like that of *Aequorea* but is triggered by a luciferase-like enzyme, ATP, Mg^{2+} and O_2.

Figure 6.9: Firefly squid in the Toyama Bay (Japan) [25].

Prawns such as krill also have photophores. They are situated on the stalks of the eyes and their luminescence is stimulated hormonally and serves for *internal* communication. Additionally, the luminescence can be used *externally* to recognise a mate and to control their shoaling behaviour. Crown jellyfish are inhabitants of the deep ocean. They scare their attackers by making their transparent umbrella or "bell" glow blue on touching. The glow of these *Periphylla* jellyfish also depends on oxidation of a luciferin involving coelenterazine in a similar reaction to that of the firefly squid.

While the ocean creatures described above generate their light by chemiluminescence, phytoplankton do so by mechanical stimulation exploiting the principles of triboluminescence (see Chapter 1.1). The breaking waves of the surf generate shear forces that act on the cells of the dinoflagellates (0.002–2.0 mm). The resulting electric field converts the luciferin of these microorganisms to its electronic excited state and the blue-green emission scares the predators. Their luciferin consists of a linear tetrapyrrole bearing a resemblance to chlorophyll (Figure 6.2). Unlike the luminescence of the organisms described earlier, no decarboxylation occurs.

In contrast to the examples of primary bioluminescence described above, some fish glow through the action of symbiotic bacteria – this is secondary luminescence. For example, the lantern fish of the Pacific Ocean contain bacteria in a type of tear duct under the eyes (Figure 6.10). These fish sleep during the day at a depth of 1200 m and migrate vertically upwards at night following shoals of zooplankton to within 10 m of the surface. They can switch "bio-headlights" on and off by turning their eye. Since this happens several times per minute, the fish swim around flashing their lights and use the pulses for orientation in finding food. The pulse frequency plays an essential role in the behaviour of the fish shoal. In the morning, they return to the deep. Bacterial bioluminescence depends on the oxidation of long chain aldehydes to the corresponding carboxylic acid with molecular oxygen in the presence of reduced flavin mononucleotide (FMNH2) and an enzyme called Lux luciferase resulting in light emission from the excited state of the flavin product (Scheme 6.5) [26].

Secondary bioluminescence also occurs in fish of the deep ocean. Anglerfish inhabit this midnight zone – they have a characteristic lantern on their heads that is

Figure 6.10: Lantern fish (ca 30 cm long) with light organ under the eye [27].

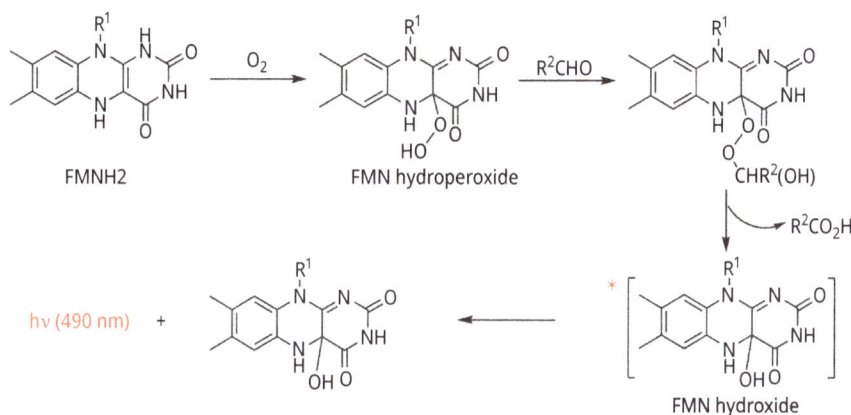

FMNH2 — O_2 → FMN hydroperoxide — R^2CHO → CHR^2(OH)

→ R^2CO_2H

FMN hydroxide

hv (490 nm) +

Scheme 6.5: Basis of bacterial chemiluminescence catalysed by Lux luciferase. $R^1 = CH_2(CHOH)_3CH_2OPO_3H_2$ [26].

stimulated to glow by the action of luminescent bacteria. The lantern acts as a lure both for mating and hunting. The female of the black angler (Figure 6.11) attracts prey straight into its open gullet and also attracts the male for pairing. The male of some species bites onto the much larger female[22] and doesn't let it go. His body grows together with hers and the male gives up all bodily functions other than sperm production. One female can even be bound to several males in this strange manner. An adaptation to the unusual habitat could be responsible: very few conspecifics swim in the deep and they are hard to find. Once a mate is found, it is beneficial to hold on tightly [28].

The sepiolid squid *Euprymna scolopes* camouflages itself in calm coastal waters in a cunning way. When it's hunting in bright moonlit nights, its photobacteria glow exactly as brightly as the moonlight: it becomes a hunter without a shadow. Moreover, its symbiotic bacterial cells generate light in a circadian rhythm.

---

22 The female is 10–60 times bigger than the male.

Figure 6.11: The black anglerfish attracts its prey with its light lure. Reproduced with kind permission © J. Muralt [28].

Secondary bioluminescence through bacteria can also be observed in rotting foods such as eggs, meat and fish. If a salted herring, for example, is kept too long, it begins to glow weakly in the dark. Bacterial bioluminescence can even be observed in humans on wound tissue. This explains an observation made during the American Civil War: of more than 16,000 wounded soldiers at the Battle of Shiloh on the Tennessee River in 1862, those whose wounds showed a blue glow were most likely to survive. This story was named Angel's Glow from then on. Thanks to the curiosity of a high school student and his microbiologist mother, this turned out be more than a legend. This photobacterium *Photorhabdus luminescens* lives symbiotically with nematodes that attack insect larvae. The photobacteria secrete toxins that kill the insect larvae and have antibacterial properties, such that the spread of dangerous bacteria is prevented [29].

6.5.2 Bioluminescence in fungi and plants

Aristotle's treatise *"On the Soul"* written about two and a half thousand years ago includes the words: *There are certain things which are, indeed, not seen in light, but which produce a sensation in darkness, such as those which burn or are luminous. Such are the fungi of certain trees,"*

Presumably, this referred to the edible honey fungus *Armillaria mellea*, that grows on wood. The depth gauge of the first submersible of the American navy, the *US-Turtle* built in 1775, was lit by a piece of cork on which a bioluminescent fungus grew! Edison and Swan had not yet invented the incandescent light at this time – that came 100 years later.

Of the hundred thousand known fungus species, only about 1% are bioluminescent. For a few, such as *Neonothopanus gardneri* (Figure 6.12), it has been shown that the glow is controlled enzymatically: the glow is dimmed by day and brightened in the dark. This adaptation ensures the attraction of the maximum number of insects which disperse the spores through the wood. Bioluminescent fungi act as cold light for submersibles and attractants for insect dispersers! The luminescent compounds of these fungi are substantially different from those encountered earlier: they use a cyclic structure called an α-pyrone. Nevertheless, they work by reaction with oxygen and decarboxylation, probably via a peroxide intermediate (Scheme 6.6) [30].

Figure 6.12: The fungus viewed by day and by night. Reproduced with kind permission © Cassius V. Stevani, IQ-USP, Brazil [31].

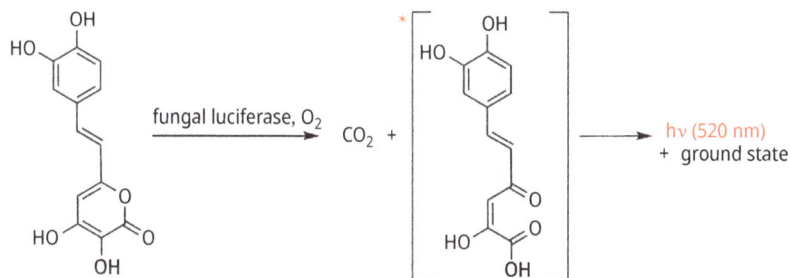

Scheme 6.6: Chemiluminescence reaction of fungi. The peroxide intermediate is not shown.

The concept of glowing plants is not that big a step from glowing fungi. In fact, you can already buy a chemiluminescent tobacco plant. The gene of a marine photobacterium has been successfully spliced into the genes of the chloroplasts of the plant. The plant is then able to produce the luciferin and luciferase. The negative image of genetic engineering threatens, however, to suppress the dream of biological light sources. A method free of genetic engineering would be to populate trees with glowing fungi – then the self-lighting Christmas tree will no longer be science fiction.

6.5.3 Overview of biological chemiluminescence

The preceding sections have shown that biological chemiluminescence is observed in organisms from bacteria through jellyfish and squid to insects and fungi. Moreover, higher organisms such as fish use symbiotic bacteria to achieve their glow. However, there is no common method of generating the luminescence. Unlike ATP, amino acids or nucleotides, no universal system has evolved. There are numerous different luciferases and luciferins. The luciferins range from sulfur-nitrogen heterocycles to flavins and pyrones as summarised in Figure 6.13 [32]. Reaction with O_2 is an essential part of the cycle in all of them and it is likely that most, if not all involve a peroxide intermediate. However, the chemiluminescence of aequorin is triggered by calcium ions alone – the reaction with O_2 is required only in the slow recovery process. Many of these reactions, but not all result in loss of CO_2 (decarboxylation).

6.5.4 Fluorescent proteins

The jellyfish *Aequorea victoria* contains the chemiluminescent protein with its coelenterazine prosthetic group described above (Chapter 6.5.1) that hands its excitation to another protein by Förster energy transfer, resulting in light emission centred at 509 nm. This is Green Fluorescent Protein (GFP), a protein that ushered in transformative discoveries. GFP contains no prosthetic group but includes a hexapeptide that has been modified to form a conjugated chromophore. The small (27000 Da), soluble protein has a beautiful structure with a cylinder of β-sheets[23] on the outside and an α-helix on the inside that encloses the chromophore as part of its primary structure (Figure 6.14). As a result, the fluorescence depends on its environment within the protein – the denatured protein does not fluoresce. The modifications principally involve a serine-tyrosine-glycine tripeptide that has been cyclised, dehydrated and oxidised [34]. The oxidant is molecular oxygen and the by-product of oxidation is hydrogen peroxide. GFP absorbs at 397 and 475 nm and emits with a quantum yield of about 0.8. It is also found in numerous other marine organisms. The chromophore is a photo-acid (see Chapter 2.2.8) in which the OH group of the tyrosine is deprotonated in the excited state prior to emission, thus lengthening the conjugation (Scheme 6.7a). The emission spectrum of analogues of GFP can be tuned right across the rainbow from about 420 to 650 nm by change of organism and by genetic engineering methods providing a remarkable range of fluorophores, including far-red emission with its greater depth penetration. There are even some analogues that switch on their fluorescence in response to

---

23 Beta-sheets contain two or more polypeptide chains arranged alongside each other and are linked by hydrogen bonds between the main chain C=O of one polypeptide chain and the N-H group of the neighbouring polypeptide.

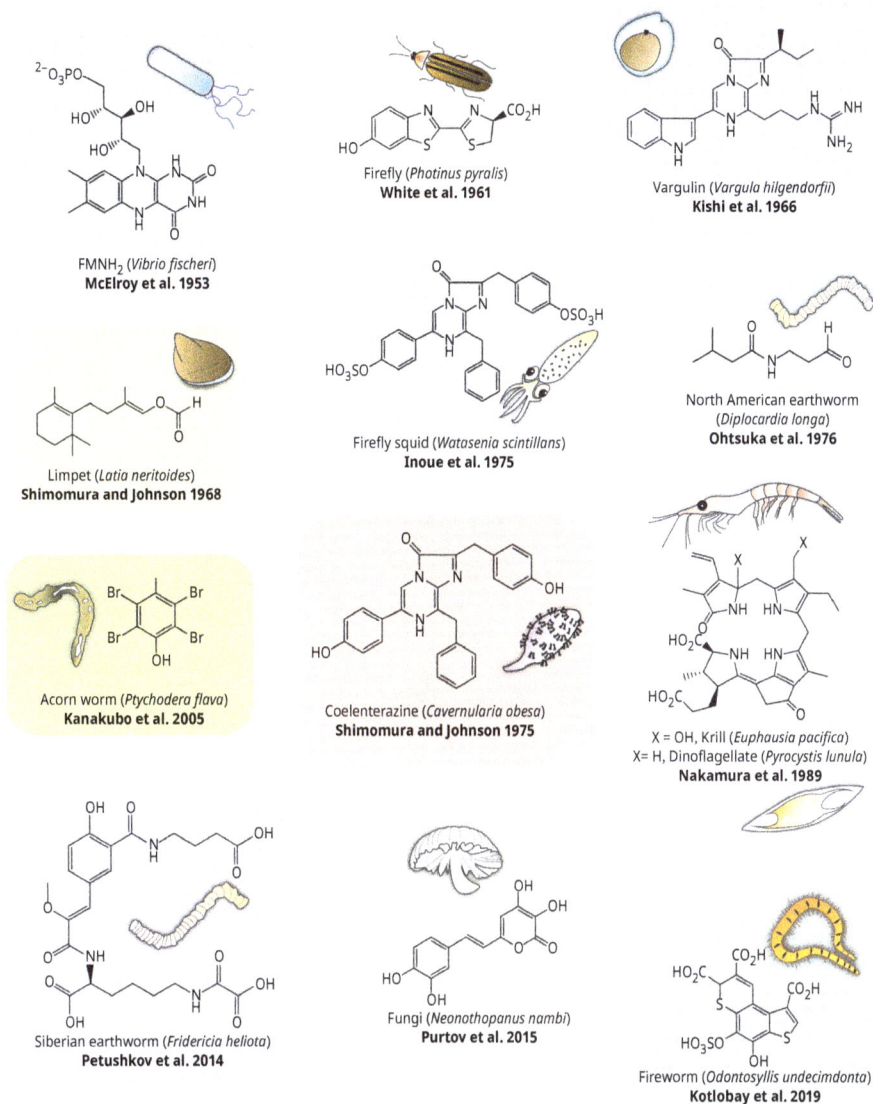

Figure 6.13: The wide variety of chromophores that form the basis of bioluminescence. Reproduced with kind permission from *Journal of Photochemistry Photobiology C: Photochemistry Reviews* **2024**, *58* 100654, © Elsevier 2024 [32].

irradiation at the appropriate wavelength, enabling spatial and temporal control (Scheme 6.7b). In the example shown, the protein normally fluoresces green, but on UV irradiation, a peptide bond breaks, a new C=C double bond is formed and a phenolic OH is deprotonated, resulting in a shift to long wavelength and red emission [19].

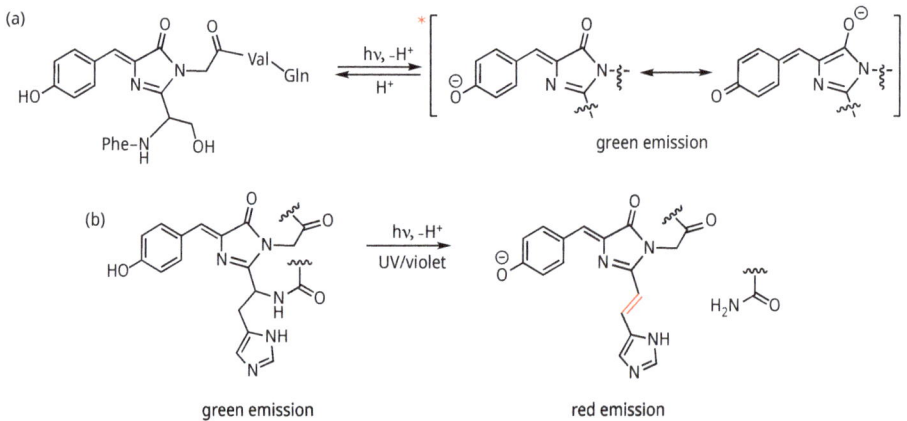

Scheme 6.7: (a) chromophore of wild-type GFP from *Aequoria victoria* with deprotonation on excitation; (b) GFP from stony coral activated by UV to switch fluorescence from green to red. Phe, Val, Gln correspond to phenylalanine, valine, and glutamine.

The success of GFP (and analogues) as fluorescence markers and sensors depends on the ability to incorporate its gene into organisms of all three different domains of life, and the fact that no additional enzymes are needed to achieve the modifications to amino acid residues. It is sufficiently small and unreactive that it does not perturb other organisms and can be fused with other proteins. Unlike small molecule fluorescence markers, it is heritable. It is typically detected by fluorescence microscopy and can be used on live organisms or live cells. The innumerable applications include imaging at super-resolution magnifications, localisation of proteins, tracing the role of cations especially Ca^{2+}, investigating cell division, growth of neurones, gene expression, spread of viruses between cells, for sensing toxins, etc. Calcium ions play an enormously important role in signalling with many functions, including fertilisation and cell division. By fusing GFP to the calcium-binding protein calmodulin and its target peptide, it is possible to build a system that responds to Ca^{2+} via FRET (see Chapter 2.2.3). Images of cell division (Figure 6.15) were obtained with the aid of two different fluorescent proteins CFP and YFP, the former emitting blue-green (476 nm) and the latter yellow (527 nm) light. In the absence of Ca^{2+}, the two emitters are too far apart for FRET, so excitation of CFP results in blue-green emission. On Ca^{2+} binding the two emitters become close enough for FRET and excitation of CFP results in yellow emission from YFP. These fusions that produce multicoloured images of Ca^{2+} signals have been nicknamed *cameleons.*

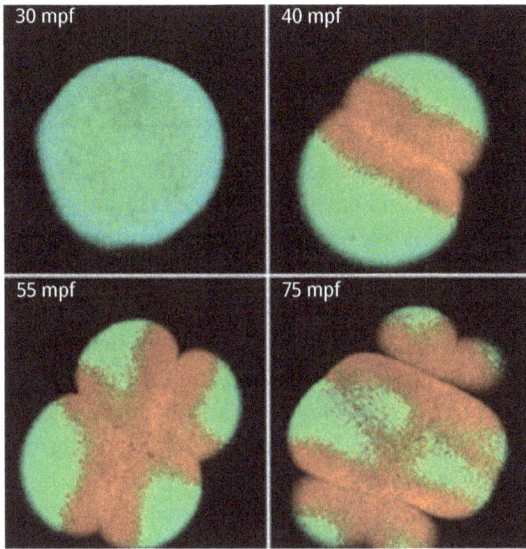

Figure 6.15: Cell division of zebrafish embryo expressing "yellow cameleon". mpf = minutes post-fertilisation. The orange colour shows the wave of Ca^{2+} that appears every time the cell divides. Published with kind permission from Nobel Prize Outreach 2024 copyright © The Nobel Foundation 2008, Roger Tsien Nobel lecture 2008. {Les Prix Nobel, The Nobel Prizes 2008, Ed. K. Grandin, Nobel Foundation Stockholm, 2008} [35].

GFP and its analogues were the first fluorescent proteins to be discovered that rely on their primary structure for their fluorescence. More commonly, fluorescent proteins incorporate prosthetic groups that fluoresce, for example, flavoproteins or phycobiliproteins. The latter is a light harvester for photosynthetic cyanobacteria that contains a chain of four pyrroles. Flavoproteins have also been used as fluorescence labels when the requirement of GFP for oxygen is an issue.

6.5.5 Fluorescence labels

There are a wealth of uses of small fluorescent molecules in biological applications, several of which have been mentioned elsewhere in the book – see textbox. There are so many different small molecules available that there are entire catalogues full of them. Here we will mention two further examples that link to what has been discussed previously.

Fluorescence Labels

	Chapter
Single molecule fluorescence - DNA sequencing	1.5
Fluorescence microscopy	1.5
Super-resolution imaging	1.5
Förster Resonance Energy Transfer (FRET)	2.2.3
Lanthanide antenna complexes for analytical application	2.2.3
Luminescence sensors for analysis of cations in blood	4.2.6

Calcium ions have a wide range of roles as "second messengers", enabled by the difference in concentration between the interior and exterior of the cell (10^{-7} M inside, 10^{-2} to 10^{-3} M outside). In the previous sections, we encountered proteins that can be used to monitor calcium ions by chemiluminescence and fluorescence. However, there are numerous applications of calcium waves and oscillations that require small molecule labels. One example is the involuntary nerve response of epileptics that can be monitored in model organisms via spikes in Ca^{2+} concentrations. The fluorophores incorporate a dye unit (for example resembling fluorescein) and a binding unit, typically an analogue of EDTA, and are designed to be selective for calcium over magnesium ions. They exhibit little or no fluorescence in the absence of calcium but light up in its presence. An example called Fluo-4 is shown in Figure 6.16, together with a time plot taken with it.

Ruthenium polypyridyl complexes have featured several times in this book. In the biological context, they are of interest both for their ability to cause photo-induced electron transfer and for their luminescence properties. Photo-induced electron transfer can cause oxidation of guanine and hence DNA damage leading to potential for PDT. The MLCT luminescence of some derivatives is quenched in water but activated in the presence of DNA – the so-called light-switch effect. Since both the complexes and DNA are chiral, these effects exhibit enantiospecificity. With suitable variation of the ligands and the DNA, specificity for particular DNA structures can be observed. In addition to the study of luminescence, key insights into the binding of metal complexes to short self-complementary DNA strands have been obtained by X-ray crystallography and by time-resolved infrared spectroscopy (see Chapter 1.8).

The key to obtaining high-quality crystals to elucidate the light switch effect was to use $[Ru(NN)_2(dppz)]^{2+}$ (NN = phenanthroline, dppz = dipyridophenazine) in the presence of double helical-DNA with Ba^{2+} to balance the charge of the DNA phosphate groups

(a)

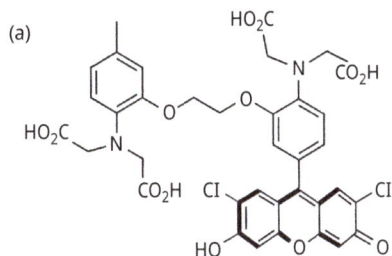

(b)

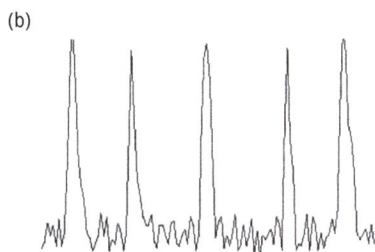

Figure 6.16: (a) Fluo-4, a calcium selective fluorescent dye, (b) time sequence of calcium spikes in the resting neurone over a period of 100 s [36].

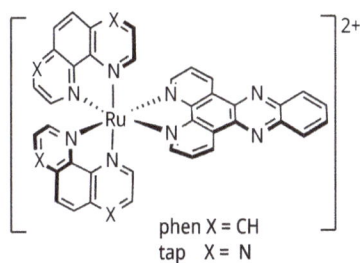

phen X = CH
tap   X = N

Figure 6.17: Molecular structure of Λ-[Ru(NN)$_2$(dppz)]$^{2+}$ (NN = phenanthroline or tetraazaphenanthrene).

(Figure 6.17). Dppz intercalates into the minor grove of the DNA double helix so that the dppz ligand forms π-stacking interactions with purine bases and distorts the DNA chain. The difference in angles of intercalation into the same DNA sequence between L and D enantiomers of [Ru(LL)$_2$(dppz)]$^{2+}$ results in different hydrogen bonds of the nitrogen atoms of dppz with the water environment. DNA damage at guanine is characteristic of complexes with LL = TAP = tetraazaphenanthrene. The damage results from photoinduced electron transfer from guanine to TAP. The effect of intercalation can be followed by TRIR spectroscopy both in crystals and in solution through the characteristic vibration of the guanine radical cation and can be correlated with crystal structures. It is also possible to determine the structure of different DNA structures bound to ruthenium complexes including a single-stranded telomeric sequence of 21 bases forming a chair-shaped G-quadruplex (a telomer reads the same forwards and backwards) [38].[24]

---

24 Much of the research associated with these complexes has been led by women scientists: Jacqueline Barton (USA, discovery of the light switch effect, 1990), Andrée Kirsch-De Mesmaeker (Belgium,

6.6 Photoacoustics

Photoluminescence and time-resolved spectroscopy are not the only ways of characterising excited states. They can also be investigated by measurement of radiationless processes with the help of acoustic detection. The tissue is irradiated with a pulse of laser light lasting a few nanoseconds. The heat generated through absorption leads to a short-lived thermal expansion of the tissue that generates an acoustic pressure wave (see Chapter 3.1.6). This wave is detected with ultrasound receivers on the surface of the skin. A layer profile of the skin is obtained by altering the wavelength and pulse length of the laser. This method gives significantly more precise three-dimensional images of the conspicuous components of the tissue than are obtained by conventional fluorescence imaging because of the reduced scattering of acoustic waves compared to light scattering. Investigations aimed at better recognition of prostate carcinomas are in progress [39].

When photoacoustic methods are used in conjunction with pulsed lasers, they can be used to determine the energetics of excitation of biological chromophores such as retinal and flavoproteins [40].

6.7 Electron transfer through proteins by long-range tunnelling

Biological energy transduction, whether in photosynthesis or respiration, depends on electron transfer from a substrate to a protein or from one protein to another. These electron transfer proteins store the electrons at prosthetic groups that may be distant from the substrate or protein supplying the electron. The prosthetic groups are often copper centres as in azurin (a bacterial protein), or iron-porphyrins (cytochromes) as in cytochrome c. Alternatively they may be organic groups based on quinones. How do the electrons travel from one centre to another and how fast do they go? It was Harry Gray who spotted how to answer this question – tag a histidine on the exterior of the protein with a ruthenium(II) bipyridyl group, forming a [Ru(bpy)$_2$(Imidazole) (His)]$^{2+}$ group. Laser excitation of this centre generates an excited state that is oxidised to Ru(III) by an external quencher. The metal centre of the protein transfers an electron to the Ru(III) intramolecularly providing the required kinetic signal. Finally, the reduced quencher reduces the metal centre of the protein back to the starting state (Figure 6.18A). The protein can be modified by site-directed mutagenesis to place histidine groups at different positions on the surface. This method allows collection of data for the distance dependence of the rate of electron transfer, k_{ET}. As an example,

photo-oxidation), Christine Cardin (UK, crystal structures of DNA-Ru complexes), Susan Quinn (Dublin, TRIR spectroscopy), Eimer Tuite (Sweden and UK, intercalation into minor groove from photoluminescence data). The roles of numerous women scientists in photobiology and the obstacles that they encountered has been described by Silvia Braslavsky [37].

the rate of electron transfer in azurin varies from ca. 10^6 s^{-1} over a distance of 17 Å to 10^2 s^{-1} over a distance of 26 Å. A plot of the logarithm of the lifetime τ ($\tau = k_{ET}^{-1}$) against distance for numerous metalloproteins yields an approximate straight line (Figure 6.18B). The slope of this line, β, lies in the region 1.1 $Å^{-1}$. The observation of scatter on this graph indicates that different protein structures differ in their ability to support tunnelling. Proof that electron transfer occurs by tunnelling comes from the lack of temperature dependence even over a wide temperature range either in solution or in the crystal.

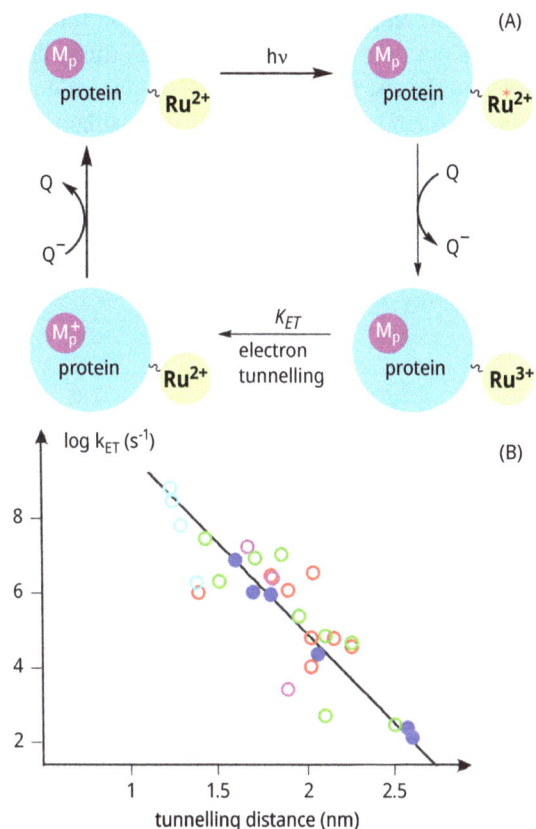

Figure 6.18: (A) Reaction scheme for flash quench measurement of electron tunnelling rate. M_p = protein prosthetic group, Q = quencher, Ru^{2+} = ruthenium tag. The diagram shows oxidative quenching. Alternatively, reductive quenching may be used. (B) Dependence of the rate constants for electron tunnelling through proteins modified with a ruthenium complex tagged onto the surface (azurin, blue; cytochrome c red; myoglobin, magenta, cytochrome b_{562}, green, high-potential iron protein with Fe_4S_4 centre, cyan). The black line represents the best straight line and its slope is β. Adapted with kind permission from *Chem. Rev.* 2010, 110, 7024–7039, Copyright 2010, American Chemical Society.

The interpretation of the electron transfer rates depends on Marcus theory which shows that the rate of electron transfer depends on the driving force and the reorganisation energy (Chapter 2.2.6 and Equation 2.24). More detailed analysis of the electron transfer rates in proteins yields the reorganisation energies. The maximum distance for this tunnelling is ca. 25 Å, yet electrons transfer over much larger distances in some proteins. In these proteins, there may be a series of stations en route so the electron tunnels from one station to the next before reaching its final destination. This role can be fulfilled by iron-sulfur centres or aromatic amino acid residues such as tryptophan. Examples include three of the proteins that we mentioned, photosystem II, DNA photolyase and cryptochrome. According to the data in Figure 6.18B, if an electron tunnels over 10 Å twice, the tunnelling rate can be ten thousand times faster than tunnelling once over 20 Å. This concept has been put to the test by binding a $Re(CO)_3$(phenanthroline) unit to a histidine on the outside of azurin (see Chapter 5.1.3). A neighbouring amino acid residue is mutated to tryptophan so as to reduce the electron transfer distance from Re to Cu of 19 Å. Laser excitation of the Re centre generates an excited state that is quenched by hole transfer from Re to Cu(I) via the tryptophan. The hole is found to transfer in 31 ns instead of the expected microsecond timescale. The return transfer from Cu(II) to Re takes place directly, not stepwise, in 3 µs [41].

6.8 Optical magnetoreception and DNA repair

Many organisms navigate with the Earth's magnetic field during their spectacular migrations. The experimental evidence (see textbox) [42] that some birds can orient with the Earth's magnetic field poses extraordinarily challenging questions. Amazingly, it appears that the birds "see" the direction of dip of the Earth's magnetic field with their eyes via a photochemical reaction in the protein cryptochrome (Chapter 6.3)! How can there be a chemical response to such a small magnetic field? The initial thought is to search for a ferromagnet but it transpires that magnetic sensitivity can proceed without.

Most chemical reactions are unaffected by magnetic fields; even those that are, require fields ten thousand times greater to elicit a significant response. Photo-induced electron transfer initially produces a radical pair $\{D^{+\cdot}A^{-\cdot}\}$. If the radicals remain in close proximity, their electron spins are correlated to one another. Initially, they are produced with opposed spins (spin singlet), but the pair undergoes a quantum mechanical oscillation between singlet and triplet state millions of times a second (see Chapter 2.1.1). The radical pair has two competing pathways, recombination to regenerate the starting material or reaction to form product. Since only the singlet state can recombine, the formation of product depends on the spin states. Consequently, this quantum mechanical oscillation is altered by application of a weak magnetic field in a way that is dependent on the field direction. Cryptochrome is expressed in the photoreceptor cells of the eyes of the European robin. Irradiation of the isolated protein at 450 nm induces formation of a pair of radicals $\{FAD^{-\cdot} \; Trp^{+\cdot}\}$ where the donor is a tryptophan residue of the

The European robin can see the dip of the Earth's magnetic field

(Photo © Robin Perutz)

As long ago as 1972, husband and wife team Wolfgang and Roswitha Wiltschko published their seminal paper on the European robin (*Erithacus rubecula*), a bird that migrates by night. The birds exhibit migratory restlessness that reveals an orientation with the inclination (dip) of the magnetic field – they find North with the dip, not the polarity of the field. In the experimental set-up, the birds are in a cage where their feet leave a mark on blotting paper. This birdcage is enclosed in a Faraday cage to eliminate the Earth's magnetic field but magnets within the Faraday cage control the field direction and inclination. How does this perception work? At 50 °N in 2022, the horizontal and vertical components of the Earth's magnetic field were ca. 20 and 45 microtesla, respectively. This tiny field must be sufficient to provide the requisite signal.

protein and the acceptor is the chromophore, FAD (Figure 6.19). A chain of up to four tryptophan residues enables the hole to hop along the chain and so increase the distance between cation and anion, slowing down the back electron transfer. The radical pair has a characteristic absorption spectrum that lasts for ca. 0.1 to 1 microsecond. The Trp$^{+\bullet}$ is acidic and protonates the reduced FAD to form FADH$^{\bullet}$ which has a different optical spectrum. Application of a magnetic field of 10 millitesla for a microsecond in the lab is sufficient to randomise the spins resulting in back electron transfer. The bird would perceive a colour change according to the magnetic field and transmit the information to the brain via the optic nerve. These experiments on the isolated protein begin to show how the magnetoreception may work, but the lifetime of the radical pair and its sensitivity to the magnetic field must be greater in vivo than observed in these experiments. Calculations of the light intensity on a starlit night show that this low intensity would be sufficient for magnetoreception. In keeping with this theory, orientation of the robin is disrupted by electromagnetic noise and by a weak oscillating magnetic field. Unlike opsins that depend on isomerisation, cryptochrome generates a photo-induced radical pair creating a suitable basis for this mechanism [43].

What do DNA repair and magnetoreception have in common? They exploit the same type of photosensitive protein. We saw in Chapter 4 (Figure 4.5), that UV-damage to DNA can induce dimerisation of adjacent thymine bases to form a cyclo-butane unit. Birds, amphibians, fish and even marsupials, but not placental mammals, possess an enzyme with a structure resembling that of cryptochrome with the ability to repair the damage. Like cryptochrome, the photolyases possess an FAD and (usu-ally) a pterin prosthetic group (Figure 6.19b). The pterin absorbs strongly in the near UV (λ_{max} 360 nm) and transfers its energy via FRET to the fully reduced FADH$^-$. Alter-natively, the FADH$^-$ absorbs directly in the same region but with a lower absorption coefficient. Following DNA binding to the enzyme, the excited FADH$^-$ transfers an electron to the thymine dimer causing cleavage of one of the cyclobutane C–C bonds (C5–C5′). The other C–C bond (C6–C6′) is broken in the next step. Remarkably, the cleavage processes following electron injection and the final back electron transfer to FADH$^\bullet$ are complete in less than 1 ns in solution, leading to the proposal of involve-

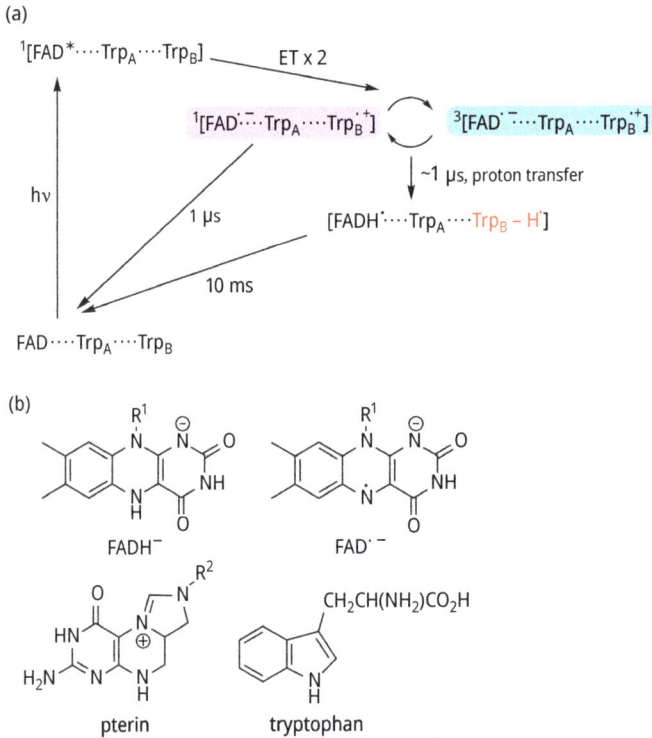

(a)

(b)

Figure 6.19: (a) Formation of radical pair in cryptochrome showing the quantum mechanical oscillation between singlet and triplet states. The lifetimes vary with species. Trp = tryptophan. Note that Trp$_B$ – H$^\bullet$ shown in red represents Trp with an H Atom removed. Adapted from ref [43a]. (b) Simplified structures of the FAD (flavin adenine dinucleotide) and pterin prosthetic groups and of the amino acid tryptophan. R^1 = $CH_2(CHOH)_3CH_2$(diphosphatoadenine), R^2 = C_6H_4-(glutamate)$_n$ (n = 3 to 6).

ment of electron tunnelling (see Chapter 6.7 above). This process has also been followed by time-resolved X-ray crystallography revealing how the $FADH^-$ adopts a butterfly-shape that switches from "wings down" to "wings up" on excitation. These DNA-repair properties have led to the incorporation of photolyases in sunscreens and other skin care products [44].

While photolyase exploits electron transfer at FAD, there is another group of proteins that exploits proton-coupled electron transfer from FAD which go by the awkward name of Blue Light Using Flavin (BLUF) domain proteins. These proteins are involved in bacterial phototaxis, synthesis of bacteriochlorophyll and other processes. The photo-reaction is thought to convert a glutamine from its keto to its enol form resulting in a change in the hydrogen-bond network and hence a conformational change [45].

As we have seen for opsins, cryptochromes and related proteins form a group with closely related structures and a variety of different functions, all involving light sensitivity – magnetoreception in birds, DNA repair in many organisms, phototaxis in plants and bacteria, and circadian rhythms in many organisms including mammals.

6.9 Photocatalysis and prebiotic evolution

The origin of life on Earth has been a subject of laboratory experiments ever since the work of Miller and Urey in the early 1950s. Their energy source was the electric discharge but light provides an attractive alternative, especially since UV radiation down to 200 nm could reach the surface of the early Earth since there was no ozone layer. The early Earth had a mildly reducing atmosphere, but meteorite impacts are thought to have brought HCN, NH_3, metal sulfides and metal phosphides.

The photo-induced conversion of HCN to adenine (formula $(HCN)_5$) was reported in the 1960s. More recent experiments have targeted UV photolysis of $[Cu(CN)_2]^-$, $[Fe(CN)_6]^{4-}$ with H_2S, SO_3^{2-}, HSO_3^- or NO_2^-. The cyanometallate anions react photochemically in solution by charge-transfer to solvent (CTTS) releasing hydrated electrons that act as strong reducing agents (see Chapter 2.3.2.5). Thus photolysis of $[Cu(CN)_2]^-$ with H_2S results in a photoredox catalytic cycle generating HSSH as a by-product. The photoreaction of HCN gives formaldehyde and glycolonitrile, $(OH)CH_2CN$. When glycolonitrile is irradiated with the cyanocuprate system in phosphate buffer, glycolaldehyde $(OH)CH_2CHO$ and acetaldehyde were generated. In the presence of HCN, these molecules can act as precursors of a series of four canonical amino acids as well as RNA precursors. Photolysis of $[Fe(CN)_6]^{4-}$ with SO_3^{2-} as the final reducing agent is an alternative photoredox system. Another photoredox cycle of $[Fe(CN)_6]^{4-}$ with nitrite can generate methylisonitrile CH_3NC leading to further pathways for nucleosides. In a related development, it turns out that precursors of purine nucleosides with C=S groups in place of C=O groups can generate the required nucleotides after photolysis followed by hydrolysis. The importance of substitution of oxygen by sulfur is demonstrated in the photoreaction of thiophosphate with glycolonitrile and HCN that leads to C4 and C5 sugars. Although this chemistry is very complex and quite unlike conventional organic synthesis, the yields are very high in many cases [46].

Prebiotic photocatalysis at minerals may have been relevant in chemical evolution (see Chapter 5.3.4.4.). Direct evidence of the viability of this hypothesis comes from laboratory experiments. Micromoles of glycine, alanine, and traces of aspartic acid and serine are formed when a methane saturated suspension of platinised titania in aqueous ammonia is irradiated with UV light. Although the mechanism is not known, intermediary $CH_3^{\bullet}$ and $NH_2^{\bullet}$ radicals may react with $OH^{\bullet}$ to form alcohols and eventually lead to amino acids [47]. Another further intermediate could be formamide. Surprisingly, upon adsorption onto a titania single crystal in high vacuum, it is converted to nucleoside bases when irradiated with UV light [48]. A rare case of C–C coupling with carbon dioxide is its addition to glyoxalate forming lactic acid through irradiation in the presence of colloidal ZnS [49].

6.10 Overview

Photo-induced charge separation, photoisomerisation, photo-induced bond breaking and making, photoluminescence, chemiluminescence, and Förster resonance energy transfer have all featured as naturally occurring phenomena in this chapter. These processes underlie photosynthesis, vision, clocks, communication, DNA damage and repair, phototaxis and magnetoreception. Biology uses small variants on the proteins and chromophores to effect different processes: for example, opsins for vision and clocks, electron transfer from FAD to tryptophan for DNA repair and magnetoreception. Labelling with organic or metal-based dyes has enabled photopharmacology and therapy, tracking and imaging, optogenetics and measurement of fundamental processes in biology such as electron transfer rates.

Questions

1. Positive and negative charges are separated following light absorption at the photosynthetic reaction centre. How does the biological system achieve efficient charge separation and minimise the back electron transfer? Are there any lessons to be drawn for biomimetic systems?
2. The light harvesting proteins transfer energy to the photosynthetic reaction centres. What are the wavelength requirements for this energy in terms of the absorption maxima of the light harvesters compared to the reaction centres? The light absorbers of the photosynthetic reaction centres form a "special pair" or excimer. How does this affect their absorption properties (see Chapter 2.3.1.2)?

3. Förster energy transfer (Chapter 2.2.3) plays roles (a) in natural biological systems, (b) in medicinal applications and (c) in fundamental studies of biological structures. Explain how it works and give an example of each.
4. Explain with diagrams the chemical relationships between: (a) free base porphyrins, (b) metalloporphyrins, (c) free base chlorins, (d) metallochlorins, (e) bacteriochlorins. What roles do each of them play in biological systems, if any?
5. How do biological systems use photoswitches based on cis-trans isomerisation of polyenes? How can it be established by experiment whether this photoswitching is best described as a prompt photoreaction or one involving an equilibrated excited state? The back-reaction usually takes place thermally. What is the significance of the rate of this reverse isomerisation?
6. The examples of bioluminescence all involve formation of peroxide groups that eliminate CO_2 or carboxylic acid leaving another product in an electronically excited state. Compare the structures of each of the compounds in this chapter containing peroxide groups. Why does elimination of CO_2 or carboxylic acid enable formation of an excited state of the other product of reaction?
7. Why is it helpful to have several analogues of GFP that emit at different wavelengths? Describe some applications other than those already mentioned in this chapter.
8. Many proteins have several iron-sulfur groups arranged in a chain. Explain the significance of this for electron transfer making use of the graph in Figure 6.18b.
9. Information may be transmitted (signal transduction) within a protein structure by mechanical changes following light absorption or by electron transfer following light absorption. Provide an example of each type.
10. It is fashionable to emphasise the role of quantum mechanical phenomena in biology. One side of this debate emphasises quantum mechanical tunnelling in electron transfer and coherence phenomena in photosynthesis. The other side points out that photochemistry is a quantum mechanical phenomenon in itself and therefore all of this chapter relates to quantum mechanics. What would you say about this to a lay audience?

References

[1] A. Griesbeck, M. Oelgemöller, F. Ghetti, *CRC Handbook of Organic Photochemistry and Photobiology, - two Volume Set*, CRC Press, Boca Raton, USA **2019**.
[2] J. B. Reece, M. R. Taylor, J. L. Simon, *Campbell Biology: Concepts and Connections*, 7th Edition Pearson Education, London **2013**.
[3] https://commons.wikimedia.org/wiki/File:Thylakoid_membrane_3.svg. Accessed on June 18 2024.
[4] https://redefiningscientists.files.wordpress.com/2015/02/eating.jpg. Accessed on June 18 2024.
[5] a), L. B. Josefsen, R. W. Boyle, *Theranostics* **2012**, *2*, 916; b) R. Alzeibak, T. A. Mishchenko, N. Y. Shilyagina, I. V. Balalaeva, M. V. Vedunova, D. V. Krysko, *Journal for Immunotherapy of Cancer* **2021**, *9*.

[6] R. Baskaran, J. Lee, S.-G. Yang, *Biomaterials Research* **2018**, *22*, 25.

[7] C.-N. Lee, R. Hsu, H. Chen, T.-W. Wong, *Molecules* **2020**, *25*, 5195.

[8] M. Overchuk, R. A. Weersink, B. C. Wilson, G. Zheng, *ACS Nano* **2023**, *17*, 7979–8003.

[9] W. J. de Grip, S. Ganapathy, *Frontiers in Chemistry* **2022**, *10*, 879609.

[10] D. Stella, K. Kleisner, *Insects* **2022**, *13*, 242.

[11] G. J. Taylor, W. Ribi, M. Bech, A. J. Bodey, C. Rau, A. Steuwer, E. J. Warrant, E. Baird, *Current Biology* **2016**, *26*, 1319–1324.

[12] N. W. Roberts, M. L. Porter, T. W. Cronin, *Philosophical Transactions of the Royal Society B: Biological Sciences* **2011**, *366*, 627–637.

[13] https://commons.wikimedia.org/wiki/Category:Lepomis_cyanellus. Accessed on June 18 2024.

[14] J. Fricke, *Physik in Unserer Zeit* **1990**, *21*, 159–160.

[15] L. Wang, C. Wang, W. S. Choi, *International Journal of Molecular Sciences* **2022**, *23*, 3779.

[16] K. Deisseroth, P. Hegemann, *Science* **2017**, *357*, eaan5544.

[17] D. Brinks, Y. Adam, S. Kheifets, A. E. Cohen, *Accounts of Chemical Research* **2016**, *49*, 2518–2526.

[18] J. Broichhagen, J. A. Frank, D. Trauner, *Accounts of Chemical Research* **2015**, *48*, 1947–1960.

[19] a) O. Shimomura, *Angewandte Chemie International Edition*, **2009**, *48*, 5590–5602. b) I. Yampolsky, O. Shimomura, Eds., *Bioluminescence, Chemical Principles and Methods*, 3rd ed., World Scientific, Singapore **2019**.

[20] S. Tripathi, Q. T. N. Hoang, Y. J. Han, J. I. Kim, *International Journal of Molecular Sciences* **2019**, *20*, 6165.

[21] U. Kutschera, W. R. Briggs, *Annals of Botany* **2016**, *117*, 1–8.

[22] S. M. Lewis, C. K. Cratsley, *Annual Review of Entomology* **2008**, *53*, 293–321.

[23] https://en.wikipedia.org/wiki/Lampyris_noctiluca#/media/File:Lampyris_noctiluca.jpg. Accessed on June 18 2024.

[24] Y. Nasu, R. E. Campbell, *Science* **2018**, *359*, 868–869.

[25] https://www.ana.co.jp/japan-travel-planner/area/chubu/toyama/0000015/main.jpg. Accessed on June 18 2024.

[26] E. Brodl, A. Winkler, P. Macheroux, *Computational and Structural Biotechnology Journal* **2018**, *16*, 551–564.

[27] D. F. Gruber, B. T. Phillips, R. O'Brien, V. Boominathan, A. Veeraraghavan, e. a. G Vasan, *PLoS ONE* **2019**, *14*, e0219852.

[28] J. Muralt, Ceratioidei – Tiefseeangler, Tintenkilby Verlag, Bern **2014**.

[29] G. Mulley, M. L. Beeton, P. Wilkinson, I. Vlisidou, N. Ockendon-Powell, A. Hapeshi, N. J. Tobias, F. I. Nollmann, H. B. Bode, J. Van Den Elsen, *PLoS One* **2015**, *10*, e0144937.

[30] Z. M. Kaskova, F. A. Dörr, V. N. Petushkov, K. V. Purtov, A. S. Tsarkova, N. S. Rodionova, K. S. Mineev, E. B. Guglya, A. Kotlobay, N. S. Baleeva, *Science Advances* **2017**, *3*, e1602847.

[31] https://www.zeit.de/wissen/umwelt/2015-03/bioluminesczenz-leuchtende-pilze-pflanzen. Accessed on June 18 2024.

[32] C. V. Stevani, C. K. Zamuner, E. L. Bastos, B. B. de Nóbrega, D. M. Soares, A. G. Oliveira, E. J. Bechara, E. S. Shakhova, K. S. Sarkisyan, I. V. Yampolsky, *Journal of Photochemistry and Photobiology C: Photochemistry Reviews* **2024**, *58*, 100654.

[33] https://www.flickr.com/photos/sjcockell/3507881934. Accessed on June 18 2024.

[34] D. P. Barondeau, C. J. Kassmann, J. A. Tainer, E. D. Getzoff, *Journal of the American Chemical Society* **2007**, *129*, 3118–3126.

[35] R. Y. Tsien, *Angewandte Chemie International Edition* **2009**, *48*, 5612–5626.

[36] L. Huang, Y. Liu, P. Zhang, R. Kang, Y. Liu, X. Li, L. Bo, Z. Dong, *PLoS One* **2013**, *8*, e59804.

[37] S. E. Braslavsky, *Photochemical & Photobiological Sciences* **2023**, *22*, 2799–2815.

[38] a), A. W. McKinley, P. Lincoln, E. M. Tuite, *Coordination Chemistry Reviews* **2011**, *255*, 2676–2692; b) C. J. Cardin, J. M. Kelly, S. J. Quinn, *Chemical Science* **2017**, *8*, 4705–4723.

[39] H. Kye, Y. Song, T. Ninjbadgar, C. Kim, J. Kim, *Sensors* **2022**, *22*, 5130.

[40] A. Losi, S. E. Braslavsky, *Physical Chemistry Chemical Physics* **2003**, *5*, 2739–2750.

[41] J. R. Winkler, H. B. Gray, *Chemical Reviews* **2014**, *114*, 3369–3380.

[42] W. Wiltschko, R. Wiltschko, *Science* **1972**, *176*, 62–64.

[43] a), P. J. Hore, H. Mouritsen, *Annual Review of Biophysics* **2016**, *45*, 299–344; b) J. Xu, L. E. Jarocha, T. Zollitsch, M. Konowalczyk, K. B. Henbest, S. Richert, M. J. Golesworthy, J. Schmidt, V. Déjean, D. J. Sowood, *Nature* **2021**, *594*, 535–540.

[44] a), Z. Liu, L. Wang, D. Zhong, *Physical Chemistry Chemical Physics* **2015**, *17*, 11933–11949; b) N.-E. Christou, V. Apostolopoulou, D. V. Melo, M. Ruppert, A. Fadini, A. Henkel, J. Sprenger, D. Oberthuer, S. Günther, A. Pateras, *Science* **2023**, *382*, 1015–1020; c) D. Ramírez-Gamboa, A. L. Díaz-Zamorano, E. R. Meléndez-Sánchez, H. Reyes-Pardo, K. R. Villaseñor-Zepeda, M. E. López-Arellanes, J. E. Sosa-Hernández, K. G. Coronado-Apodaca, A. Gámez-Méndez, S. Afewerki, *Molecules* **2022**, *27*, 5998.

[45] A. Lukacs, P. J. Tonge, S. R. Meech, *Accounts of Chemical Research* **2022**, *55*, 402–414.

[46] a), N. J. Green, J. Xu, J. D. Sutherland, *Journal of the American Chemical Society* **2021**, *143*, 7219–7236; b) J. D. Sutherland, *Angewandte Chemie International Edition* **2016**, *55*, 104–121.

[47] a), W. W. Dunn, Y. Aikawa, A. J. Bard, *Journal of the American Chemical Society* **1981**, *103*, 6893–6897; b) R. Saladino, C. Crestini, G. Costanzo, E. DiMauro, *Current Organic Chemistry* **2004**, *8*, 1425–1443.

[48] S. Senanayake, H. Idriss, *Proceedings of the National Academy of Sciences* **2006**, *103*, 1194–1198.

[49] a), M. I. Guzman, S. T. Martin, *Chemical Communications* **2010**, *46*, 2265–2267; b) D. Vadivel, F. Ferraro, D. Merli, D. Dondi, *Photochemical & Photobiological Sciences* **2022**, *21*, 863–878; c) L. Botta, B. Mattia Bizzarri, D. Piccinino, T. Fornaro, J. Robert Brucato, R. Saladino, *The European Physical Journal Plus* **2017**, *132*, 1–7.

7 Religion, philosophy and art

7.1 Religion and philosophy

The life-giving role of light has led people to honour or worship the sun in almost all cults and religions. The Sumerians worshipped the sun-god *Utu*, the Babylonians *Shamash*, the Hindus *Surya*, the Aztecs *Toñatiuh* and *Huitzilopochtli*, the Mayans *Itzamná* and the Incas the god *Inti*. Both the people of the Baltic and of Japan worshipped a sun-goddess: *Saulé* and *Amaterasu*, respectively. *Amaterasu* is the most important deity in the Shinto religion: she personifies the sun and light and is considered as the foundress of the Imperial House of Japan [1, 2].

These deities fought and conquered the darkness in human beliefs, turning the Earth to a place of light, life and fertility. In these sun cults, people celebrated the daily and annual cycles of the sun. Sacrifices or other rituals guaranteed the return of the sun. The pharaoh Echnaton or Akhenaten who declared the sun as the only god (Aten) expressed it in his Great Hymn to the Sun: *you rise in perfection on the horizon of the sky, living Aten who started life* [3]. This declaration is impressively displayed in the form of a wall-relief (Figure 7.1). In the unending strife between darkness and light, Echnaton decides in favour of light as the only form of being. The darkness of the night symbolises death. The sun-god Aten travels across the sky in a sunboat during the day, in the night he crosses the waters of the underworld [4]. According to the ancient Greeks, the youthful god *Helios* was driven in his chariot by four fiery stallions in time with the motion of the sun. The Delphic oracle was dedicated to the Sun-god *Apollo* pointing to a connection between the future and light. In northern mythology, it is the horses *Árvakr* (the early riser) and *Alsvidr* (the speedy) who pull the goddess *Sol* across the sky. However, *Alsvidr* is always pursued by the wolf *Sköll* who reaches him on the day that the world ends and will devour him. Sol quickly gives birth to a daughter who is much more beautiful than she is and provides the new world with life-giving light.

The Old Norse name Sol, in Old High German *Sunna*, stands in German tradition for the personification of the sun and is mentioned in the *Merseburg Spells (Merseburger Zaubersprüchen)* and the Icelandic *Edda*. This female form is transformed to a male in the Greco-Roman tradition. Apollo is not only the God of light, but also of healing, Spring, moderation and moral purity, as well as of the arts of music and poetry. The Babylonians introduced the seven-day week and dedicated one day to the sun, Sunday. The Roman emperor Constantine declared it to be a day of rest; he worshipped *Christos Helios* (*Christ, the true sun*). The word *Christ* stems from the Greek *Christos* and means the anointed one. *Christos Helios* was worshipped by the Greeks and Romans and corresponds to Osiris, one of the many Egyptian sun gods worshipped before Echnaton. The name Jesus (ΙΗΣΟΥΣ) is derived from *Iesos*, the Greek Goddess of healing, who was the daughter of Apollo. In the early church, Christians

https://doi.org/10.1515/9783111029375-007

Figure 7.1: Echnaton and Nefertiti with their daughters, Egyptian Museum Berlin. Copyright Richard Mortel from Riyadh, Saudi Arabia, via Wikimedia Commons [5a].

abbreviated the name Jesus to its first three letters, leading after translation into Latin to the acronym *IHS* which appears in many Christian works of art.[1] It is then unsurprising that the provider of light, the sun, appears as a disc or circle in many artistic objects in churches. An impressive example is the baroque altar by Bernini in St Peter's Basilica in Rome (Figure 7.2).

There is even a link with the sun hidden in the Christian designation *church* for the House of God since it stems from the name of the Greek goddess *Circe*, the daughter of *Helios*. In Christian symbolism, Christ becomes the Light of the World and God the Light of Truth. The sun, the provider of light, is the symbol for Life, justice, bliss and Christ. An example is found in St. Paul's Letter to the Corinthians:

---

[1] The abbreviation could also have been taken over from the Egyptian Trinity "Isis, Horus, Set" (the mother, child and father of the Gods).

Figure 7.2: Altar by Gian Lorenzo Bernini in St Peter's Basilica in Rome (ca. 1650 C.E.). Copyright Gary Todd from Xinzheng, China, PDM-owner, via Wikimedia Commons [5b].

For God, who commanded the light to shine out of darkness, hath shined in our hearts, to give the light of knowledge of the glory of God in the face of Jesus Christ.[2] In pre-Christian architecture, sun worship expressed itself in the columns and obelisks stretching to the sky. The obelisk in front of St Peter's Basilica in Rome is considered as a sign of the unification of sun worship with Christian belief and symbolises the birth of the universal church of Rome. In the cathedral in Aachen, Charlemagne (about 800 CE) had a narrow window added next to his throne that changes the light of the sun into a mystical pattern. A few hundred years later, the young Louis XIV, King of France, even appeared in a ballet as a personified sun – the sun king (Figure 7.3).

---

2 2. Corinthians Ch 4 v 6.

Figure 7.3: King Louis XIV (1643–1714) in Le ballet de la nuit (1653) as an elegant symbol of the sun, painted by Henri Gissey [6] (public domain).

The most beautiful manifestations of the central role of light in the Christian religion are the gothic cathedrals that can be considered as architecture of light. In most churches, the altar lies at the East end and the entrance at the West end. From the East, when the sun rises, the believers expect the Second Coming of Christ (sol invictus). For that reason, the priest and congregation pray towards the East. The believers enter the house of God from the West through the church door which prepares them for the majestic space. The entry into the light space of the cathedrals resembles the entry to heaven since the gothic stained-glass windows bring the light to the believers and enable their participation in the divine ceremony. A mystical aesthetic lightroom is generated by the filigree columns of stone stretching to the heavens enclosing windows filled with stained glass and monumental rose-windows. The cathedral is turned into a medium of true light. On the way eastward to the altar, believers traverse their life's course symbolically with their goal of meeting their redeemer. There they will meet the cosmic and heavenly, leaving behind the material world in the West. This direction by the light should bring to believers the hope of resurrection in the light of God. Abbot Suger (1081–1151 CE), the builder of the church of St Denis in Paris, the first gothic cathedral, planned to erect a building in which he wanted the "entire church to shine with the wonderful and uninterrupted light that illuminates the interior beauty in all its glory". The architects succeeded brilliantly in this aim as is shown in Figure 7.4. In the succeeding 350 years the gothic style spread across much of Europe and building techniques improved to allow the stonework to become finer and the area of glass to increase. In the fourteenth century, it was discovered that yellow or orange glass could be made by applying silver nitrate to the reverse of the glass before firing. These developments reached their apogee in the Great East Window of York Minster, built in the perpendicular gothic style and glazed in a light hue by master glazier John Thornton. At ca. 23 × 10 m, it is about the size of a tennis court with more than 300 panels of stained glass and was completed in 1408 [7].

Figure 7.4: Light shining through the triforium and clerestory windows of the cathedral of St Denis in Paris. The original design stems from Abbot Sugar in the mid-twelfth century but was substantially modified in the succeeding century [8].

Control of light also plays an important role in modern sacred buildings. One of the prime examples is the church in the Japanese town of Ibaraki (Osaka prefecture), built in 1989 to the design of the architect Tadao Ando. The minimalist design leaves the church in darkness but for the cruciform cut in the altar wall that illuminates the worshipping congregation with morning light from the East (Figure 7.5).

Figure 7.5: Church in Ibaraki (Osaka prefecture) [9].

An equally impressive example is the renovated chapel at the St Joseph Cemetery in Cloppenburg (Lower Saxony, Germany) with its purist simplicity. A cross of light in the centre of a dark concrete front wall appears almost as if it had been cut out. This impressive effect is created by embedding light-conducting fibre-optic cables into the conventional concrete slabs. This "light concrete" collects the external daylight and directs it to the surfaces of the interior.

In contrast to the example above, the light-artist James Turrell employs coloured artificial light to symbolise God as filling the interior with light to overflowing. In his chapel of the Dorotheenstadt Cemetery in Berlin, the light becomes part of the building, since the church interior is not illuminated conventionally, but is filled with coloured light [10]. It is as if the walls of the building have become weightless; the believer experiences the pure light that draws him to the Christian hope of resurrection. The effect is similar to that of a painting by Caravaggio[3] where the source of the light is hidden (see below).

The word light has an important function as a metaphor in philosophy just as it does in the science of photosynthesis and photonics. The changing use of the light metaphor reflects the development of human understanding in the course of history. Already in the eighth century BCE, Homer wrote in *The Iliad* that life means seeing the light of the sun. The darkness of Hades as the sphere of death was the opposite of this Olympic brightness. Hesiod (seventh century BCE) saw light as a power that does not only see through itself but through its opposite. Parmenides followed Hesiod and expressed the view that light penetrates everything and can therefore be equated with knowledge. Plato drew the general conclusion in around 400 BCE: light illuminates its opposite (darkness) and overcomes whatever is placed in its way. Philosophy should render clarity to the darkness of everyday through the *true light* as *idea of the good* – the origin of all that is *right and beautiful*. Almost 800 years later, Augustine argues that it is only possible to reach God through spiritual perception of light: God is the *light making* light. In the time of the Enlightenment, the metaphorical role of light came to the fore once more. Goethe expressed his interest in the science of light and the poetry of light in *Faust* at the start of the nineteenth century, as we highlighted in the Prologue. The Light of the Enlightenment encompasses and illuminates the darkest corners; reason is a power of the day, not of the night. Later, Friedrich Nietzsche expressed the connection between light and human striving in his poem *Ecco Homo* in the late nineteenth century.

> Yes! I know whence I come!
> Like a flame, unsatisfied
> I glow and consume myself.
> All that I touch, turns to light,
> All that I leave behind, is coal:
> Assuredly I am a flame.[4]

7.2 Art

Just as in architecture the role of light goes well beyond its purely physical effect. In religious works of art, halos or glories around the bodies of the saints together with the surrounding light-beams indicate holiness. In books of the Middle Ages strong colours and sparkling jewels point to the shining divinity of God. Bright lights were achieved with gold leaf. At the beginning of the Renaissance, a time of light and en-

---

3 Michelangelo Merisi, named Caravaggio (1571–1610).
4 Translation by Scott Horton, Harper's Magazine, March 28, 2010. https://harpers.org/2010/03/nietzsche-ecce-homo/

lightenment, the idea that *truth shines out* became a characteristic of this epoch. This metaphorical, mystical light was soon humanised and transferred from God to Man; soon light became an actor in the painting [11]. The recognition of the potential of oil painting by the Flemish painters of the early fifteenth century enabled lighting to be far more effectively depicted than in tempera painting on panels or frescoes for wall painting [12]. Oil painting allowed a translucent layer to be painted over an opaque layer, whereas frescoes had a matt finish. With these techniques Jan van Eyck became a virtuoso in depicting reflections in a mirror, on armour or on skin. In the second half of the century, Geertgen tot Sint Jans was able to depict a night scene with directed illumination. Somewhat later at the time of early baroque, this idea came to expression in the paintings of Caravaggio. His paintings show subtle lighting in which the central scene seems to be illuminated by mild sunlight. The message of the artist is transmitted through a background that is mostly dark strongly contrasting with the light foreground. The painting of *Narcissus* illustrates the legend of the incredibly beautiful narcissistic young man (Figure 7.6). According to Ovid's *Metamorphoses*, he repeatedly refused the advances of numerous young women. One day when standing by a spring he fell in love

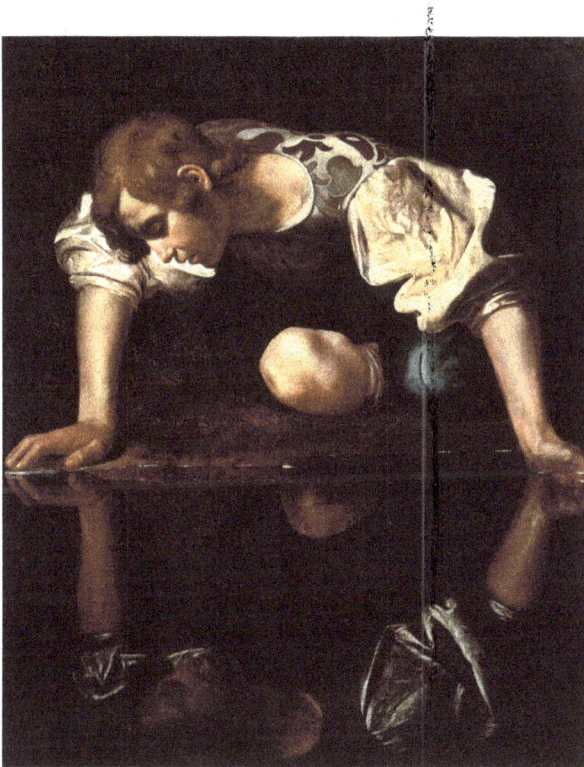

Figure 7.6: Narcissus by Caravaggio (1595). Galleria Nazionale d'Arte Antica, Rome [13]. Caravaggio, Public domain, via Wikimedia Commons.

with his own reflection and couldn't turn away from it. Then a leaf fell into the water and the resulting waves distorted his picture into a hideous grimace, whereupon he died. His body turned into a narcissus, the flower that is still named after him. The fascinating tension of the painting can be understood as a warning of the darkness of life that can sometimes emerge from the beauty of youth.[5] The reflections on the water represent the origin of painting as a mirror of reality.

The contrast between light and shade in painting of the early baroque, exemplified by Caravaggio, became a novelty that gave inspiration and fame to later artists. A celebrated example is the Night Watch by Rembrandt (Figure 7.7) in which the rich details of alternating light and shade create the effect on the onlooker of a group of people in movement. The hand of the captain in the centre illuminated by light coming in from the left of the picture, even casts a shadow on the golden coat of the lieutenant standing to his right. We can still see Caravaggio's influence in the *Discovery of Phosphorus* painted by Joseph Wright (Chapter 4.1) more than a century later.

Figure 7.7: Detail of The Night Watch by Rembrandt (1642). Rijksmuseum [14], Amsterdam (public domain).

---

5 This corresponds to the vanity motif that is also a favourite of baroque painting: the Judaeo-Christian concept of the transience of all earthly things.

Artistic interest in the myriad effects of light on the sky and water took a very different turn in the works of J. M. W. Turner (1775–1851). Turner was fascinated by the light on mist and storm clouds, sunsets and sunrises. For him, the ever-changing light on clouds inspired a completely new way of depicting a landscape, anticipating the work of the French Impressionists. He was an inveterate traveller, watching the light on the sea from Margate on the coast of south-east England, on Scottish lochs, on Venetian canals, and on Lake Lucerne in Switzerland on his many trips to the Alps. His sketchbooks and watercolours (Figure 7.8) have a spontaneity and naturalism that is much less evident in his large oils painted in the studio. Not only was Turner a brilliant observer, but he also lectured and wrote about theories of colour including the colour wheel (Chapter 1.1) and followed scientific developments of the time. He was very aware of Newton's discoveries that white light could be split into the colours of the rainbow and recombined.[6] Goethe's theories of colour influenced him greatly, though he did not always agree with them [16].[7] The polymath John Ruskin (1819–1900) was inspired both by Turner and by the underlying science. Both in his writing and his painting, he was up to date in the science: meteorology, classification of clouds, effects of light on the eye (Figure 7.9). He measured

Figure 7.8: J. M. W. Turner, Sarner See (Switzerland) (watercolour, ca. 1842) [19]. (public domain).

---

6 Until Newton, the general understanding of colour followed Aristotle's theory that the colours arose from the mixing of light and dark.

7 For general information on vision, colour, perception and culture see e.g. ref. [15].

Figure 7.9: John Ruskin, Champagnole (watercolour, 1846, Yale Centre for British Art) [20]. (Wikimedia, Commons license).

the colour of the blue sky using a device called a cyanometer and drew intricate geometric patterns to assist in modelling the patterns of clouds. His five-volume book, *Modern Painters*, showed his own acute observational skills in the numerous effects of light that he pointed out. He also defended the accuracy of Turner's revolutionary paintings from his detractors.

In Monet's studies of light, the reflections of the low sun on the snow are yellow and pink, while the shadows of the fences take the complementary colours. In the knowledge of Newton's colour wheel, Monet later restricted his palette to the prismatic colours and their mixtures, banishing black, grey and brown. Seurat read up on the theory of colour and light and also on the way in which neighbouring colours affect one another. His pointillist technique of combining dots of different colours was inspired by his reading [17].

The modern science underlying the effects of light on the sky and water, from mirages to rainbows and caustic curves, is a reminder of the worlds of Turner and Ruskin [18].

Photochemistry brought a new revolution, starting in the late 1820s, that changed art and culture, entertainment and exploration, journalism and war, science and medicine: photography. The technology of recording black-and-white images with silver halides (see Chapter 4.2.6), started by Niépce, Daguerre and Fox-Talbot, changed our perception of the world. By the early 1900s, the initially primitive methods were extended to slide projection, stereo photographs, studies of locomotion, electric-powered movie cameras, X-ray photography and colour photography. Although competing methods such as cyanotypes (see Chapter 2.3.2) did not take off, they resulted in some key landmarks such as the first book of photographs by Anna Atkins (1843). The way was open for new generations of artists working with light and landscape through photography, for example Herbert Ponting who recorded Antarctica for Robert Scott's expedition, and Ansel Adams whose evocative photos reveal the landscapes of the American West. The art of still and movie photography was all pervasive throughout the twentieth century and remains dominant even if digital methods have largely supplanted silver halides.

Epilogue

Life on Earth rests on three pillars: light, water and carbon dioxide, for without photosynthesis by green plants there would be no food to eat or oxygen to breathe. In the earliest cultures, the sun was worshipped as the highest God; in Christianity Christ is the Light of the World and the sun is a symbol for righteousness and bliss. The columns and obelisks that reach for the heavens are the architectural signs of the worship of holy light. Since the Renaissance the physical and chemical effects of light have become prominent. The interaction of light with materials can be well described with the help of quantum theory, facilitating the explanation of the most diverse phenomena. The chemical nature of substances is changed through the absorption of light, enabling the highly selective processes of life in both fauna and flora. Among them are our internal clock, our hormonal balance and the germination and growth of plants. In other substances such as chlorophyll and semiconductors, light absorption generates opposite electrical charges that enable photosynthesis and photovoltaics. Making use of the reverse principle, application of an electrical potential to modified semiconductors generates the cold light of LED-lamps. Artificial photosynthesis on semiconductor surfaces (photocatalysis) also rests on the principle of charge generation with its promising methods for chemical storage and utilisation of solar energy. The biggest challenges are the generation of hydrogen from water and fertilisers from nitrogen in the air. New fundamental science underpins research both in photovoltaics and photocatalysis with developments that cannot yet be foreseen. Since the present models and mechanisms may change, the Erlangen poet Friedrich Rückert (1788–1866) had it spot on when he wrote:[8] [21]

> Wer Schranken denkend setzt,
> die wirklich nicht vorhanden
> und dann hinweg sie denkt,
> der hat die Welt verstanden.

---

8 The translation was assisted by Dr. Antonia Ruppel and Prof. Dr. Bernhard Forssman.

He whose thoughts do limits set
Which truly don't exist
And then thinks them away
Has understood the world.

Imagination and experimental art will always ensure the birth of new knowledge.

References

[1] *The Cambridge History of Religions in the Ancient World*, Ed. M. R. Salzman Cambridge University Press, Cambridge, UK, **2013**.

[2] https://en.wikipedia.org/wiki/Solar_deity. Accessed on June 19 2024.

[3] W. K. Simpson, *The Literature of Ancient Egypt*, Yale University Press, **2003**.

[4] U. Berner, *Licht. Religiöse und literarische Gebrauchsformen*, Ed. W. Gebhard, Peter Lang Group AG, Frankfurt am Main, **1990**.

[5] a) https://commons.wikimedia.org/wiki/File:Akhenaten,_Nefertiti,_and_three_daughters_beneath_the_Aten;_from_Amarna;_18th_dynasty;_ca._1345_BCE;_Pergamon_Museum,_Berlin_(2)_(26350816948).jpg. b) https://commons.wikimedia.org/wiki/File:Chair_(Throne)_of_St._Peter,_St._Peter's_Basilica_(48466425951).jpg. Accessed on June 19 2024.

[6] https://picryl.com/media/ballet-de-la-nuit-1653-67ecae. Accessed on June 19 2024.

[7] https://www.vam.ac.uk/articles/stained-glass-an-introduction, *Victoria and Albert Museum*. Accessed on June 19 2024.

[8] https://upload.wikimedia.org/wikipedia/commons/b/b1/SaintDenisInterior.jpg. Accessed on June 19 2024.

[9] https://hkang7.wordpress.com/2011/01/16/church-of-the-light/. Accessed on June 19 2024.

[10] https://encrypted-tbn0.gstatic.com/images?q=tbn:ANd9GcQUVdo6uT8kLukynmD8wN0L_P3ePgMpVvz-aRkap4vboJzvSGJ8&s. Accessed on June 19 2024.

[11] C. Lechtermann, H. Wandhoff, *Licht, Glanz und Blendung*, Peter Lang Group AG, Frankfurt am Main, **2008**.

[12] J. Dunkerton, *Giotto to Dürer: Early Renaissance Painting in the National Gallery*, Yale University Press, New Haven, USA, **1991**.

[13] https://commons.wikimedia.org/wiki/File:Narcissus-Caravaggio_(1594-96).jpg. Accessed on June 19 2024.

[14] https://picryl.com/media/rembrandt-the-nightwatch-wga19147-f82630. Accessed on June 19 2024.

[15] M. Livingstone, *Vision and Art – The Biology of Seeing*, Revised edition, Abrams, New York, **2014**.

[16] J. Gage, *Colour in Turner, Poetry and Truth*, Studio Vista, London, **1969**.

[17] W. Homer, *Seurat and the Science of Painting*, MIT Press, Cambridge, Mass, USA, **1964**.

[18] M. G. J. Minnaert, *Light and Colour in the Outdoors*, Springer, New York, **1993**.

[19] https://garystockbridge617.getarchive.net/media/jmw-turner-the-sarner-see-lake-sarnen-evening-c1842-watercolor-bcaf3c?action=download&size=original. Accessed on June 19 2024.

[20] https://commons.wikimedia.org/wiki/File:John_Ruskin_-_Champagnole_-_B1997.14_-_Yale_Center_for_British_Art.jpg. Accessed on June 19 2024.

[21] F. Rückert, *Die Weisheit des Brahmanen, ein Lehrgedicht in Bruchstücken. Werke, Band 2*, Bibliographisches Institut, Leipzig und Wien, **1897**.

Index

https://doi.org/10.1515/9783111029375-008

.

www.ingramcontent.com/pod-product-compliance
Lightning Source LLC
Chambersburg PA
CBHW080933220326
41598CB00034B/5771